MW01358101

The Science and Art of Simulation

Michael M. Resch · Nico Formánek ·
Ammu Joshy · Andreas Kaminski
Editors

The Science and Art of Simulation

Trust in Science

Editors
Michael M. Resch
High Performance Computing Center
University of Stuttgart
Stuttgart, Baden-Württemberg, Germany

Nico Formánek
High Performance Computing Center
University of Stuttgart
Stuttgart, Baden-Württemberg, Germany

Ammu Joshy
High Performance Computing Center
University of Stuttgart
Stuttgart, Baden-Württemberg, Germany

Andreas Kaminski
Department of Philosophy
TU Darmstadt
Darmstadt, Germany

ISBN 978-3-031-68057-1 ISBN 978-3-031-68058-8 (eBook)
https://doi.org/10.1007/978-3-031-68058-8

Cover illustration: istock.com/VadimShechkov

This Springer imprint is published by the registered company Springer Nature Switzerland AG
The registered company address is: Gewerbestrasse 11, 6330 Cham, Switzerland

Synopsis of Contributions

Philosophy of Trust

Johannes Lenhard—Heresy and Honor. A Historical Perspective on Trust in Science
A historical consideration of how and why trust became a pressing issue in contemporary philosophy of science. Four factors of science are singled out that contribute to trust being a pressing issue. Conclusion: Trusting science is affected by factors outside from science—this is the heresy.

Britanny Gentry—Trusting Science: Is There Reasonable Distrust of Reputable Scientific Authority?
Highlights a symmetry between the formation of distrust and trust. Argues that distrust in science can be rational if one has a history of negative experiences with science.

Jörn Wiengarn—Can There be an Epistemic Authority?
Defuses a skeptical argument against epistemic authorities by ascribing to them the ability to give robust reasons. Under this definition of authority three popular objections are shown to be inadequate.

Elizabeth Stewart—Trust Science With What? Trust-Building Dialogue between Scientists and the Public
Proposes a framework of trust domains, that is domains of expectations of truster and trustee. If these domains are misaligned trust is broken. But it can be fixed by making explicit and negotiating the content of trust domains. It is suggested that a trust building dialogue between scientists and the public in the case of COVID-19 vaccines will involve such negotiations of trust domains.

Trust in Science

Pierluigi Barrotta, Roberto Gronda—Scientific Experts, Epistemic Wisdom and Justified Trust
Argues that the type of trust that laypeople should place in experts is epistemic-pragmatic trust. This amounts to experts having epistemic wisdom, thereby extending the traditional two component account of trust in experts.

Lara Huber—Confidence: Calibrating Trust in Science
Argues that the concept of calibrated trust can be elucidated by two case studies, namely the IPCC and IPBES reports. These reports tune confidence in scientific findings according to intersubjectively adjusted metrics. Calibrated trust in science is thus found to have three components: adequate assessment of information, consensus formation and reliance on actual scientific outcomes.

Petar Nurkić—$#*! scientists say: monitoring trust with content analysis
Analyzes trust enhancing rhetoric strategies of scientists during COVID-19 when engaging the public with qualitative content analysis.

Trust and Policy

Vlasta Sikimić—Trust in science during global challenges: the pandemic and trustworthy AI
Trusting science and trusting AI can be achieved with similar measures. Epistemic and ethical layers of science need to be connected to achieve trustworthy science.

Sarah Malanowski, Nicholas R. Baima, Ashley G. Kennedy—Science, Shame, and Trust: Against Shaming Policies
Argues that non-communal shaming policies to motivate the public are unethical and ineffective. They are ineffective because they are non-communal, that means the shamed individual doesn't trust the originator of the policy in the first place. They are unethical for a variety of reasons, but mostly because they view humans as morally deficient from the outset, i.e. as beings that must be shamed into action. Recommends that shaming policies should be avoided.

Elena Popa—Decision Making, Values and (Dis)Trust in Science: Two Cases from Public Health
Discusses two cases were injustice in public health decisions led to distrust in science and argues that in such cases distrust is warranted. Concludes that trust in science is not only an epistemic function but also depends on justice and equity inherent in decisions that are informed by science.

Jamie Shaw—Trust and Funding Science by Lottery
Discusses four different hypotheses how allocating science funds by lottery might affect trust in science and suggests that in two cases lotteries might be beneficial

for fostering trust in science—the other two cases being detrimental. Calls for more empirical research on the topic.

Sociological, Communicative and Media Aspects of Trust in Science

Paula Muhr—Establishing Trust in Algorithmic Results: Ground Truth Simulations and the First Empirical Images of a Black Hole
Argues that the concept of *transcription* from media science explains how epistemic trust in black hole imaging was established. Epistemic trust is found to be strongly context and media dependent.

George Zoukas—Trust and science communication in the Internet era: The case of mainstream climate blogging
Science communication case study of exchange between climate scientists running climate science blogs and their lay audience. Argues that climate science blogs are central for lay audiences to obtain and evaluate primary source knowledge, thereby increasing the trustworthiness of climate science.

Birte de Gruisbourne—Emancipatory Data Literacy and the Value of Trust
Trust has an affective component which is at odds with the idea of explicitly increasing trust by teaching data literacy. Simply teaching data literacy as a skill also creates the danger of shifting the responsibility of trusting "correctly" to the individual. The concept of empowering data literacy is proposed to remedy the situation.

Andrija Šoć, Monika Jovanović—Only a Theory? Substantive and Methodological Strategies for Regaining Trust in Science
Discusses social correlates of trust in science and their measurements in surveys. Argues that the notion of science in such studies is prone to misunderstanding and in many cases doesn't latch onto what scientists would view as science. As a remedy suggests that improved science communication could at least partly lead to a better public understanding of the scientific process.

Yujia Song, Maciej Balajewicz—Undermining Trust in Science: No Fraud Required
Reminds us that trust in science is not only a relation between individuals but to a large extent shaped by institutionalized practices like peer review and grant funding. Looking at pressures emanating from such institutions argues that naive trust in individualistic science can be easily undermined. Voices the hope that a more nuanced view of scientific practice will lead to a more realistic assessment of public trust in science.

Contents

Contributors

Nicholas R. Baima is Associate Professor of Philosophy at Harriet L. Wilkes Honors College, Florida Atlantic University. He received his Ph.D. in philosophy from Washington University-St. Louis and specializes in ancient philosophy and ethical theory. His work in ancient philosophy has appeared in journals such as *Ancient Philosophy*, *Phronesis*, and *Journal of the History of Philosophy*, while his work in ethical theory has appeared in *Ethical Theory and Moral Practice* and *Journal of Value Inquiry*. He is the coauthor, with Tyler Paytas, of *Plato Pragmatism: Rethinking the Relationship of Ethics and Epistemology* (Routledge, 2021). With Sarah Malanowski, he is writing a book on the ethics of video games, *Why It's OK to be a Gamer* (Routledge).

Maciej Balajewicz is an independent researcher, receiving his Ph.D. from Duke University (North Carolina, USA) in the area of fluid dynamics. His research interests include physics-informed machine learning and his current work focuses on rank algorithms in large-scale industrial applications.

Pierluigi Barrotta is full Professor of Philosophy of Science and holder of the Galileo Galilei chair at the University of Pisa. He graduated in Philosophy at the University of Pisa and Scuola Normale Superiore, and studied Economics at the University of Cambridge, where he earned the M.Phil. degree. After obtaining the Ph.D. in Philosophy of Science from the University of Genoa, he was Visiting Fellow at the Centers for Philosophy of Science at the University of Pittsburgh and the London School of Economics. He has published in philosophy and history of science, epistemology of economics and the social sciences, epistemology of environmental sciences, and dialectical rationality in science. He is currently carrying out research on the relationships between science and democracy, and the role played by scientific experts in democratic societies. On this topic, he has published the book Scientists, Democracy and Society: A Community of Inquirers, New York and Heidelberg 2018.

Birte de Gruisbourne studied philosophy with a focus on critical theory, post-structuralism and philosophy of science in Berlin and is now a research assistant at the chair of "Digital Cultures/Digital Humanities" at the Institute of Media Studies at the University of Paderborn. Her dissertation deals with the relation of autonomy and care. Here, she is not only interested in classical intersubjective caring relationships, but also in other types of relationships that are characterised by caring attitudes or patterns of care. In the project DataLiteracySkills@OWL she furthermore works on ways to make data literacy accessible to university students. The aim here is to establish a power-critical and context-sensitive concept of data and its social interrelations.

Brittany A. Gentry is Postdoctoral Fellow at Utah State University. Areas of research include philosophy of science, epistemology of measurement, and metaphysics with specialties in philosophy of time, physics, mathematics, and epistemology of trust.

Roberto Gronda is Assistant Professor of Logic and Philosophy of Science at the University of Pisa. He received his Bachelor's and Master's degrees from the University of Torino, and his PhD from Scuola Normale Superiore of Pisa. He is the author of Dewey's Philosophy of Science (Springer, 2020), and of several essays in philosophy journals. He is the editor of Esperti scientifici e complessità (Pisa University Press, 2020) and Pragmatismo e filosofia della scienza (Pisa University Press, 2017). His fields of research are the philosophy of scientific expertise, American Pragmatism, and the philosophy of humanities.

Lara Huber is a Senior Researcher and Lecturer at the Institute of Philosophy of Kiel University. Much of her research and teaching in philosophy of science and research ethics has revolved around the coevolution of scientific methods and technical devices. She has published broadly about standardisation in the sciences, the cultural and technical history of measurement and related questions, for example, how normality is conceptualised in medicine, psychology and every-day life. Lately, her work focuses on exploring scientific criteria that are key in prioritising ends, evaluating methods, or assessing outcomes of research.

Monika Jovanović is an assistant professor at the Department of Philosophy, Faculty of Philosophy, University of Belgrade. Her dissertation centered on the topics of aesthetic concepts, objectivity and evaluation (published in 2019 as Epistemology of Taste: Sibley and Contemporary Aesthetics). Monika Jovanović has taught primarily courses related to Aestetics, but also took part in courses on Ethics and Methodology of Science. Her main areas of interest are traditional and contemporary aesthetics, and philosophies of art (especially literature, painting and film). Her interests also pertain to experimental aesthetics, where she is interested in the ways one can experimentally discuss the possibility of aesthetic anti-relativism. Among other publications, she recently authored articles titled: 'Beyond Internalism/Externalism Dispute on Aesthetic Experience: A Return to Kant', 'Response-Dependence and Aesthetic Realism: Zangwill and Pettit', and 'Crises, Thought Experiments and Fiction: Moral Intuitions Between Theory and Practice'.

Andreas Kaminski is a Professor of Philosophy of Science and Technology at TU Darmstadt. His research areas include: Social Epistemology: The philosophy of trust and testimony, Technical Epistemology: The role of technology in scientific processes (computer simulation & machine learning), Politics of Technology: Modeling for policy. Andreas Kaminski is co-editor of the Yearbook of Philosophy of Technology. He is a senior scientist at the High-Performance Computing Center of the University of Stuttgart (HLRS), where he established and led a department for the philosophy of computer-based sciences. Previously, he was a visiting professor at RWTH Aachen.

Ashley G. Kennedy is Associate Professor of Philosophy at Harriet L. Wilkes Honors College, Florida Atlantic University. She received her Ph.D. in philosophy from the University of Virginia and she specializes in applied issues in medicine, science, and global justice, and her work has appeared in various journals such as *Bioethics, Journal of Global Ethics, Journal of Medicine and Philosophy*, and *Studies in History and Philosophy of Science.* She recently published *Diagnosis* (Oxford, 2021), the first philosophical and ethical guidebook for medical trainees, and has a second book, *An Introduction to Science and Public Policy* forthcoming with Routledge. She is currently working on a third book, *Child Labor in the Global Context*, which is under contract with Oxford.

Johannes Lenhard is a senior researcher at RPTU Kaiserslautern, Germany. Since 2020, he is located in the Laboratory for Engineering Thermodynamics where he is establishing philosophy in science and engineering. He has received his doctoral degree in mathematics from the University of Frankfurt (1998), long before he wrote his habilitation thesis in philosophy at Bielefeld University (2012). How does using the computer change the methodology and epistemology of the sciences? How does computational modeling transform the use of mathematical tools? Lenhard's research aims at tackling these questions in a way that speaks to philosophers, historians, and scientists alike. Readers find the preliminary peak of his achievements in: J. Lenhard: Calculated Surprises. A Philosophy of Computer Simulation, New York: Oxford University Press, 2019.

Sarah C. Malanowski is Instructor of Philosophy at the Harriet L. Wilkes Honors College, Florida Atlantic University. She received her Ph.D. in philosophy from Washington University-St. Louis and specializes in philosophy of cognitive science and biomedical ethics. Her work in philosophy of cognitive science has appeared in journals such as *Frontiers in Neuroanatomy* and *Synthese*, while her work in biomedical ethics has appeared in *Bioethics*, *Journal of Medicine and Philosophy*, and *Neuroethics.* With Nicholas R. Baima, she is writing a book on the ethics of video games, *Why It's OK to be a Gamer* (Routledge).

Paula Muhr is a Postdoctoral Researcher at the Institute for History of Art and Architecture, Karlsruhe Institute of Technology (KIT) and a visual artist. She is also a member of the History, Philosophy, and Culture Working Group of the Next Generation Event Horizon Telescope collaboration. She studied visual arts, art history, theory of literature, and physics before receiving a Ph.D. in Visual Studies from the Humboldt-Universität zu Berlin and a Postgraduate Diploma in Fine Arts (Meisterschülerin) from the Academy of Fine Arts Leipzig. Her transdisciplinary research is at the intersection of visual studies, image theory, media studies, science and technology studies (STS), and history and philosophy of science. She examines knowledge-producing functions of new imaging and visualisation technologies in natural sciences, ranging from neuroscience over medicine to black hole physics.

Petar Nurkić is a Ph.D. candidate and research assistant at the Institute of Philosophy (Faculty of Philosophy, University of Belgrade). His areas of expertise are social epistemology and philosophy of science. He is currently working on an epistemic network analysis of institutional communication between experts and the general public, with special attention to epistemic elements such as authority, reliability, and trust. The above activities are carried out within the project funded by the Faculty of Philosophy in Belgrade, called Man and Society in the Time of Crisis.

Elena Popa works as postdoctoral researcher at the Interdisciplinary Centre for Ethics, Jagiellonian University, Krakow. She is principal investigator for the project 'Values, Trust, and Decision Making in Public Health', funded by the European Commission and the Polish National Science Center under a Marie Skłodowska-Curie COFUND grant. Her areas of specialization are general philosophy of science, philosophy of psychology, and philosophy of medicine. Her research focuses on interventionist theories of causation and causal inference, values in science, with special emphasis on cultural and social issues in medicine (particularly psychiatry and public health), and the philosophy of AI. She has published in journals such as *Synthese Studies in History and Philosophy of Science,* and *Philosophy and Technology.*

Jamie Shaw is currently a postdoctoral fellow at Leibniz Universität, Hannover on a project entitled "Biases in Scientific Inquiry" working alongside Torsten Wilholt and Manuela Fernandez-Pinto. Before this, I was a SSHRC sponsored postdoc at the University of Toronto after completing my Ph.D. dissertation "A Pluralism Worth Having: Feyerabend's Well-Ordered Science" at the University of Western Ontario under the supervision of Kathleen Okruhlik. My research focuses on methodological and politics issues embedded in science funding policy as well as the history of 20th century philosophy of science (with a focus on the thought of Paul Feyerabend). I am the co-editor of *Interpreting Feyerabend: Critical Essays* (with Cambridge University Press) and have published several papers in philosophy and science policy journals on these topics.

Vlasta Sikimić is an Assistant Professor at the Philosophy and Ethics group of Eindhoven University of Technology. Her research focus is on Philosophy of Science, Philosophy of AI, Empirical Philosophy, Science Policy, and Animal Ethics. In her research, Vlasta promotes the idea of increasing knowledge acquisition through an inclusive and supportive environment. She is also active in professional organizations and policy advising bodies. For example, she was the Chair of the Organizing Committee of the European Philosophy of Science Conference in Belgrade in 2023 and a member of the Advisory Committee of the Serbian Government on the Ethical Use of AI.

Andrija Šoć is a Research Fellow at the Institute of Philosophy, Faculty of Philosophy, University of Belgrade. His main research interests pertain to areas in which Political Philosophy, Political Science and Research Methodology can fruitfully intersect. He is currently working on topics related to democratic backsliding, civic participation, political deliberation, problems of defining the concept of trust and ways in which normative positions can be informed by and in turn themselves inform empirical research of civic attitudes to current societal challenges. Another major area of his interest is Kant's philosophy and ways in which different aspects of Kant's tought can be brought to bear upon contemporary philosophy. Some of his recent publications are: 'From Deliberation to Participation: Democratic Commitments and the Paradox of Voting', 'Towards a Comprehensive Concept of Trust: Empirical Research and Theoretical Analysis', 'Deliberative Education and Quality of Deliberation: Toward a Critical Dialogue and Resolving Deep Disagreements'.

Yujia Song is Assistant Professor of Philosophy at Salisbury University, Maryland, USA. Her research focuses on the nature and value of interpersonal understanding, and topics related to it, including empathy, appreciation, thinking, intellectual virtues, and emotions. Her current project explores the aesthetic dimensions of our ethical life, such as the role of imagination in moral perception, the importance of expression in interpersonal relations, and aesthetic appreciation of disabled bodies

Elizabeth Stewart is currently a Visiting Assistant Professor at Howard University in Washington, D.C. Her work focuses on theorizing issues related to trust and trustworthiness, specifically with respect to Artificial Intelligence and scientific practice. She has recently published a paper, Negotiating Domains of Trust in *Philosophical Psychology*, in which she develops the concept of a trust domain. She is currently working on a project that details the connections between feminist conceptions of vulnerability and trust relationships.

Jörn Wiengarn is a research associate at the Institute of Philosophy at TU Darmstadt. His research focuses on the social-eptemological side of scientific and technological developments, with particular emphasis on questions of trust and mistrust in this context.

George Zoukas is a postdoctoral research fellow at the Department of History and Philosophy of Science, National and Kapodistrian University of Athens (NKUA), currently working in the research project *BIO-CONTEXT—Contextualizing biobanking in Greece: histories, practices, discourses*, funded by the Hellenic Foundation for Research and Innovation (HFRI). He has a Ph.D. in Science, Technology and Innovation Studies (University of Edinburgh, 2019), an M.A. in Society, Science and Technology (Maastricht University, 2012), an M.Sc. in History and Philosophy of Science and Technology (NKUA/NTUA, 2012), and a B.Sc. in History and Philosophy of Science (NKUA, 2010). His research interests include the sociology and history of social media and the internet, the communication and public image of science and technology, and the sociology of scientific knowledge (SSK). He has held research and teaching positions at the University of Edinburgh, the National and Kapodistrian University of Athens, and the University of West Attica.

Trust in Science. A Systematic-Historical Introduction

Andreas Kaminski

Abstract The phrase 'trust in science' raises the question of how trust can have a place here, especially when it is not 'merely' about public trust in science, but about trust in science itself. Does this not constitute a category mistake? Or perhaps a fact that nonetheless violates central epistemic norms? An intuitive strategy to address these concerns is to epistemologize trust itself. Accordingly, trust in science would itself be a matter of knowledge. However, epistemic trust (in science) presents both a conceptual and a practical problem. Normative positions that seem to offer a solution to these difficulties like the assurance view face their own conceptual and practical problems. This paper demonstrates how theory development leads to a virtue theory of trust, which, by unifying the epistemic and normative dimensions, allows for solving the conceptual problems. However, it too is confronted with challenging practical issues.

Torben finds himself in a waiting room where he encounters a newspaper article that claims coffee has protective effects against cancer; however, he recalls that a recent study could not detect any protective effect. When the doctor invites him in, she informs him that the lab results are ready: The abdominal pains he suffers from are not explainable by the detailed blood picture in front of him. Therefore, an ultrasound is now performed. Torben can see on the large screen what the doctor sees, but he recognizes nothing but vague bodies in different shades of gray. The doctor concludes that there are no signs of a serious cause that could explain his pain. She suggests that he reduces stress and drinks less coffee.

As Torben later walks to his car, he feels the enormous heat of the summer day and thinks that these are signs of the changing climate. He is glad to at least drive an electric car equipped with a driving assistant. On the radio, he listens to a science podcast about recent studies evaluating the effectiveness of masks to prevent or at least slow the spread of Covid-19. Arriving at the office, he opens his emails and finds a message from a colleague. She sends him access to the measurement data he has

A. Kaminski (✉)
TU Darmstadt, Darmstadt, Germany
e-mail: andreas.kaminski@tu-darmstadt.de

M. M. Resch et al. (eds.), *The Science and Art of Simulation*,
https://doi.org/10.1007/978-3-031-68058-8_1

been waiting for, so that his research group can perform a computer simulation on fluid dynamics, intended to reduce the formation of turbulence—which helps reduce the fuel consumption of airplanes.

This description of half a day in Torben's life contains various episodes where his trust in science plays a role (or several). As artificial as this story might sound, many of us will probably find ourselves in similar situations over the course of a day where this trust in science is significant. Assuming Torben trusts a scientific study, he—like us—would not think specifically about it or even discuss it. For instance, Torben trusts his colleague that the measurement data are correct. Of course, it matters that they don't deviate significantly from his expectations. But this still leaves room for a lot of things that could go wrong: The measurement data could be incorrect in many ways, without appearing blatantly false. Torben also believes the doctor who examines him—and (somewhat) the laboratory that has analyzed his blood, although he knows none of the people working there or the procedures involved. His decision to buy an electric car was also significantly influenced by his belief that it's ecologically beneficial to drive such a car; he trusts the testimonials of experts here, even though there are differing positions among them. However, he treats the studies about the effects of coffee differently; the discrepancy between the two studies makes him suspicious, and he initially believes neither of them.

When trust becomes an explicit topic, it is often because it has become a problem, this is a common diagnosis.[1] Those who trust typically do not think or talk about this trust. Current, controversial topics such as debates about climate change, Covid-19, or AI seem to support this. Trust is usually discussed in the context of these topics because it has become problematic for individuals or groups.[2]

These debates are intertwined with another contemporary development: the rise of political disinformation, which exploits and amplifies existing polarization. One interpretation of polarization is to understand it as polarization of trust and distrust relationships. Disinformation campaigns, on the other hand, also target topics such as climate change, the pandemic, or AI. This creates the image of a societal Achilles body, that from head to toe is vulnerable to doubt and mistrust. In this sense, trust has become a problem that is discussed daily in public discourse.

If we take a step back, it becomes apparent that the problem of trust in science distinguishes two issues. On the one hand, trust in science poses a theoretical problem, and on the other hand, it presents a practical problem. It is worthwhile to separate

---

[1] See for example Böhme (1998). An argument is rarely given for this diagnosis. It appears to be more of a rule of observation. However, Lagerspetz has developed a theory of trust, according to which it is inherent in the form of trust to be 'tacit' (1998, p. 28–46, 2001).

[2] Empirical studies paint a nuanced picture of the development of trust in science, which changes locally, contextually, and based on group membership, such as education level. For Germany, see the surveys by "Wissenschaft im Dialog", known as the "Wissenschaftsbarometer" (Science Barometer), available online https://wissenschaft-im-dialog.de/suche/?q=barometer (last time visited on January 9, 2024). For the U.S. context, see Pew Research Center, November 2023, "Americans' Trust in Scientists, Positive Views of Science Continue to Decline", available online at https://www.pewresearch.org/science/2023/11/14/americans-trust-in-scientists-positive-views-of-science-continue-to-decline (last accessed on January 9, 2024).

these at first, as they are interrelated in the following way: the answer given to one problem has consequences for the answer that can be given to the other problem.

The *theoretical-conceptual* problem involves the question of whether and in which sense one can have trust in science. It may seem as if this constitutes a category mistake. Trust would thus be a form of relationship that has no place or, more precise, *should* have no place in science. The problem concerns the question of what conceptual relationship exists between trust and science. Here, too, a further distinction is initially suggested: trust in science and trust within science. The former refers to the relationship of the public, the political or legal system to science. The latter, on the other hand, denotes a relationship within science, i.e., between scientists. Even if it may be easier to accept that there is trust (or distrust) in science and this does not constitute a category mistake, it is also claimed that there is trust within science; not just factually, which could easily be dismissed as a violation of an epistemic norm, but as a meaningful epistemic relationship.

The *practical* problem is how to assess the trustworthiness of science, scientists, studies, models, or results. Assuming that trust in science or even within science represents a meaningful form of relationship: How can a layperson or another scientist solve the practical problem of whether trust should be granted? How will they know whether it is reasonable to trust?

This also immediately raises the need for clarification: Henceforth, the practical problem of trust in or within science does not mean the aim would be to increase trust or enhance the acceptance of studies or the like. Rather, it pertains to what characterizes well-founded trust or distrust in or within science. In other words, what is the reasonable basis for judgments about trustworthiness?

This points towards the connection between the theoretical-conceptual and practical problems as mentioned above: The answer to the question of what the conceptual relationship between trust and science is prepares the answer to the question on what basis practical judgments about trustworthiness can be made.

1 The Conceptual Problem

We have become accustomed to understanding 'trust in science' as a description of a societal problem. However, one should not take it for granted; it may even sound paradoxical: Was science not supposed to be based on evidence, which through insight, demonstration, and explanation should render trust unnecessary? A modern assumption is: Just as trust and knowledge seem to exclude each other, trust and science are also two mutually exclusive forms.

1.1 *First-Hand Experience and Logical Reasoning as Sources of Knowledge*

Modern epistemology begins with Descartes rejection of inherited doctrines to find an "unshakeable" foundation (Descartes, 2008 [1641]: 17). To this end, Descartes explains, it is beneficial to enter seclusion, to withdraw from everyday life and other people, as well as their opinions: "The moment has come, and so today I have discharged my mind from all its cares, and have carved out a space of untroubled leisure. I have withdrawn into seclusion and shall at last be able to devote myself seriously and without encumbrance to the task of destroying all my former opinions." (Descartes, 2008 [1641]: 17) It is important to note that the isolation Descartes seeks here is not merely a narrative accessory. Certainty could only be found in radical independence from others—through pure and wholly self-sufficient thinking.

Francis Bacon, on the other hand, challenged the dependency of the epistemic self on everyday language in his doctrine of the Idols. The "idols of the market-place" consist of the inherited words that designate and thus distort objects: „Plainly words do violence to the understanding, and confuse everything; and betray men into countless empty disputes and fictions." (Bacon, 2000 [1620]: 42) The "idols of the theater" represent the teachings of others, which, for Bacon, have also "created false and fictitious worlds" (ibid.). Robert Hooke, a contemporary of Descartes and Bacon, therefore advised that the natural philosopher should imagine himself as a person who has entered a completely foreign continent, to view the objects of his research as if seeing them for the first time (Hooke, 1666: 62).

These ideas characterize the image that the public will have of science and the self-perception that scientists will have of themselves; they determined the epistemology and the philosophy of science for a long time. According to this image:

(1) Other individuals are seen primarily as a source of disturbance, irrespective of traditions such as Rationalism or Empiricism. Sociality is primarily understood as a disturbance, distraction, distortion from the true state of things.
(2) Therefore, it is necessary to detach and isolate oneself from others as much as possible to let the evidence speak for itself.

Modern epistemology is firmly founded on the idea that there can only be *two sources of knowledge*: first-hand experience and first-hand logical reasoning. Leibniz refers to them as truths of fact and truths of reason (Leibniz, 1710: 71, 1714: § 5), Hume calls them "matters of fact" and "relations of ideas" (Hume, 2007a, 2007b: Sect. IV), and Kant eventually states: "Our cognition arises from two fundamental sources in the mind, the first of which is the reception of representations (the receptivity of impressions), the second the faculty for cognizing an object by means of these representations (spontaneity of concepts)" (Kant, 1998 [1781]: B74)—The crucial point is:

(3) Knowledge (and evidence) can only be gained through one's own experience or through one's own reasoning.

It can only be considered knowledge if it is *my* knowledge—and for that, it must be traceable back to my own experience and my thinking. Others are *only* understood as influences that disturb and distort my experience and my thinking.

1.2 Science as "Organized Skepticism"

This concise historical reconstruction elucidates how and why the expression "trust in science" was historically unexpected and indeed seemed contradictory. If (1) others are a source of disturbance, if (2) the aim is to gain evidence in a self-sufficient manner, and if (3) only one's own experience or one's own reasoning could be considered for this purpose, then knowledge could not be gained through others, but only by isolation from them and by focusing on knowledge processes that are isolated and purified from social influence as much as possible.

Against this historical backdrop, the sociologist Merton characterized science as "organized skepticism" (Merton, 1973 [1938]: 264). While most institutions require unreserved trust, in science, doubt is a virtue (ibid., 265). "[S]cientific research is under the exacting scrutiny of fellow experts [...], the activities of scientists are subject to rigorous policing, to a degree perhaps unparalleled in any other field of activity" (Merton, 1973 [1942]: 276).

Merton did not assume that scientists behave virtuously in an epistemic sense due to a special character trait, but rather because their practice is so normatively structured that discovered misconduct has immense consequences for them. In other words, while Merton corrected the notion that science begins with individual autarky, he assigned a crucial role to the social in scientific practice. This role, however, is limited to norming the behavior of scientists, where organized skepticism plays a significant role. This skepticism is directed at other scientists, but also at oneself.

Daston and Galison have shown how the epistemic virtue of objectivity emerged in the middle of the 19th century (and thus surprisingly late for scientific self-understanding). At its core, objectivity is doubt about the role of the subject in the knowledge process (Daston & Galison, 2007: 44); objectivity as a virtue does not negate subjectivity in the research process in an abstract way, rather it should guide the formation of a scientific self that does not distort the objectivity of the research process. Thus, it becomes part of the scientific 'pedagogy' to *practically* exercise objectivity. The aim is "willed willessness": the rigid determination for an epistemically selfless self (Ibid.: 53). Trust in science, therefore, appeared to be once again, albeit in a different way than at the beginning of modern times, a category mistake. The scientist should first mistrust himself. Thus, it also seemed questionable for others to trust a scientist.

1.3 Trust in Science and Trust Within Science

If science is organized (self-)doubt, then trust within science indeed seems to be a misguided notion. However, might this apply only to the internal perspective and not the external one? *Within* science, i.e. between scientists, it is argued, there is no trust—and if there is, it is misplaced. Trust towards science, however, might be a comprehensible term when it refers to the relationship of others to science. Laypeople, politicians, in short, the public, can have trust in science.

This interpretation is also supported by the examples mentioned at the beginning, namely that issues such as climate change and COVID-19 measures, including vaccinations or masks, became problematic primarily in the relationship between the public and science. The problem was not the trust or lack of trust of a scientist in other scientists, but the public's trust in the statements of scientists.

But does this eliminate the conceptual difficulty of speaking of trust in science? Can trust within science only be viewed as a violation of epistemic norms?

2 Trust in Science

That trust *factually* plays a role within science is likely an uncontroversial claim. The questions start with whether it is *avoidable*? And whether it *should be avoided*?

Scientific processes are and must be highly specialized in many areas which no single individual could carry out alone. This is clearly seen in the work at research centers like CERN, where hundreds of scientists, organized into various domains, conduct experiments together. Another example is the publication of the first visual representation of a black hole by a telescope; the paper names 106 research institutions involved (Akiyama et al., 2019). Such cases exemplify the *synchronous* dependence of scientists on one another. There is also a *diachronic* dependence on individuals who have not only never met but who can also be separated by vast historical distances: data that was collected a long time ago; mathematical models that were formulated centuries ago; programs, programming languages, software libraries that generations of developers have contributed to exemplify this. These brief remarks indicate how far scientific practice has moved from the notion Hume wanted to assert: an individual mind having the experience of a thing. What is referred to here as 'experience' is mediated in various ways through the achievements of others, on which each person who experiences the black hole is dependent upon.

This argument might not seem entirely convincing. It could be countered that, in the end, in Hume's sense, one person will have had an experience—their own experience. A scientist observes an experiment or follows in a simulation the formation of a tornado. Research is not possible without such 'experiences'. Yet, even if we neglect that the experiment or computer simulation are based on necessary technical prerequisites, which many others have prepared, even the experience itself remains dependent on the achievements of others: the experience remains dependent on the theories that guide the observation of the object (Fleck, 1935, 115–121; Hansson, 1958, ch. 1; Sellars, 1963; for the question of what this implies for the justification of scientific statements, see, for instance, Adam, 2002); but also on the way explanations are given, or assumptions that have entered into the model, the type of visualization (Daston & Galison, 2007; Richter, 2013).

Husserl even invoked such an idea for the history of mathematics in his *Crisis*-text. There could be no scientific progress if every scientist had to repeat the discoveries made thus far, on which their own work is based. Beyond a certain progress in mathematics, a researcher's life would then only offer sufficient time for such repetitions (Husserl, 1992 [1938], §9).

2.1 Dependence on Others Without Trust?

Acknowledging the fundamental dependence on others within the scientific community, an aspect deeply entrenched in the philosophy of testimony, does not inherently imply the incorporation of trust into science. It is conceivable that one may and one must depend on others without *trusting* them. This notion could be further explored through the following argument: A scientist relying on the work of peers does not operate on blind trust or trust at all; instead, she seeks substantiating evidence that such reliance is warranted. Addressing this contention, a scientist would meticulously evaluate dependencies rather than accepting them uncritically. Proceeding only after scrutiny affirms that reliance is justifiable. Further, it illustrates that while scientists may depend on their colleagues, they do so selectively, based on evidence and not trust. This approach underscores the nuanced decision-making process regarding on whom they choose to rely upon.

Consider, for instance, a scientist preparing to simulate a tornado. This individual must ascertain the reliability of the numerical solver, assess the integrity of the data to be used, and validate the model of the phenomenon in question. Similarly, when aiming to date a human bone, a scientist scrutinizes the chosen method for age determination. In neither case is the approach haphazard or unconsidered. Particularly in the former scenario, the availability of multiple datasets and solvers mandates a thorough critique. It seems unlikely that a conscientious scientist would employ any method without a preliminary confirmation of its dependability.

However, the act of verifying and evaluating a method merits further examination. In simulation, one might envision a scientist benchmarking the numerical solver against established analytical solutions. Tests could extend to compilers, assessing whether the manufacturer's compiler outperforms alternatives in speed or accuracy. Yet, we soon encounter inherent limitations. Successful approximation in test cases by the numerical solver does not guarantee similar success in untested scenarios. More to the point though, the provenance of the so-called gold standards themselves is often external. Although a significant portion can be independently verified—a practice both possible and sometimes necessary—much cannot be fully validated without depending on others. For example, the replication of data collection is frequently impractical, and the benchmarks for tests are usually sourced from peers. Climate models are typically products of communal effort, reflecting the input of countless predecessors.

This notion is equally applicable to the scientist employing radiocarbon dating.[3] The method relies on an array of discoveries from physics, such as the disparate half-lives of isotopes C12 and C14, alongside measurement technologies like the Geiger counter, calibration techniques such as dendrochronology, and metabolic theories concerning the uptake and post-mortem behavior of isotopes. It is virtually unfeasible for an individual to independently replicate these foundational discoveries. Moreover, it begs the question of the scope of such replication: To what extent should the development of a Geiger counter presuppose or incorporate external contributions, be it the metal, the power source, the loudspeaker, or the inert gas? This illustration highlights the stark limitations of epistemic independence.

Once more, the objection might be posited that an undue burden is being imposed—specifically, the conflation of reliance on others with trust in them is unwarranted. Must scientists possess an understanding of the reasons why a method is applied in a particular manner, or of the operational intricacies of the technology they employ?

2.2 Technical Reliability Instead of Trust?

Scientists are not required to replicate every discovery and development made by their predecessors. Instead, they benefit from accessing these advancements in the form of *technology*. Colleagues across various fields gather *data*, establish *databases*, create *models*, construct *measurement instruments*, invent clever *simplifications*, develop *algorithms*, write programming *code*, and design *visualization methods*, such as the CAVE system. This epistemic division of labor is facilitated by the transmission of results through technological means, which notably alleviates the burden of acquiring foundational knowledge. This alleviation encompasses understanding both within the context of discoveries and the rationale behind them. Utilizing technology does not necessitate knowledge of its origins or operational principles; what is essential is knowing how to employ it effectively. This perspective on technology was foundational to Husserl's exploration of the technicization in the history of the mathematical sciences (Husserl, 1992 [1938], §9; see also Ihde & Kaminski, 2020; Kaminski, 2024). For example, the Pythagorean theorem can be applied technically without knowledge of its discovery, the identity of its discoverer, or the proof behind it. Husserl's notion for technology, though not exclusive to mathematics, is aptly applied to it as it embodies a broader characterization of technology: it liberates us from the necessity of understanding the reasons behind discoveries and justifications. The critical knowledge required is the methodology of application, applicable equally to technologies ranging from automobiles and computers to mathematical

[3] This example is taken from Carrier (2008: 70–73).

theorems. This detachment from foundational knowledge not only facilitates technological but also scientific progress, allowing scientists to leverage the work of others without delving into its origins or justifications *to some degree*.[4]

In essence, the acquisition of such foundational knowledge is neither possible nor necessary. Nevertheless, the efficacy of technology remains evaluable. It is not imperative to comprehend the mechanisms through which technology achieves specific outcomes (these mechanisms may remain the secret of the Black Box). What is crucial is the ability to assess the desirability, intentionality, and suitability of the results, as well as the technology's reliability in producing beneficial effects. This leads to a nuanced, yet prevalent concept of trust defined as reliability. Under this interpretation, reliability is not antithetical to trust but embodies rather its core, and this understanding seems to be particularly apt to the scientific self-image. This conceptualization views the issue of trust through an epistemic lens. Thus, the question arises: Can reliance on reliability substitute for trust? Or, rephrased: If trust is equated with reliability, does this not present a model of trust that is inherently aligned with the ethos of scientific inquiry?

3 Epistemic Theories of Trust

By the late 1980s, research on trust, previously a marginal topic, began to accelerate rapidly. Since then, two paradigms[5] of trust have been established: *Epistemic* approaches that understand trust as a problem of knowledge, and *normative* approaches that primarily see it as a relationship of mutual recognition between two individuals. Theories conceptualizing trust in terms of reliability represent a significant form within the epistemic paradigm.

The term reliability was introduced into the philosophy of trust by Annette Baier as this field began to develop. In her seminal essay "Trust and Antitrust", Baier posits a polar difference between trust as a specific form of reliability and general reliability. Her distinction is based on the type of motives attributed to others. These different motives present various reasons for expectations that lead either to trust or to (general) reliability. One might *rely* on the recognized habits of another person, their

[4] The difficulty lies in knowing when one can dispense with this knowledge and when one cannot. Especially in scientific contexts, justificatory knowledge and practical knowledge can significantly overlap.

[5] The term paradigm has been used somewhat vaguely for some time, typically without considering Kuhn's reflections. When I refer to a paradigm here, I specifically wish to highlight a point that plays a central role in Kuhn's work: The emergence of a new paradigm is often attributed to the current paradigm raising a series of problems and appearing unconvincing (especially to a younger generation of researchers) in its ability to solve these issues. See, for example, Kuhn, particularly Chap. 12 et passim.

fear of punishment, or their self-interest. *Trust*, however, is based on the presumed benevolence of the other person towards the potential trustor (Baier, 1986: 234).[6]

Most subsequent approaches refer to these considerations, whereby the difference between trust and reliability itself is usually only maintained by normative theories. The latter claim to deal with actual trust as opposed to mere reliance.[7] In contrast, epistemic approaches often simply speak of trust while conceptualizing reliability. Therefore, from the perspective of normative approaches, which present very good reasons for this, they do not address trust at all. Naturally, there are numerous variants within each paradigm, but the fundamental differences are surprisingly clear and distinct. The difference between epistemic and normative theories particularly concerns the nature of the reasons (in Baier's terms: motives) on which trust is based:

> Epistemic approaches assume that the problem of trust is a problem of knowledge. Consequently, it can be resolved through the right kind of knowledge. What is known serves as the epistemic ground for trust (or distrust).

This framing of the problem predates the current surge in trust research. In the mid-18th century, David Hume suggested that the question of trust in others must be decided by no other method than that used to investigate nature: induction, based on experience (Hume, 2007a, 2007b [1748]: 80 f.). His basic idea was that we make experiences about how often the reports of others coincide with reality. We can then form track records that indicate, based on past experiences, how reliable their attestations are. This general method can be applied to individuals, groups, or types of speech situations. The same can be done with forecasts or promises.[8]

$$\text{Reliability} = \frac{\text{Number of True Assertions}}{\text{Number of Assertions}}$$

Track Records: One way to operationalize Hume's approach

---

[6] Baier writes: "What is the difference between trusting others and merely relying on them? It seems to be reliance on their good will toward one, as distinct from their dependable habits, or only on their dependably exhibited fear, anger, or other motives compatible with ill will toward one, or on motives not directed on one at all. [...] Trust is often mixed with other species of reliance on persons. Trust which is reliance on another's good will, perhaps minimal good will, contrasts with the forms of reliance on others' reactions and attitudes" (Baier, 1986: 234).

[7] Illustrative of this are (Hertzberg, 1988, 2010; Lahno, 2002). The distinction between trust in a narrower normative sense and trust as reliability also re-appears in other terminologies, such as the difference between affective and predicative trust (Faulkner, 2007), or between the Assurance and Evidential view of trust (Moran, 2006).

[8] Hume also discusses trust in another section, specifically in the *Treatise of Human Nature*, III.2.5. There, he explores the question of how the obligation of promises arises, which, at least for Hume, mirrors the question on what basis it can be trusted that a promise will be kept. His example discussed there, the promise of mutual help during harvest, is revisited by James Coleman, the founder of sociological rational choice and game theory—though without mentioning Hume (Coleman, 1990: 93–94). Moreover, Hume's response, that the obligation is based on the self-interest of the individuals or society if cooperation is to continue, forms the core idea not only in Coleman's approach but also among other rational choice theorists, particularly Russel Hardin and his concept of trust as "encapsulated interest" (Hardin, 2002).

In his *Enquiry*, Hume also hints at a second method: We can select *indicators* that act as a specific type of sign, namely, signs that connect *observables* with the value of trustworthiness that interests us but is not directly observable. For instance, we might choose as the observable the way someone speaks or their facial expression as an indicator to assess their credibility (Hume, 2007a, 2007b [1748]: 81). Which track records (of individuals, specific personality traits like age, groups, or types) and which indicators to choose are ultimately *empirical* questions and cannot be decided a priori.[9]

$$\text{Reliability (of an indicator)} = \frac{\text{Instances in which a Feature is Observed given True Assertions}}{\text{Number of True Assertions}}$$

Indicators: A second way to operationalize Hume's approach

Hume's approach represents the prototype of a reliability theory of trust. Trust as reliability is a central form of epistemic approaches. Its characteristics include:

1. Epistemic theories of trust make *predictions* about others' future behavior.
2. These predictions are presented in the form of *probabilities*. A track record can be formed based on the ratio of true reports or fulfilled promises to the total number of given reports or promises. An acquaintance who has made promises 15 times but only kept them 3 times, therefore, has a poor track record, meaning low trustworthiness. If the track record is understood as a probability, it would be 1/5. Similarly, indicators can have a probabilistic basis: How likely is it that the occurrence of an observable indicates the presence of the indirectly inferred value (trustworthiness)?
3. *Trust* thus represents a *positive expectation value*, referring to a high probability based on past experiences. Mistrust, conversely, represents a negative expectation value.
4. *Reliability* thus appears twice within this framework; this becomes especially clear when considering competing indicators: Firstly, as the reliability of the person in question, secondly as the reliability of the indicator or track record to evaluate the person's reliability (reliably).
5. This shifts the *subject of trust*. Trust is primarily based on *one's own epistemic abilities*: One first trusts oneself to correctly assess the situation. This results from the dependency relationship between first and second-order reliability mentioned in point 4. Initially, a person relies on the reliability method and their own epistemic abilities to apply it correctly. Trust or mistrust in the other person is then merely a conclusion. For this reason, epistemic approaches can also be applied to non-personal objects of trust without major changes. The object of trust, in terms of its reliability, can also be an instrument, a system, an AI model.

---

[9] Empirical studies can determine which indicators are reliable and to what extent. Hume's program has been implemented empirically in the history of psychology. Around 1900, track records were initially established to examine the trustworthiness of witnesses of different ages; particularly in the second half of the 20th century, indicators were tested for how reliably they could predict the credibility of witnesses (Kaminski, 2019).

6. *Epistemic reasons* for trust are based on knowledge that can be empirically obtained in Hume's sense.[10]

Later developments in rational choice and game theory represent extensions of the epistemic approach. With Coleman, the magnitude of potential gain and loss is considered alongside the probability of gain or loss (which, even for him and subsequent theorists, is primarily interpreted in terms of trustworthiness). Thus, the question of whether it is rational to trust is interpreted as a special case of a risky decision under uncertainty.[11] However, there are no fundamental changes regarding the six characteristics mentioned; rather, expansions and refinements are made within its theoretical space. Thus, epistemic reasons and expectations within the framework of rational choice and game theory can now also be formed through analyses of preferences and agents' action options.

4 The Practical Problem of Epistemic Approaches

At first glance, the *epistemic* concept of trust seems ideally suited for application in the sciences. If this were true, then scientists would create track records regarding the accuracy of other scientists' reports, forecasts, or explanations. Or they might assess trustworthiness based on indicators (such as reputation or report style). Certainly, this plays a big role in assessing the reliability of measurement devices and models. The reliability of AI applications that classify objects in image data is assessed in a similar manner.[12]

$$\text{Reliability} = \frac{\text{Number of True Classifications}}{\text{Number of Classifications}}$$

The Humean approach in AI

The same applies to vaccines or Covid-19 tests for example (Fig. 1).[13]

In journalism, it is not uncommon for track records to be made of politicians. A similar approach is sometimes taken with the statements of scientists. The newspaper Ubermedien published an article about a German virologist, Hendrik Streeck, who became known during the COVID-19 pandemic. The article documents 12 statements

[10] However, it is not the only way; an alternative is presented by Coady's logical, namely transcendental-pragmatic argument for the generic trustworthiness of others, which can be understood a priori solely under the condition of the existence of a language and language community. See Coady (1992, especially Chaps. 8–9).

[11] Coleman thus compares the problem of trust to the question of when it is rational to place a bet. The betting model provides the analytical structure for the formula that should guide rational decisions for trust (Coleman, 1990: 99).

[12] In AI research, the term (robust) accuracy is sometimes used for this purpose.

[13] In this context as in AI, reliability is called accuracy.

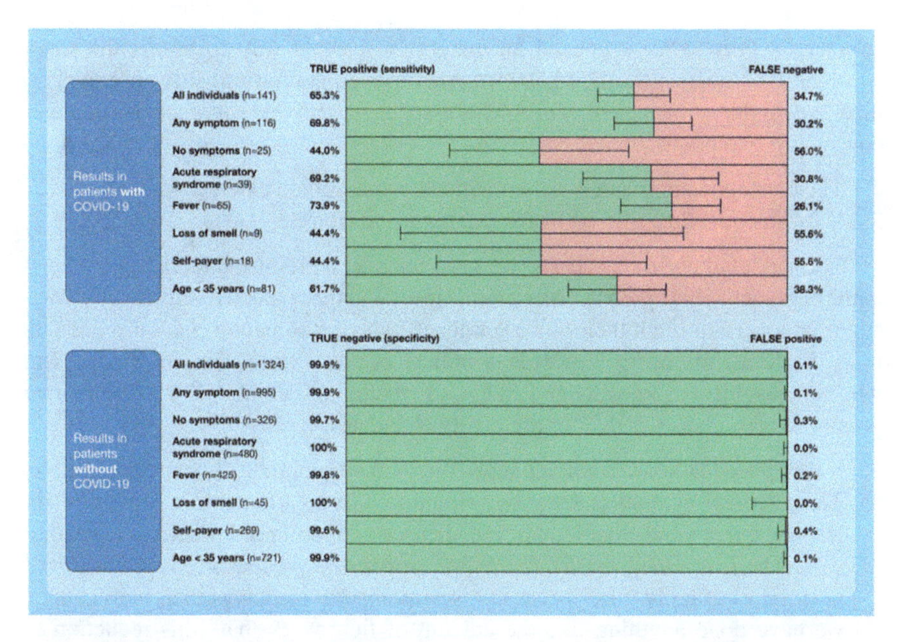

Fig. 1 The graphic shows the reliability of a specific test for Covid-19 (Jegerlehner et al., 2021: 121)

Fig. 2 Journalistic track record of a scientist about his twelf covid-19 statements

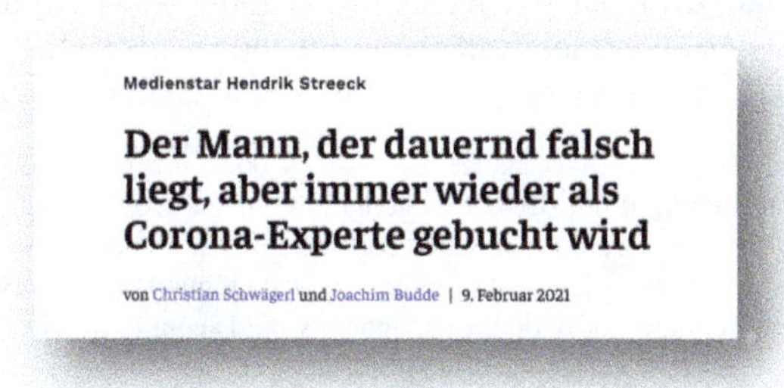
Medienstar Hendrik Streeck

Der Mann, der dauernd falsch liegt, aber immer wieder als Corona-Experte gebucht wird

von Christian Schwägerl und Joachim Budde | 9. Februar 2021

(forecasts and the like) made by Streeck, for example, about the number of deaths and whether there would be a second wave, to demonstrate that in these 12 cases, he was either incorrect or flip-flopped between positions (Fig. 2).

Thus, the article aims to show that Streeck has a poor track record regarding COVID-19, and then questions why he is still frequently heard in the media (why he is still considered trustworthy).[14]

Thus, it seems that Humean track-records are a standard tool for measuring the reliability of scientists, measurement tools and models. Therefore, the epistemic

---

[14] "Der Mann, der dauernd falsch liegt, aber immer wieder als Corona-Experte gebucht wird" (ÜberMedien, 9.2.2021, https://uebermedien.de/57343/hendrik-streeck-der-mann-der-dauernd-falsch-liegt-aber-immer-wieder-als-corona-experte-gebucht-wird, last time seen 17.7.21).

approach seems to have opened an extremely feasible way to assess the trustworthiness of science through knowledge. However, contrary to first appearances, there are fundamental *practical* problems associated with this approach, particularly concerning two points:

1. The epistemic approach is only practicable to a limited extent.
2. If its practicability is to be increased, it ends in a circle.

To understand these objections, we must uncover a premise underlying the epistemic approach. This approach is epistemic insofar as the respective subject of knowledge can assert, through their *own* experience, whether someone else's statement is true. The formation of track records or indicators presupposes this knowledge. But who is this subject of knowledge that knows whether a statement is true? Here, two possibilities exist, as Coady[15] impressively elucidates:

1. If it implies that it's an individual person, then this approach is scarcely feasible. This becomes evident as testimony would lose its significant role. If we already knew whether what others say is true, then we wouldn't need to question whether we can believe what they say. In Coady's words: "Now I characterized this sort of position as 'plainly false' because it seems absurd to suggest that, individually, we have done anything like the amount of field-work that" this reduction to individual experience would necessitate (Coady, 1992: 82).

This might seem like an objection to the *general* feasibility of epistemic trust from an individual perspective, but what about a more limited, local application? Isn't an expert in a completely different situation to assess the accuracy of scientific processes and results in their area of expertise compared to a layperson? Coady correctly notes:

> [M]any of us have never seen a baby born, nor have most of us examined the circulation of the blood nor the actual geography of the world nor any fair sample of the laws of the land, nor have we made the observations that lie behind our knowledge that the lights in the sky are heavenly bodies immensely distant nor a vast number of other observations that [the reduction to individual experience] would seem to require. (Coady, 1992: 82)

But does this objection also apply to an astronomer evaluating the work of another astronomer? Or a mathematician reviewing the proof of another mathematician? Clearly, they are in a different situation compared to laypeople in their field of expertise. For them, certain errors are evident (peculiarities in the results that raise doubts, significant methodological issues in a study) that remain hidden to laypeople. Nonetheless, a non-individual residuum remains. I will discuss this using the example of mathematics, as this case likely gives rise to the most persistent expectations that the epistemic program can be implemented individually. It should be noted that this example naturally extends the concept of experience in an unusual way, but in my view, this does not pose a problem: here, experience would mean having independently understood a mathematical proof.

---

[15] Coady (1992: Ch. 4, especially pp. 80–82) shows this with regard to an ambiguity in Hume's text, which on the one hand speaks of experience in the sense of the individual person, such as David Hume's experiences, and on the other hand of experience in the collective sense of humanity.

Three reasons significantly limit the individual feasibility of the epistemic strategy, even though the limit remains flexible:

a. A sample, such as the replication and verification of *one* mathematician's mathematical proof, does not yet constitute a track record. For that, numerous replications and verifications would be necessary.

If our mathematician were to verify numerous proofs of one or more other mathematicians to build such a track record, additional problems that limit the feasibility of this approach would emerge:

b. Reconstructing numerous proofs of another mathematician would provide a better epistemic basis for evaluating this one mathematician. A track record could then be established for him. However, this would still be a very limited applicability of this epistemic approach. One would not gain anything about the work of other mathematicians.
c. If a mathematician were to undertake the immense task of creating a comprehensive archive of other mathematicians' track records by establishing *numerous individual* track records, the work of this mathematician would consist of little more than replication.
d. Yet, even if we were to imagine, counterfactually, a world in which a person undertakes such work, they would remain dependent on the achievements of others. This is partly because the premises of proofs must also be independently developed. According to the individual epistemic approach, the person must not show any dependence on others; that also means the form of what a proof is, on which other axioms or theorems the respective proof is based on, must not be taken over from others.

This demonstrates how we encounter a fundamental limit to the feasibility of such an approach. This was precisely Husserl's point, why for him the history of mathematics is a history of mechanization. Husserl, driven by the idea of a complete foundation of philosophy, viewed that any further progress in mathematics would be confronted with the limit of a lifetime (and the individual's capability regarding memory performance, overview, etc.).

The considerations above would more clearly delineate the feasibility in the case of astronomy, where the manifold dependence on others regarding the used data, measuring instruments, models, etc., would become evident. In light of this, an additional argument that Coady puts forth concerning the fundamental dependence on others at the core of what is considered personal experience could be supplemented. Coady discusses the case where an individual sees the Queen during a state visit in Melbourne. The individual could claim to have had their own experience of the Queen's presence in Melbourne. However, this personal experience relies on the achievements of others, which made this experience, namely seeing the Queen, possible. How does the individual know, for example, that it is the Queen? From photos, film recordings, and reports given to them by others. How do they know it is a state visit? From newspaper reports, descriptions of how these are conducted, etc. This point could be applied to the dependence on others within the framework of scientific experience; it is usually referred to as theory-ladenness, which has already been a topic of discussion. In the context of the current argument, it concerns the dependence on others, which thus intrudes into one's own individual experience.

Given the objections, the path of establishing individual *track records* may be discarded. Yet, do *indicators* not offer a more promising alternative? Within their framework, one might limit oneself to samples that are used indicatively instead of conducting extensive tests: As signs of the abilities and virtues of the mathematician or astronomer: how careful, precise, knowledgeable he or she is. However, this also does not lead to a reduction to individual experience. For the reliability of the indicator would still need to be individually verified; for this, the individual scientist must independently know what, for example, constitutes a good study, a good proof, a good explanation. But this is learned socially. Yet, that the formation of an idea of a good proof is socially learned does not mean that the later application of this idea cannot provide autonomous evidence. An analogy: A physician may learn in social processes what a certain bacterium looks like when viewed through a microscope. Once learned, he can independently determine whether a sample shows this bacterium. He may use certain indications for this; but then, isn't his evidence independent of others? To a certain extent, this is not the case: the morphological indications may be identified independently (spherical, rod shape, etc.) after training, but that these indications are indications for this or that bacterium is not part of this observation. For this, other socially developed, optimized, and trained indications (staining tests, biochemical tests) are usually required.

Although it is certain that the epistemic trust strategy, understood *individually*, is only very limited in its feasibility, no one would sensibly dispute that evaluating reliability plays a significant role in the sciences. As said before, AI models are tested for their reliability in correctly classifying objects in image data—for many years, for example, improvements in reliability in object recognition have been pursued using the ImageNet database. Furthermore, the reliability of diagnostic performances is also compared between doctors and AI applications (Liu et al., 2019).[16]

How, then, does it reconcile that the epistemic trust strategy, understood individually, is only limitedly feasible when there are numerous approaches that formally or informally evaluate the reliability of scientists and scientific models or instruments? These evaluations are possible, but only as *collective* achievements. To illustrate using the comparison of how reliable AI models are in diagnosing compared to doctors: Many individuals are involved in conducting this comparison. This is already shown by the data collection. Correct image data must be labeled with a ground truth in

---

[16] Beyond purely reliabilistic approaches, epistemic indicators such as the affiliation of scientists play an informal yet significant role in assessing their potential trustworthiness. Measures like the H-index, while not allowing direct assertions about their trustworthiness (though it may be understood to suggest it), the RipetaScore aims to enable a direct statement about trustworthiness: "Many who use these metrics make assumptions that higher citation counts or more public attention must indicate more reliable, better quality science. While current metrics offer valuable insight into scientific publications, they are an inadequate proxy for measuring the quality, transparency, and trustworthiness of published research. Three essential elements to establishing trust in a work include: trust in the paper, trust in the author, and trust in the data" (Sumner et al., 2022: 1).

order to carry out this comparison. This 'ground truth' will mostly need to be determined by individuals beyond those conducting the test (experienced doctors or lab data based on different testing methods than the image data). No individual conducts the measurement of reliability alone, no individual can usually do this on their own.

This situation introduces the aforementioned *circle*: *We trust others in order to evaluate the trustworthiness of others.* Trust is placed in the reliability of an AI model or of doctors because it is believed that the evaluation of their reliability has been conducted in a trustworthy manner (by the authors of the study and by those who have reviewed the study; as well as by those who have determined the ground truth through lab tests). Naturally, a review can again be conducted to verify whether this trust in the evaluation, that the trust in AI models or doctors is justified, can take place. This review, however, will typically be a collective process itself, where the assessment of the trustworthiness of those contributing to this review plays a role.

This circle is therefore inevitable in most cases. It primarily poses a problem in light of the claim of the epistemic trust strategy: This strategy is aimed at ensuring that trust in the sciences occurs only in a special kind, namely as epistemic trust, which does not prevent autonomous assessment by the individual scientist. While scientists are highly dependent on others, they can manage this dependency in an autonomous epistemic way—so goes the fundamental idea and the seductive suggestion of the epistemic strategy. One may not know everything that colleagues know, but one can independently know whether the colleagues are worthy of our trust or not. This exact autonomy is undermined by the circular trust relationships. A can epistemically evaluate the trustworthiness of B, but only by trusting C and D in the process, without typically being able to place fully autonomous epistemic trust (Fig. 3).

Let us consider the track record of the virologist Streeck mentioned above. Determining whether his statements (forecasts, recommendations) were true or accurate presupposes this very knowledge: whether they have been true or not. And this knowledge is typically acquired by placing our trust in others. Is his claim about how many people in a specific city (Heinsberg in Germany) died from Covid-19 accurate? Was

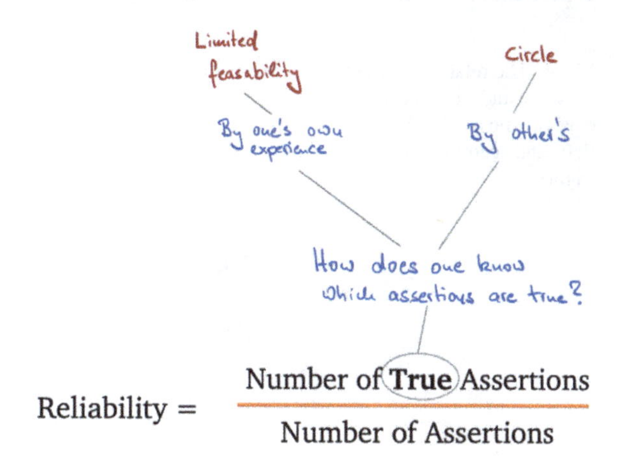

Fig. 3 The two dead ends of Hume's epistemic strategy

there a second wave? It might seem trivial to judge the latter (there was!). But can we know this *without* others?

The philosopher Lahno, in his critique of Rational Choice Theory and epistemic approaches, has developed an analogous argument. In epistemic approaches, trust is to be based on evidence (for or against the trustworthiness of another). Thus, evidence forms the foundation on which trust or distrust is concluded.

Lahno's point is this: In numerous cases, this evidence is provided to us *by others*. It is these others whom we choose to trust or distrust in the first place. Therefore, evidence cannot be the *independent foundation* of our trust or distrust. For evidence itself is deemed trustworthy or untrustworthy *based on our trust* in others who present this evidence to us (Lahno, 2002: 132) (Fig. 4).

It is possible that the arguments used so far are preaching to the converted. Yet, I am skeptical as to whether we all know what we were converted to: Trust enters the domain of science. And it cannot be reduced to epistemic trust. The arguments have shown that this is impossible on an individual level to justify scientific results in general. However, if one were to respond, 'Then let us place our trust collectively, but in an epistemic sense,' this does not seem a clear answer to me. For what does it mean to trust epistemically in a collective sense? It does not mean that *I* can know epistemically on my own when this trust is justified—and in this sense, a fundamental, although not universally definable, boundary line is indicated as to how one can trust purely epistemically.

It is sensible to add two caveats.

The aforementioned does not imply that A cannot have epistemic insights into the trustworthiness and reliability of B, C, and D, which are autonomous in the sense of an individual strategy, namely through sampling. However, these do not constitute (reliable) track records or developed indicators. The possibilities for this, as demonstrated, are practically very limited. In a small environment, there may be many opportunities to obtain such samples or to build up track records. But it cannot be a *universally* applicable strategy, it remains limited to the immediate environment. Where this boundary lies cannot be generally stated.

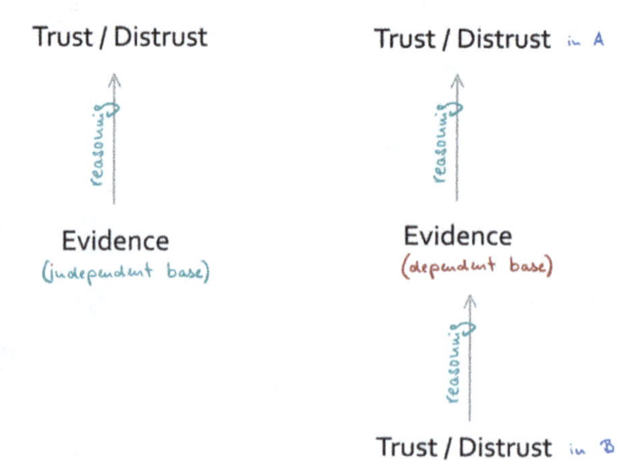

Fig. 4 The relation between evidence and trust according to epistemic approaches (left) and according to Lahno (right)

It also does not mean that A is blind in terms of evaluating B. A can use the 'eyes' of C and D to see how trustworthy B is. And perhaps A can also epistemically assess the reliability of C and D, but given the limited individual feasibility of the epistemic strategy, it remains the case that it is not a universally applicable strategy that can solve the problem of trust.

5 Normative Theories of Trust

Before I discuss the normative approaches that have been developed, I must preface this with a remark on understanding what normative approaches within the theory of trust entail: The formulation of scientific norms or the establishment of institutional sanctions do not constitute contributions to a normative theory of trust. The reason is that such norms are intended to change motivations and interests or assist in developing a habitus that enhances the *reliability* of good scientific practice. In this context, norms are consequently seen as epistemic reasons for fulfilling expectations of science. Therefore, they belong to the field of epistemic approaches.

Normative theories of trust have primarily emerged as a response to principal *conceptual* problems of epistemic approaches. These conceptual problems are closely related, such that from the first problem, the subsequent ones arise.

1. *Reasons and Trust*: There are epistemic reasons that increase the certainty of expectations, but they are incompatible with trust. Such epistemic reasons cannot, therefore, be the foundation of trust. Epistemic theories do not exclude these reasons, nor are they capable of doing so, because they would need to develop a theory of the right kind of reasons for trust, which would result in abandoning the purely epistemic approach (see Kaminski, 2017). Examples of epistemic reasons that increase the certainty of expectations include evidence arising from surveillance. Surveillance and trust are mutually exclusive. This does not mean that, for example, laboratories cannot be monitored; it simply means that they are not trusted if they are. Similarly, the evidence one has that others fear the deterrents one is ready to use against them is a good epistemic reason to predict their behavior, but does not provide a reason to trust them. Threats of punishment (like control in general) are not issued out of trust, but rather due to a lack of trust.
2. *Trust and Forecasts*: Forecasts are based on knowledge. The better the knowledge regarding the respective process, the more reliably future events can be predicted. Imagine someone possessing vast knowledge that would allow them to predict processes with unfailing reliability. Such a person would hardly claim to trust the object of their forecasts. Their cognitive expectation would not be a trust expectation.
3. *Trust and Relationship*: Those who perceive themselves to be in a trusting relationship are deeply disappointed when it turns out that the other person does not believe *their* words, does not believe *them* but instead believes evidence that is

independent of their own words. The consequence is the same: A believes that p. But the relationship is different: A does not believe B that p. Thus, it is not the evidence independent of B that trust is based on, but rather it is bound to the form of relationship that A and B have (Moran, 2006).

4. *Trust and Disappointment*: Epistemic theories are incapable of articulating the specific nature of disappointment in trust relationships. Due to their theoretical structure, they can only interpret disappointments in trust as cognitive surprises or, if they are probability-based, as unavoidable possibilities inherent to the game, akin to the improbable green zero in roulette. However, disappointed trust results not merely in a cognitive surprise or a sober disillusionment. When trust is betrayed, it is hurtful, it causes pain, it outrages (Faulkner, 2007: 881; Lahno, 2002: 139, 173 f). It leads to anger, contempt, moral indignation. Even if proponents of rational choice theory or other adherents of epistemic theories of trust might agree, they are nonetheless unable to elaborate this *within* their theory. Their theoretical foundation does not provide them with the means to do so.

To address these fundamental conceptual issues, normative approaches make two changes: on the one hand, they emphasize the role of the form of the relationship between two individuals who trust each other; on the other hand, they attempt to create a space for reasons beyond the epistemic realm. Notable examples of this approach include the works of Moran (2006) and Ross (1986).

While normative theory does resolve the conceptual problems mentioned, its solution introduces its own conceptual and practical issues. Both problems stem from the way normative approaches attempt to solve the relationship between trust and reasons. Since epistemic reasons lead to the problems outlined above, Moran, for example, tries to uncover a realm of reasons beyond evidence. In his view, this consists of the 'assurance' one person gives to another. Although this assurance can itself be viewed evidentially, doing so does not capture it in the manner intended by the speaker. In other words, there is no harmony between the two. Such harmony is only present when the assurance is understood in the terms intended by the one who offers it: as an intended, voluntary assumption of responsibility (this also highlights how normative concepts replace epistemic ones).[17]

This change aims, among other things, to solve the problem that the speaker can be hurt when believed for their evidence but not for their words. The issue with this solution attempt is that the speaker could also be hurt by Moran's approach: Now the listener cannot provide reasons for believing her. This might not be immediately apparent, but it becomes clear when we consider the following: Moran no longer allows any epistemic reasons.

This is exemplified in the statement by Kristin Urquiza, whose father Mark Urquiza died from the Covid-19 virus in June. Her father believed the words of the then US President Donald Trump, as she explains, when he claimed in the spring of 2020 that the coronavirus posed no further danger and that social restrictions should be lifted. He subsequently went to a karaoke bar and died of the virus in the

[17] It is not an assurance if it is not intended and addressed to be told to someone (Moran, 2006: 277).

following weeks. His daughter summarized this in the words: 'His only preexisting condition was trusting Donald Trump'.[18]

Both a trustworthy friend and colleague as well as a deceiver can give 'assurance'. How, then, can one practically distinguish between trustworthy and untrustworthy individuals if no epistemic reasons can be named? Both offer their assurance. Trust is therefore necessarily blind, it threatens to conflate conceptually with naivety—and ultimately, no good reasons can be given for why I trust my long-time friend with whom I have been 'through thick and thin'. Such reasons would appear epistemic, leading to a track record and therefore missing the essence of trust. In this respect, Moran's approach, contrary to his intention, also hurts the speaker –just in a different way than epistemic theories.

6 Trust and Trustworthiness as Virtues

In light of the mentioned conceptual and practical problems, it was natural to develop approaches that understand the epistemic and normative dimensions of trust in their unity. The necessity of this unity was demonstrated, among other things, by the fact that betrayed trust can provoke both epistemic learning to trust more appropriately and moral indignation. Or also, that trust requires epistemic reasons but cannot be reduced to epistemic reasons alone. In other words, it is about recognizing and acknowledging as unified acts in relation to others.

The advantage offered by virtue theory in this context lies in avoiding a separation from the outset. Consider the conditions under which a person A can attribute a virtue to person B:

1. For A, something about B's behavior must *manifest* in a way that matches the behavioral descriptions of this virtue, for example, being honest, friendly, just, or indeed trustworthy. A can then refer to B's *experienced* behaviors.
2. However, A must also have reasons to believe that B's behavior is not merely coincidental or strategic but falls under the description characteristic of such a virtue. A must have reasons to think that B is guided by this virtue in their behavior, motivated by it. For instance, someone who shows friendliness only while the supervisor is in the room, losing all friendliness once they leave, displays merely a strategic orientation. They are not motivated by friendliness as a virtue but by career advancement, for which friendliness is a means from the perspective of the person behaving in this way. In other words, B must recognize friendliness as a virtue to be oriented and motivated by it, allowing A to reasonably attribute this virtue.
3. Now, A must also recognize the particular behavior as virtuous. This is illustrated by a problem that can be called the imposter or trickster problem. If A does not recognize honesty or trustworthiness *as virtues* themselves, this opens the

[18] Cf. https://edition.cnn.com/2020/08/17/politics/kristin-urquiza-democratic-national-convention-coronavirus-father-trump/index.html (last visit: 12.03.2024).

possibility that A understands the honesty or trustworthiness of B merely as a good mean for making predictions about B's behavior. Honesty or trustworthiness then are good predictors for A. And a trickster, for whom these are not virtues, can exploit the honesty of another to better deceive them. If A is such a trickster, they cannot attribute honesty to B as a virtue (or only in an ironic sense).

4. Therefore, for both to understand and trust each other as trustworthy, they must recognize and acknowledge each other as trustworthy. And thus, facilitating the realization of their initially merely possible relationship.

Such a virtue theory differs from existing trust theories that reference virtue ethical considerations by taking the inseparability of recognition and acknowledgment, of beliefs and values, as it starting point (see Kaminski, 2017, 2021). In my view, it resolves the *conceptual* problems discussed above of the one-sided epistemic or normative approaches. It no longer poses a problem to understand how individuals can learn and react with moral injury when their trust is betrayed. It also allows for the exclusion of problems like that of control (as an inappropriate epistemic reason *for* trust).

However, this virtue-theoretical approach comes with its own *practical* problem. For virtues to be recognized and acknowledged, they must be practiced in practical enactments. Yet, it is questionable whether a person can recognize if another is virtuous in a specific sense under the following conditions: (1) Interactions are only occasional, often even unilateral. One might read an article or watch an interview. (2) The life world of the involved parties is highly functionally differentiated. For example, how can a layperson determine whether a scientist is trustworthy when the signs by which this can be determined are tied to highly differentiated practice areas? The core idea of the objection suggests that what indicates trustworthy scientific practice may differ among radiology, Renaissance art history, sociology of friendship, or mathematical topology. How, for example, can a layperson recognize whether a doctor discussing their radiation therapy for cancer treatment is trustworthy in the sense of virtue theory?

Here, a particular aspect of the virtue theory of trustworthiness and trust may come to our aid, at least *partially*. Trustworthiness and trust, as virtues, are directed towards other virtues. When we attribute trustworthiness to someone, it pertains to other virtues that may be relevant in the given situation—such as being honest, brave, friendly, or just. In other words: trustworthiness and trust represent the unity of virtues because we cannot comprehend them as fulfilled if other virtues are not met. We cannot attribute the virtue of trustworthiness to a person if we do not consider them to be honest or just, especially if this is relevant in the current situation for determining whether they are trustworthy or not. Thus, a partial solution emerges to the question of how the virtuousness of others can be assessed: Do they exhibit other virtues? Do they prove to be honest or brave in the sense of free speech?

However, this seems to open a series of potential errors, especially in the scientific field. I will explain this using an example from a research project in radiation oncology in which I participated. The patients were suffering from cancer. Their situation presented them with trust issues. Were they receiving the right treatment? Were they being treated well? Not only is this an existential situation, as their lives might depend on this therapy, but also the effectiveness of the radiation therapy is not directly perceivable. Even though the treatment is loud: patients do not see, hear, feel, or smell whether the rays are hitting the tumor cells. They rely on the assessment and tests of the medical staff. But how can they judge if they are being treated well and correctly? A psycho-oncologist said in an interview: It can happen that patients look at the doctor's shoes and then become suspicious: The shoes are not clean. How can I then trust him? The situation can be understood here as signs (virtues) such as the doctor's apparent tidiness being taken for the tidiness of his treatment. Clearly, however, this is a questionable correlation.

This is why it remains a partial and fallible solution. (1) It is unclear to what extent we are capable of assessing these other virtues, and (2) it is unclear how relevant the respective virtues may be for the given situation—another example, a virologist can be trustworthy even if she is not friendly, and so forth.

Despite these limitations, virtue theory has another advantage: it takes the trust issue from the individual level to the social. We have seen, especially given the practical problems of epistemic approaches, that individual solutions are not practically feasible. The circle of trust cannot be avoided. The virtue ethical perspective has the advantage here that virtues are learned within the framework of a social practice. And scientific practice is fundamentally a social practice, as seen above. But modern societies differentiate these fields of practice, making it challenging to judge across different fields.

7 Where to Go from Here?

What emerges from the discussion of these three paradigms of trust for the issue of trust in science and trust within science? Three points seem noteworthy to me:

1. We only gain a theoretically adequate approach to trust by avoiding the abstract one-sidedness of either epistemic or normative foundations. In my view, virtue theory achieves this.
2. All three approaches lead to fundamental practical problems. This also means there is no easy fix for the problem of trust; in particular, it cannot be solved individually. However, viewing the trust problem on the social level also presents complex difficulties.
3. The existence of three practical approaches to the topic can also be seen differently, namely as interconnected but complementary strategies for dealing with the trust problem. Epistemic approaches to the reliability of scientific studies are immensely useful. Where they can be applied, they should not be neglected. However, they do not eliminate the need for trust, nor do they resolve the corresponding practical problems.

In this edition, epistemic approaches to the topic predominate. Given the subject matter, trust in *science*, this is not surprising, as we have seen at the beginning.

However, the synopsis of individual contributions reveals the aforementioned expansion. While the different approaches theoretically exclude one another, in practice they complement each other, provided one understands how epistemic approaches must build on other frameworks. This includes considering the role of individual scientists, institutions, and science communication, to briefly mention diverse aspects of the contributions here.

References

Adam, M. (2002).Theoriebeladenheit und Objektivität: Zur Rolle der Beobachtung in den Naturwissenschaften. De Gruyter.

Akiyama, K., et al. (2019). First M87 Event Horizon Telescope Results. VI. The Shadow and Mass of the Central Black Hole. In *Astrophysical Journal Letters, 875.*

Bacon, F. (2000 [1620]). *The New Organon.* Cambridge University Press.

Baier, A. (1986). Trust and Antitrust. *Ethics, 96*(2), 231–260.

Böhme, G. (1998). Trau, schau, wem! In *Die Zeit.*

Carrier, M. (2008). *Wissenschaftstheorie zur Einführung.* Junius.

Coady, C. (1992). *Testimony: A philosophical study.* Clarendon Press, Oxford University Press.

Coleman, J. S. (1990). *Foundations of social theory.* Belknap Press of Harvard University Press.

Daston, L., & Galison, P. (2007). *Objectivity.* Zone Books.

Descartes, R. (2008 [1641]). *Meditations on first philosophy: With selections from the objections and replies.* Oxford University Press.

Faulkner, P. (2007). On telling and trusting. *Mind, 116*(464), 875–902.

Fleck, L. (2021 [1935]). Entstehung und Entwicklung einer wissenschaftlichen Tatsache: Einführung in die Lehre vom Denkstil und Denkkollektiv. Suhrkamp.

Hanson, N. R. (1958). *Patterns of discovery: An inquiry into the conceptual foundations of science.* Cambridge University Press.

Hardin, R. (2002). *Trust and trustworthiness.* Russell Sage Foundation.

Hertzberg, L. (1988). On the attitude of trust. *Inquiry, 31*(3), 307–22.

Hertzberg, L. (2010). On being trusted. In A. Grøn & C. Welz (Eds.), *Trust, sociality, selfhood* (pp. 193–204). Mohr Siebeck.

Hooke, R. (1705 [1666]). A General Scheme, or Idea of the Present State of Natural Philosophy and How it Defects may be Remedied By a Methodical Proceeding in the making Experiments and Collecting Observations. In *The Posthumous Works of Robert Hooke. Herausgegeben von Richard Waller* (pp. 1–70). S. Smith and B. Walford.

Hume, D. (2007 [1739]). *A treatise of human nature: A critical edition.* Oxford University Press UK.

Hume, D. (2007 [1748]). *An enquiry concerning human understanding.* Oxford University Press.

Husserl, E. (1992 [1938]). Krisis der europäischen Wissenschaften und die transzendentale Phänomenologie: Eine Einleitung in die phänomenologische Philosophie. Bd. 8. Gesammelte Schriften. Meiner.

Ihde, D., Kaminski, A. (2020). What is postphenomenology? A controversy. In A. Friedrich, P. Gehring, C. Hubig, A. Kaminski & A. Nordmann (Eds.), *Unheimlichkeit und Autonomie. Jahrbuch Technikphilosophie 6* (pp. 261–287). Nomos.

Jegerlehner, S., Suter-Riniker, F., Jent, P., Bittel, P., & Nagler, M. (2021). Diagnostic accuracy of a SARS-CoV-2 rapid antigen test in real-life clinical settings. *International Journal of Infectious Diseases, 109,* 118–122. https://doi.org/10.1016/j.ijid.2021.07.010

Kaminski, A. (2017). Hat Vertrauen Gründe oder ist Vertrauen ein Grund? Eine dialekti-sche Tugendtheorie von Vertrauen und Vertrauenswürdigkeit. In J. Kertscher und J. Müller (Eds.),

Praxis und, zweite Natur'. Begründungsfiguren normativer Wirklichkeit in der Diskussion (pp. 121–139). Mentis.
Kaminski, A. (2019). Begriffe in Modellen. Die Modellierung von Vertrauen in Computersimulation und maschinellem Lernen im Spiegel der Theoriegeschichte von Vertrauen. In N. J. Saam, M. Resch und A. Kaminski (Eds.), *Simulieren und Entscheiden. Entscheidungsmodellierung, Modellierungsentscheidungen, Entscheidungsunterstützung* (pp. 167–192). Springer.
Kaminski, A. (2021). *Die verwickelte Einfachheit von Vertrauen – und seine spekulative Struktur* [angenommene Habilitationsschrift, Universität Marburg].
Kaminski, A. (2024). Phänomenologie. In M. Gutmann, B. Rathgeber, K. Wiegerling (Eds.), *Handbuch Technikphilosophie.* Metzler [in press].
Kant, I. (1998 [1781]). *Critique of pure reason.* Cambridge University Press.
Lagerspetz, O. (1998). *Trust: The tacit demand.* Springer, Netherlands.
Lagerspetz, O. (2001). Vertrauen als geistiges Phänomen. In M. Hartmann und C. Offe (Eds.), *Vertrauen: Die Grundlage des sozialen Zusammenhalts* (pp. 85–113). Campus.
Lahno, B. (2002). *Der Begriff des Vertrauens.* Mentis.
Leibniz, G. W. (1985 [1710]). Die Theodizee von der Güte Gottes, der Freiheit des Menschen und dem Ursprung des Übels. In Philosophische Schriften, Bd. II.1-2. WBG.
Leibniz, G. W. (1985 [1714]). Die Prinzipien der Philosophie oder die Monadologie. [„Les Principes de la Philosophie ou la Monodologie"]. In Philosophische Schriften, Bd. I (pp. 439–483). WBG.
Liu, Y., Kohlberger, T., Norouzi, M., Dahl, G. E., Smith, J. L., Mohtashamian, A., Olson, N., Peng, L. H., Hipp, J. D., & Stumpe, M. C. (2019). Artificial Intelligence-based breast cancer nodal metastasis detection: Insights into the black box for pathologists. *Archives of Pathology & Laboratory Medicine, 143*(7), 859–868. https://doi.org/10.5858/arpa.2018-0147-OA
Merton, R. K. (1973 [1938]). Science and the social order. In *The sociology of science: Theoretical and empirical investigations* (pp. 254–266). University Press.
Moran, R. (2006). Getting told and being believed. In J. Lackey & E. Sosa (Eds.), *The epistemology of testimony* (pp. 272–306). Oxford University Press.
Merton, R. K. (1973 [1942]). The normative structure of science. In *The sociology of science: Theoretical and empirical investigations* (pp. 267–278). University of Chicago Pr.
Richter, M. (2013). *Scientific visualisation: Epistemic weight and surpluses.* Lang.
Ross, A. (1986). Why do we believe what we are told? *Ratio, 1*, 69–88.
Sellars, W. (1963). Empiricism and the philosophy of mind. In *Science, perception and reality, Atascadero* (pp. 127–196). Ridgeview Pub.
Sumner, J. Q., Vitale, C. H., & McIntosh, L. D. (2022). RipetaScore: Measuring the quality, transparency, and trustworthiness of a scientific work. *Frontiers in Research Metrics and Analytics, 6.* https://www.frontiersin.org/articles/10.3389/frma.2021.751734

Philosophy of Trust

Heresy and Honor. A Historical Perspective on Trust in Science

Johannes Lenhard

Abstract This chapter examines the discourse on trust from a historical and philosophical perspective. When and under what conditions did it arise? What are the salient features that drive this discourse? The argument revolves around the observation that trust is an ambivalent concept that can mean both honor for and heresy against science. It will be argued that the discourse on trust in science is relatively recent but has (some) historical roots. Four factors are identified that act on different time scales and that together contribute to the recent discourse on trust.

1 Introduction

If the reflection on science were performed as a concert, one theme would advance to the ear of any attentive listener: Why science deserves trust, and why scientific knowledge is a trustworthy product. Many and various instruments are involved in this theme, from books by prominent authors, like Naomi Oreskes' *Why trust science?*, to interdisciplinary networks like the newly founded *Trust In Information Network*. This paper does not enter the discussion on whether and when something can be trusted or should not be trusted, but rather takes one step back and examines the *discourse on trust* from a historical and philosophical perspective. When and under what conditions did it arise? What are the salient features that drive this discourse? This chapter revolves around two themes.

(i) Trust is an ambivalent concept; and
(ii) the discourse on trust in science is relatively recent. The following enquiry identifies four contributing factors that act on different time scales.

One salient feature of trust is that it values the person or thing that is trusted. In this very common usage, a finding of science deserves trust whereas a mere rumor does not deserve trust. The value of trust can also be seen from the efforts made to gain

---

J. Lenhard (✉)
Department of Social Sciences and Laboratory of Engineering Thermodynamics, RPTU Kaiserslautern, Kaiserslautern, Germany
e-mail: johannes.lenhard@rptu.de

M. M. Resch et al. (eds.), *The Science and Art of Simulation*,
https://doi.org/10.1007/978-3-031-68058-8_2

trust—think of Ka's "Trust in me" song to seduce Mowgli in Disney's *Jungle Book*. The thrill for the audience results from a classical suspense effect—Ka repeatedly succeeds in making Mowgli trust, but only almost. The audience is fully aware that Mowgli's trust would be unwarranted, even worse: would be deadly, whereas Mowgli himself is not aware of the danger. Ka's repeated attempts testify how valuable trust is. In short, the label "can be trusted" is like a badge of honor.

However, the reputation of trust is not unequivocal. That is clearly expressed in the Ka-Mowgli scene. It is this second side that seems to be in flagrant conflict with scientific knowledge. According to the common meaning of trust, it signifies a relationship that is not based on evidence. In the words of philosopher John Hardwig:

> After all, trust, in order to be trust, must be at least partially blind. And how can knowledge be blind? Thus, for most epistemologists, it is not only that trust plays no role in knowing; trusting and knowing are deeply antithetical. (Hardwig, 1991, p.693)

Saying that science deserves trust thus is not a purely appreciative statement but entails a potential devaluation insofar it suggests that science needs trust because it is short in evidence. From this perspective, speaking about trust is a heresy against science. Thus, vacillating between honor and heresy, trust is an ambivalent concept. Depending on how the situation of science is perceived, different actors can highlight different facets of the trust concept. Therefore, a historical perspective on the discourse about trust in science has diagnostic power.

A first finding is almost obvious. Today, trust is an issue of growing prominence in the philosophical discussion of science. One generation ago, however, this issue was virtually non-existent. Looking for hits in bibliometric data shows that, as a topic mentioned explicitly in the literature on philosophy of science, trust had a slow start around 1990.[1] The straightforward explanation is that "trust" was absent because trust went unquestioned. The arguably basic stance is that trust in truth is warranted under all definitions of trust and truth. As long as science is seen as approximating truth, trust in science is a non-issue. Surely, scientific knowledge is not absolutely certain, but even competing approaches that challenge some given account and promise a better approximation to truth would belong to science. The prominence of trust, and the motivation to write this paper, go back to the basic stance loosing its commonsensical validity.

Different actors disagree on the functional role of trust in science. The paper proposes a simple categorization into two guiding pictures, suggesting a relatively simple opposition that reflects the ambivalent nature of trust. In the substitution picture, trust replaces truth (Sect. 2), whereas the foundation picture identifies trust as a prerequisite for finding truth (Sect. 3). Examining the discourse on trust faces the complication that trust is closely related to truth and to (ascribed) expertise. The discourses on these notions cannot be fully separated. Hence the following examination will include positions on expertise and truth—or their perceived crises—where they are helpful to analyze the positions on trust—or the perceived crisis of trust.

---

[1] My data analysis is so superficial—just scanning a couple of prominent journals—that it cannot count for more than anecdotal evidence, which is fine for the present context.

Section 4 very briefly discusses some instances in the philosophical reasoning about science that indicate trust was not absent but rather invisible. By no means this account is exhaustive, much more can be done. Pointing out a couple of instances where trust is an invisible but not absent topic is merely sufficient for the claim. Finally, Sect. 5 identifies four factors on different time scales. It is argued that these factors together motivate the recent perceived need for moving trust up on the philosophical agenda. Moreover, these factors foster the ambivalent character of trust between heresy and honor.

2 Trust Trumps Truth: The Substitution Picture

The first picture displays the position of trust relative to truth in a radical way, namely as trust substituting truth. The substitution picture underlies recent and prominent public debates. The reference to "alternative facts" made by the Trump presidency is an iconic example. It is certainly not the only example. There is a widely perceived loss of self-evidence regarding the truth or credibility of facts in general and of scientific findings in particular. This debate was covered by the news media throughout the presidency of Donald Trump, but it reaches back much longer and also covers more ground than Mr. Trump did.

Already in 2005, the comedian Steven Colbert invented the word "truthiness" to make fun of an anti-science attitude in US politics and public discourse. He expressed this attitude concisely: "I don't trust books. They're all facts, no heart." Colbert argues to the point: Trust is something precious. But the acting first person is aware that lending his or her own trust need not be based on facts or evidence. Since trust has a subjective facet, it can be based on emotion. According to the truthiness attitude, emotions provide a basis that is even preferable to facts or truth. It is an appeal to feelings at the expense of facts. In this way, trust trumps truth. The term truthiness for this attitude hit the nail; the notion made it into wikipedia: "Truthiness is the belief or assertion that a particular statement is true based on the intuition or perceptions of some individual or individuals, without regard to evidence, logic, intellectual examination, or facts." (Wikipedia, looked up Oct 20, 2021: https://en.wikipedia.org/wiki/Truthiness).

A second and related example is from the British politician Michael Gove: "I think the people of this country have had enough of experts (…)."[2] He said this with the Brexit referendum coming, expressing that trust in a decision should not depend, or according to the people does not depend, on evidence sanctioned by experts. Experts play a role, providing their opinion with some authority. The group of experts is not identical with the group of scientists, but when scientists are involved in some policy issue, they normally act as experts. Thus, the verdict against experts includes a verdict against scientists. These might be journalistic examples, providing accidental

[2] The sentence continues "with organisations with acronyms saying that they know what is best and getting it consistently wrong."

evidence at best, but the perceived crisis of truth has received wide attention also in terms of books like Kakutani (2018), including academic books like McIntyre (2018). Qua title, this literature is about truth, but qua content, it is about trust that replaces truth.

The substitution picture of trust is nourished by an anti-scientific movement. According to the established view, trust should be based on facts. The anti-establishment picture proceeds in two steps. First, trust is freed from its dependence on objective evidence. As an act of subjective valuation, a person can choose or decide in what to trust. And since trust can be awarded independently of factual evidence,[3] one can substitute facts by some emotional access and jump directly to trust. The reliance on facts gets substituted by reliance on trust. To base trust on scientific evidence requires an extra step, namely the valuation of science as authority in matters of evidence. This step is what adherents of the substitution picture are not willing to go. Thus, there is an element of heresy against the authority of science: If *trust* is the pivotal issue, there are many ways to reach it, not just the one via scientifically sanctioned pathways.

Throughout history, all types of propaganda have relied on the ability to convince people of something regardless of facts. This is hardly big news. The present debate is particular because it is not (only) about persuading people, but rather because it works with an outspoken opposition against science. At the same time, this opposition constitutes itself against the background of a politically relevant science. Scientific experts might state that this and that is the scientific evidence and therefore they see these possible ways of acting and their consequences. Then, the substitution picture makes room for a modus tollens: If one prefers a course of action that looks highly disadvantageous according to the expertise, one can pull off trust from scientific evidence and confer it upon a different source.[4]

In modern times, science has acquired a very high authority to the extent that matters of truth and of acting right are intermingled in the institution of science. Doubting the epistemic authority of science for many looks like a mistake. In this regard, the institution of the church was in a somewhat similar situation in pre-modern times. Hence, dubbing the substitution picture of trust as "heresy" seems fitting.

3 No Truth Without Trust: The Foundation Picture

The second picture places trust and truth in a very different relationship. Neither does trust substitute truth as in the first picture nor is trust founded on truth as in the basic stance. According to the second picture, it is the other way round. True

---

[3] However indirect such evidence might be related to truth.

[4] By the way, this is interestingly different from a skeptical position. A skeptic would distrust a recommendation by experts if it takes certain knowledge for very highly certain. However, the option of substitution would not be open to a skeptic. In real life examples, like climate skepticism, boundaries are blurred because many so-called skeptics seem to adhere to a substitution standpoint, i.e., being skeptic because they don't like the picture.

knowledge is based on trust. This point of view originates not so much from political discussions, but rather from studies of science that have philosophical, historical, or sociological character. The base line is that science is a communal enterprise. When science claims to attain facts, this results from an organized effort of a community. No single scientist can carry the epistemic burden, but has to trust in many pieces that others contribute. Philosopher John Hardwig describes how trust is the basis for scientific knowledge:

> Modern knowers cannot be independent and self-reliant, not even in their own fields of specialization. In most disciplines, those who do not trust cannot know; those who do not trust cannot have the best evidence for their beliefs. In an important sense, then, trust is often epistemologically even more basic than empirical data or logical arguments: the data and the argument are available only through trust. If the metaphor of foundation is still useful, the trustworthiness of members of epistemic communities is the ultimate foundation for much of our knowledge. (Hardwig, 1991, p. 693/4).

This description can be called a (pragmatic) foundation picture because truth can be attained only on the basis of trust. When Hardwig wrote his essay, he could firmly assume that an individualistic epistemology occupies a hegemonic position. And he found the time is ripe to attack such individualist epistemology head on. For him, the foundational function of trust for scientific knowledge proves the social and community-oriented nature of epistemology.

While Hardwig is an early source for a discourse that uses the concept of trust *expressis verbis*, there are important forerunners that pointed out the social character of science and epistemology. One prominent instance is sociologist of science Merton (1973) who saw the essential features of science as rooted not so much in individual scientists but in science as an institution. One characteristic feature of science, according to Merton, is its organized skepticism. Tellingly, Merton does not speak of organized trust. Arguably, he granted trust is present as a shared basis and abundant resource whereas the part in question, the one that must be organized, is skepticism. Another forerunner is Ludwik Fleck who argued in his 1935 essay on the "Genesis and Development of a Scientific Fact" (Fleck, 1979) that some observation or statement reaches the status of an established fact through a social process that takes place in, and is controlled by, a "thought collective".

Admittedly, speaking of trust in science is somewhat vague. In the most common case, trust refers to a relationship—like "I trust in x"—where the variable x stands for a person. In the present case, however, this variable may stand for a scientist, a scientific community, a scientific result, or even a method. Consequently, trust might be bestowed or denied through targeting different of these options. The substitution picture works with a basic notion of trust while, in the foundational picture, scientists pragmatically work in an organized way. According to the foundational picture, the scientific method is embedded in a social process that is characterized by mutual trust. And trust is foundational because when it comes to knowledge there is no serious alternative to scientific knowledge (in contrast to what the substitution picture says).

4 Invisibility, Not Absence

As indicated above, prior to the 1990s, trust is a rare find in philosophical examinations of science. But even when "trust" does not play a prominent role, this does not imply that the issue of trust was non-existing. Without doubt, the issue has deeper historical roots. Trust is not only virtually absent in the classical accounts of philosophy of science, epistemology has been deliberately purged from trust.

Philosophical epistemology in the enlightenment times put the individual in the center. John Locke, to give a typical example, articulated the basic rationale: Anybody who wants to secure real knowledge should rely on the evidence that she or he directly gets. Everything that is merely reported from others, or even reported via some tradition, is of second rank, epistemologically speaking. Thus, first rank knowledge is a matter that involves only the relationship between a knowing individual and the world. Consequently, if you trust something, you do not really know it.[5] Trust is alien to traditional epistemology because trust has social aspects and therefore does not fit into an epistemology that focuses entirely on the individual. Roughly speaking, in philosophical epistemology, the flight of fancy of the knowing individual subject lasted from Locke to Kant (or even to Husserl). Things were less clear in the earlier 17th century. Obviously, Descartes was an important source for the individualist tradition, though his rationalistic approach was very different from Locke's empiricism.

Even if philosophical epistemology centered on the individual knower, scientific knowledge has always been more than the sum total of single knowledge seekers. Science is a socially organized endeavor, an institution. Importantly, trust played a significant role in the formation of modern science. Steven Shapin has made a number of contributions that elucidate the role trust played in early science. The Royal Society, for instance, was founded as a community of gentleman and the trustworthiness of scientists was an important basis for claims about the (perceived) validity of scientific knowledge. It is not the point here to summarize the already existing rich literature in studies of science. Rather, the modest point merely is to indicate that the issue of trust was not absent.

Here is another piece of evidence. Over the 19th century, social and communal features of scientific knowledge received more attention from the philosophical side. Pragmatism underlines that scientific knowledge relies on a process that happens in a community.[6] The scientific character then originates from the ways in which the community of scientists controls the process. Truth is not produced by the single individual researcher but, rather, it is the endpoint of the research process, i.e., a promise that scientific knowledge converges toward the truth. C.S. Peirce can count

[5] This is close to the account that Hardwig argues against, see Sect. 2.

[6] Robert Brandom, a contemporary pragmatist, reads Hegel's Phenomenology of Mind as an inquiry that shows the basic role of trust (Brandom, 2019). His reading confirms the hypothesis that recent perspectives bring the role of trust to the fore. Trust as an issue is recognized in more and more contexts. Still, the publisher advertises Brandom's account as a challenge for dominant (non-trust-oriented) modes of thought in contemporary philosophy.

as the most thorough proponent and pioneer of this viewpoint. Notably, even if knowledge gains a strong social aspect, the primary focus is not on something like trust, but on the *logic* of research, i.e., on something independent of the actors' attitudes.

The recognition of philosophy of science as a separate philosophical discipline is closely connected with the work of logical empiricism in the first half of the 20th century. Although there were also strong voices present in the Vienna Circle who emphasized the social nature of knowledge, such as Neurath and Zilsel, empiricism was predominant. The soon to follow focus on language makes it quite clear that the contact of the individual with the world takes place through a medium that itself has a social character. Trust, however, hardly was addressed explicitly in these considerations (according to my fallible knowledge). A last stop on this rushing journey is Karl Popper. His demarcation criterion for science is falsifiability. He never tired of renewing his appeals that scientists should put forward bold hypotheses and immediately follow up with persistent attempts at refuting these hypotheses. Again, Popper was looking at the logic of research. He was downright critical about too much trust into hypotheses. However, disparaging trust into single hypotheses does not altogether abandon trust as a component in science. On the contrary, when Popper underscored skepticism, he did so against a strong backdrop of trust. The entire technological and social machinery of producing empirical evidence and of developing theories and hypotheses was simply granted. Popper's declared high-level skepticism actually is swimming in a bath of trust.

Thus, the absence of trust in the philosophical discourse does not imply the absence of trust in scientific practice. Throughout the development of science as an institution in broad terms, trust built an important resource without which dissemination and institutionalization would have been hardly possible at all. Steven Shapin, in good accordance with Hardwig (1991, see Sect. 1), underlines that trust was not absent but rather "invisible" (1994)—and I signal my agreement by using this term in the heading of the current section.

Today, trust has become visible. Moreover, talks about a public crisis of trust in science have become a visible topic in public as well as in academic debates. Among the factors that contribute to this (perceived) crisis are cases where scientific experts have been shown not to deserve trust. The following section examines more principled factors.

5 Four Factors Put Trust on Display

The obvious question to ask is what factors contributed to trust becoming visible and even a focal point in the discourse on science? My thesis is that different factors work together. I discern four of them; each one contributes, but none of them can be singled out as the main culprit.

(i) Institutional organization: ongoing specialization and division of labor

(ii) Positioning of Science and technology in society
(iii) Social nature of science as a topic for study
(iv) Science in a social network.

I do not harbor the belief that these four factors provide the complete picture. Furthermore, my summary does not aim at anything like a full fledged historical analysis of any of these factors. What follows is a brief explication of what these factors stand for.

Factor (i): Institutional organization: ongoing specialization and division of labor

The 19th century saw a broad institutionalization and tremendous growth of academic science. Scholars became experts in special disciplines or sub-disciplines. This dynamic continued over the 20th century. The resulting fragmentation was often observed. An early example is provided by Romantic poets' complaints about the alienation and fragmentation to which a person is subjected. Today, science seems to be way more specialized. However, the situation is not only marked by (sub-)disciplinary specialization but, additionally, by a networked infrastructure that connects many disciplines but does not integrate them. An exemplar is software usage. Users of software packages often can employ them like a commodity that they know how to use although the inner workings remain opaque. For example, the enormous upswing of computational quantum chemistry is based (not exclusively, but significantly) on researchers' competence in handling software that does not require to be an expert in quantum theory (cf. Lenhard, 2014; Wieber & Hocquet, 2020).

Of course, it is not impossible to critically check models, or software, or other tools—even under the conditions of opacity. But the normal way of science is to build on existing knowledge, instruments, thought styles, and the like. Critical examination then is directed to local questions, like to competing models or parameter settings. In short, since more than a century, the institutional organization of science has seen increasing levels of specialization and division of labor. Single scientists thus must increasingly rely on the work of others. Hence this development fosters the role of trust.

Factor (ii): Positioning of science and technology in society

For a long time, doing scientific research was an affair where a very limited number of researchers addressed a limited audience. The industrial revolution was surely an epochal transformation that affected societies in a fundamental manner. However, technological feats like steam engines and steel bridges were more products of engineering than of scientific research. Science was closely related to philosophical thinking about epistemology, often a source inspiring the philosophers, but was of no great significance for society. In this respect, the character of science has changed fundamentally. Science has long occupied a place in the midst of society (as has engineering), occupying a prominent place and also being relevant to ordinary citizens in manifold ways.

A plausible choice for locating the turning point is the Manhattan project. Then, science became "big science"—calling for numbers in researchers and budgets that required substantial engagement from the government (Price, 1963; Weinberg, 1967). Famously, US president Eisenhower feared that big science was getting so firmly established in a military-scientific complex that it was out of reach from (parliamentary) political influence.[7] Oreskes and Conway (2010) tell a tale that is also related to this factor. Their prominent study "Merchants of Doubt" (Oreskes & Conway, 2010) documents how (a small number of) scientific experts are paid by companies and work toward undermining public trust in scientific findings. Influence through scientists is so significant that they can sell doubt to a government and society. Climate is another obvious example, among very many examples, where scientific results and methods are of great concern to society.

When science is of concern, it must also be eligible for deliberation in (more or less) public discourse. A recent direction in artificial intelligence (AI) is so-called explainable AI that tries to frame ways that make methods and results of machine learning, especially deep learning, accessible. Arguably, this movement is motivated by the need to make science accessible to a wider audience. Conversely, without such access, societal trust in science is hard to maintain.

According to Carl Sagan "We live in a society exquisitely dependent on science and technology, in which hardly anyone knows anything about science and technology." (1990, p. 264). The urgency expressed in Sagan's verdict comes not so much from the barely existent knowledge (who did know anything about science in early modernity?) but from the importance of science and technology to society, the exquisit dependance.

Since more than half a century, science and technology approach the center, or better: marketplace, of society. They affect the lives of laypersons in important ways. This turns trust from an affair internal to science into an issue for the general public. By the same token, scientists are more involved with societal interests, even governmentality, and can count less as disinterested creators of knowledge.

Factor (iii): Social nature of science as a topic for study

The third factor sets in the most recent. It is a turn in the perspective on science that can aptly be called social turn. Pivotally, Thomas Kuhn described science with the help of community oriented paradigms (1962). He acknowledged Ludwik Fleck as a forerunner (Fleck, 1979 <1935>). Of course, pioneering works in the sociology of science already existed, including Mannheim (1936) and Merton (1973).

Today, there is a great variety of topics and thinkers whose work elaborates on this turn. Examples include the debate around science and values, the feminist philosophy of science, and surely a constructivism à la Latour. Philosophers like Longino (1990) or Oreskes (2019) make pluralism a criterion for valid science. According to this sort of pluralism, only dissenting voices and the combination of different perspectives can

[7] Weinberg who coined the term "big science" in 1961 was scientific advisor to Eisenhower.

lead to valid knowledge.[8] Recent work by Nancy Cartwright, to name just one more example, on "Why trust science?" (2020) takes a similar approach, highlighting the plurality of scientific outputs on different levels (the "tangle of science") as the source of overall reliability. Here is not the space to enter a meaningful discussion of these very valuable contributions. The point here is rather simple. When philosophical studies of science scrutinize levels where plurality, dissent, tangle, etc. play a role, the issue of trust becomes a subject too, because the production and maintenance of trust is located at these levels.

Factor (iv): Science in a social network

Science has acquired a social nature in a further sense. Today, and mainly due to digital media, chunks of knowledge are readily available to a wide audience of scientists and laypeople without mediation by experts, or gate-keepers. That means these chunks are easily untied from interpretations and explanations but can be embedded in a variety of contexts. Shapin sees the problem of trust as a problem about identifying the experts: "The present-day problem is not mistrust in scientists but, rather, a problem in deciding who the scientific experts really are." (Shapin, 2004, p. 46) The crisis of trust thus goes back to a crisis of expertise. While expert knowledge still counts as trustworthy knowledge, it is not clear how to determine who the expert is. In *Crisis of Expertise*, Gil Eyal briefly reviews a number of surveys that address trust (Eyal, 2019, ch. 3) and finds little evidence that there is a trust crisis in science per se.[9]

I argue that different parts of science take part in very different social networks. Scientists working in computational quantum chemistry may disagree on what is the best modeling approach in a certain context. This does not undermine (mutual) trust. On the contrary, trust is the working basis for assuming that dissenting voices do not make up false evidence, but have reasons for their opinion that are worthy for expert evaluation. A sort of pluralism and mutual criticism functions as a ferment for scientific epistemology. Large swaths of science, but not all of it, are of this kind. When science is more closely connected to politics and to the public arena, it is part of a very different social network. This is what Eyal calls "regulatory science". There, pieces of expert knowledge are demanded to support (or defy) policy measures. Pluralism can easily support different sides of conflicting political opinions. Everyone knows instances where expertise was made up to back political agendas or economic goals.[10] Under these conditions, trust in science is not a working condition but rather becomes the object of strategic action.

---

[8] Any sort of pluralism and perspectivism challenges the concept of truth. McIntyre, for instance, argues that postmodernism (which he vaguely identifies with perspectivism) is a factor leading to a crisis of truth (McIntyre 2018, Chap. 6).

[9] There are other voices. Nichols (2017), for instance, drastically diagnoses the *Death of Expertise* through computer networks and the erosion of respect for facts.

[10] Rampton and Stauber (2001), for instance, have a point when they warn against trust in experts who have connections to industry. The already mentioned Oreskes and Conway (2010) make a related point when they argue that expertise was (and can be) targeted at undermining trust.

6 Do These Factors Make Trust in Science Waning?

First of all, trust is not a given, natural state of affairs. Trust needs to be maintained. Science is an institution that attracted extraordinary trust over more than two centuries—comparable arguably only to the church (in earlier times). It is hard to preserve this level of trust under changing conditions—the church again illustrates the point. And the finer the time scale of factors, the quicker the conditions are changing. This is not an argument that trust cannot be maintained; it is just a reminder that a high level of trust is something remarkable.

All four factors from Sect. 5 contribute to making trust an issue. They are not exhaustive; more factors are present. An obvious one is the current so-called reproducibility crisis that has gripped medicine and psychology—and likely is not contained to these disciplines (Ioannidis, 2005; Open Science collaboration, 2015). When scientific research false short of its own standards, this is surely not raising the level of trust. However, the discourse on trust is related to the discourses on expertise and on truth, both of which are topics that attract growing attention in sociology and philosophy. I expect that the four factors listed above provide only a preliminary picture.

The four factors do not necessarily lead to a crisis of trust. The arguably most critical dynamic results from the joint appearance of factors (iii) and (iv) because pluralism is a presupposition for valid scientific knowledge (iii) and a resource for raising doubts in it (iv). However, this is a conflict only if both factors are active at the same time. This need not be the case. Plurality in a specialized circle of practitioners does not undermine expertise. On the contrary, special expertise might consist in knowing how to utilize differing results for making progress. For example, how model results at odds with each other throw light on modeling assumptions and how to modify them.

But what about the "regulatory lane" (Eyal) of science where scientific knowledge is transferred into political contexts and where expertise shall serve political goals? There, pluralism means that pieces of expertise stand in conflict with each other. As pointed out above, this is a resource that can quickly lead to mistrust. But it need not. A carefully framed use that acknowledges alternative approaches and existing uncertainties can handle conflicting pieces without loss of trust. Frankly admitting what is not known can even increase trust. The critical point is that adequate handling of uncertainty in scientific knowledge and expertise requires high standards on all sides of scientists, politicians, and the public. Scientific experts can easily destroy the balance by not communicating limits and boundaries, politicians by ignoring these limits in the service of political preferences, and the public by paying insufficient attention to these limits, uncertainties, and caveats. Maintaining an adequate balance is not easy to achieve and requires high levels of knowledge, benevolence, and judgement from all actors involved.[11]

[11] It easier to identify requirements than to find good recipes for procedures that fulfill these requirements. A potential example is climate science with its comprehensive assessment reports that aim at consensus but also discuss extant uncertainties in an accessible manner. However, these reports

This observation brings us back to my title: heresy and honor. Early scientists like Galileo were accused of heresy because they challenged the position of the church as the prime authority on knowledge. For Galileo, the authority of scientific knowledge could not be based on trust in knowledge sanctioned by the church, but must be grounded in independent evidence—a heresy. I have just argued that dealing with pluralism requires a delicate balance from scientific as well as laypersons. Because maintaining trust in science is not a question that can be answered by science alone, today's discourse on trust questions the independent authority of scientific knowledge. Thus, it commits a very different heresy—now directed against science.

References

Brandom, R. (2019). *A spirit of trust. A reading of Hegel's phenomenology.* Harvard University Press.

Cartwright, N. (2020). Why trust science? Reliability, particularity and the tangle of science. *Proceedings of the Aristotelian Society, 120*(3), 237–252. https://doi.org/10.1093/arisoc/aoaa015

Eyal, G. (2019). *The crisis of expertise.* Polity Press.

Fleck, L. (1979). In R. K. Merton (Ed.), *Genesis and development of a scientific fact* (T. J. Trenn), foreword by Thomas Kuhn. University of Chicago Press.

Hardwig, J. (1991). The role of trust in knowledge. *The Journal of Philosophy, 88*(12), 693–708.

Ioannidis, J. P. A. (2005). Why most published research findings are false. *PLOS Medicine, 2*(8), e124. https://doi.org/10.1371/journal.pmed.0020124

Kakutani, M. (2018). *The death of truth: Notes on falsehood in the age of Trump.* Tim Duggan Books.

Kuhn, T. S. (1962). *The structure of scientific revolutions.* University of Chicago Press.

Lenhard, J. (2014). Disciplines, models, and computers: The path to computational quantum chemistry. *Studies in History and Philosophy of Science Part A, 48*, 89–96.

Longino, H. E. (1990). *Science as social knowledge: Values and objectivity in scientific inquiry.* Princeton University Press.

Mannheim, K. (1936). *Ideology and Utopia.* Routledge.

McIntyre, L. (2018). *Post-truth.* The MIT Press.

Merton, R. K. (1973). The normative structure of science. In *The sociology of science. Theoretical and empirical investigations* (pp. 267–278). University of Chicago Press.

Nichols, T. (2017). *The death of expertise: The campaign against established knowledge and why it matters.* Oxford University Press.

Open Science Collaboration. (2015). Estimating the reproducibility of psychological science. *Science, 349.* https://doi.org/10.1126/science.aac4716

Oreskes, N. (2019). *Why trust science?* Princeton University Press.

Oreskes, N., & Conway, E. M. (2010). *Merchants of doubt: How a handful of scientists obscured the truth on issues from tobacco smoke to global warming.* Bloomsbury Press.

Price, D. J. (1963). *Little science, big science.* Columbia University Press.

Rampton, S., & Stauber, J. (2001). *Trust us, we're experts. How industry manipulates and gambles with your future.* Tarcher/Putnam.

Sagan, C. (1990). Why we need to understand science. *Skeptical Inquirer, 14*(3), 263–269.

seem to become so time consuming and unrewarding for many scientists that they prefer more open, adventurous, and therefore unbalanced research.

Shapin, S. (2004). The way we trust now. The authority of science and the character of the scientist. In P. Hoodbhoy, D. Glaser & S. Shapin (Eds.), *Trust me, I'm a scientist* (pp. 42–63). British Council.

Shapin, S. (1994). *A social history of truth: Civility and science in seventeenth century England*. Chicago University Press.

Weinberg, A. (1967). *Reflections on big science*. The MIT Press.

Wieber, F., & Hocquet, A. (2020). Models, parameterizations, and software: Epistemic opacity in computational science. *Perspectives on Science, 28*(5), 610–629.

Trusting Science: Is There Reasonable Distrust of Reputable Scientific Authority?

Brittany A. Gentry

Abstract Is there reasonable distrust of reputable scientific authority? This paper considers the role of experience in the epistemic process of trusting authority and argues that distrust based on experience mirrors rational processes of belief formation and so produces rational, though sometimes wrong, beliefs. Part one establishes the importance of experience in the basic process of developing trust in authority and in formal epistemologies. Part two considers four ways in which people experience scientific authorities: (1) expertise, (2) distinguishing between individual and group identity, (3) shared identities, and (4) transparency. Given the role of experience in trust formation, the paper concludes that in some cases where non-scientific communities have negative experiences of scientific authority and distrust a reputable scientific authority, those communities have, *prima facie*, a rational distrust and are likely to distrust.

Why do people distrust reputable science? Recent environmental and health debates have highlighted a large group of people who distrust scientific authority. There is a tendency among some scientific, social, and media platforms to denigrate that group of people for being anything from uneducated to stupid to science deniers (Goldenberg, 2021; Krause et al., 2021). While some people accurately fall into categories like uneducated or science deniers, research suggests that such descriptions are often inaccurate over-generalizations about why people distrust science and neglect the epistemic processes by which many people arrive at distrust. I point out something rather obvious and I hope non-controversial. It is this: once we notice the foundational role that experience plays in belief formation, at both descriptive and prescriptive levels, it becomes obvious that people's experience of scientific authority is going

I owe many thanks to Brett Sherman for numerous comments and suggestions on various drafts of this paper. I am also grateful to Nico Formánek, Elizabeth Stewart, and the attendees of the SAS: Trust in Science Conference for comments and critiques as well.

B. A. Gentry (✉)
Department of Communications and Philosophy, Utah State University, 0720 Old Main Hill, Logan, UT 84321, USA
e-mail: brittany.gentry@usu.edu

M. M. Resch et al. (eds.), *The Science and Art of Simulation*,
https://doi.org/10.1007/978-3-031-68058-8_3

to be foundational to their beliefs about it, including beliefs about trustworthiness. It is then not only predictable but even likely that where individuals and communities have negative experiences of scientific authority, they are likely to form the belief that scientific authorities are not trustworthy. The goal of this paper is not to provide a defense of what we are like or of our epistemologies but, rather, to argue that because we develop beliefs in certain ways and have epistemic accounts that justify those beliefs as rational, we have reason to recognize as rational, beliefs, even wrong ones, that are produced by the same processes. This paper will conclude that in some cases where communities distrust a reputable scientific authority because of negative experiences of scientific authority, those communities may have a rational distrust and are *likely* to distrust. It will not attempt to provide a list of such cases, but only establish that where beliefs are produced in certain ways, they qualify as rational. Moreover, an important clarification to maintain throughout this paper is that rational beliefs are distinguished from moral, praiseworthy, or true beliefs—beliefs are rational, not irrational, in virtue of various rational processes by which they are formed, not in virtue of being commendable.

The paper will consider, first, the foundational role of experience in human development and in formal epistemology and, second, provide four examples of distrusting scientific authority that are based on experience.

1 Experience as Foundational to Belief Formation

Experience is foundational to belief formation at a descriptive level—what we are like—and at a formal, prescriptive, level—our rational processes for producing beliefs. This result is not surprising but does have important implications for understanding why people have certain beliefs and for maintaining reasonable expectations of what beliefs people will hold.

1.1 What We are like

Studies in neuroscience and psychology on the development of trust in early childhood to adulthood show the importance of negative and positive experiences in trust formation and in recovery from breaches of trust.

Childhood experience is foundational and formative to our capacity to develop beliefs about trustworthiness. We rarely trust *tabula rasa*—instead, some experience or set of experiences is usually involved. The adage, "first we try, then we trust" turns out descriptive of how we develop trust in authority. The pattern of experience as a foundation for trust develops in early childhood and continues into adulthood. Initial tasks of childhood begin with a child's practice of allowing her mother out of sight and developing trust through repeated experience that her mother will return (Erikson, 1963, p. 247), which then "builds a shield against insecurities and existential fears" (Bogaert et al., 2000, 505). This development process proves foundational for the "relational trust [that] is essential for interpersonal and social functioning" (Bogaerts et al., 2000, p. 505). A child trusts parents based on repeated experience of her parents—experience that begins in infancy and continues into adulthood. These childhood experiences of attachment and bonding prepare children for and predict the sort of intimacy patterns and capacities they will have as adults (Bogaerts et al., 2000, p. 504). To the degree that such experiences proved the untrustworthiness of the parents or parent figures, the child develops into an adult who struggles not only to trust other people but to identify trustworthy persons (Bogaerts et al., 2000, p. 504). And children notoriously test and reaffirm the trustworthiness of a parent's advice and maxims by doing the exact opposite of what they are told—as when they touch hot stoves or shaving razors.

This pattern of trying before trusting continues into adulthood and shows up in most areas of life (Lankton et al., 2012, 657). For example, studies on consumer trust of websites shows that prior experience is so important a factor in whether consumer continue to use a website and purchase from a vendor that brands are well-advised to ensure customers have positive experience (Chen et al., 2010, p. 545). That finding is not new or surprising—marketing industries and customer service departments have long worked to defend and promote the trustworthiness of a company via promises to deliver positive and reliable experiences for their customers. Brand recognition becomes an important factor in the success of a company due to the reliability and trustworthiness it represents (Chen et al., 2010).

Moreover, the quality of our experiences, positive or negative, influence our ability to update our beliefs about people and future outcomes. Research suggests that our brains are relying on both initial social impressions and ensuing experiences to determine trustworthiness. In brain-imaging studies using trust games, participants demonstrated the ability to update initial impressions and future likelihoods about the behavior of a gaming partner based on their experiences during play with that partner (Fareri et al., 2012). Earlier studies also show that "people may treat partner reciprocity as a conditional probability and appear to learn the trustworthiness of a partner through PE [prediction error] driven learning" (Chen et al., 2010; Fareri et al., 2012, p. 14). These studies provide support for thinking that "trustworthiness is a belief about probability of reciprocation based initially on implicit judgements, and then dynamically updated based on experiences" (Chen et al., 2010, p. 87).

Certain experiences are more important to the development of a belief than others. Experiencing positive outcomes with a good partner has more influence on belief about trustworthiness than experience of positive outcomes with a bad partner—likewise experience of losses with a bad partner had more influence than losses with a good partner (Fareri et al., 2012, p. 14). Additionally, confirmation biasing leads to some over emphasizing of initial impressions, however, studies show evidence that "participants can successfully learn the likelihoods of their partner's reciprocation rates over time from trial-to trial feedback via a reinforcement learning process" (Chen et al., 2010; Fareri et al., 2012, p. 14). The combination of the greater weight that initial impressions play in beliefs about trustworthiness as well as the roll subsequent experience has in updating beliefs about trustworthiness provides reason for thinking that while a person's initial experience with scientific authorities is especially important, her following experiences with scientific authorities will, nevertheless, update her initial beliefs about the trustworthiness of scientific authorities. Moreover, changing beliefs to fit with evidence, is laborious and difficult when it requires abandoning preferred beliefs (Goldenberg, 2021, Ch. 2). People examine evidences that go against their belief more critically than evidences in favor of their belief, even to the point of denying evidence or reframing the issue (Goldenberg, 2021, p. 46). Rather than conclude that such resistance to updating beliefs is unreasonable, Goldenberg suggests that it is the product of risk assessment rooted in cultural cognition and points to the need to address false beliefs and distrust in a manner that works with already established beliefs rather than directly against them to avoid the increased difficulty of directly countering beliefs (Goldenberg, 2021, pp. 50–53).

The categories of 'negative' and 'positive' experience also warrant further consideration. For example, does the negative experience that a child has when given a shot or a person has when a doctor informs them of a terminal illness count as the kind of negative experience that undermines a person's trust in scientific authority?[1] If *all* negative experiences amount to breaches of trust, then it might indeed turn out that many cases of distrusting scientific authorities are cases of irrational distrust. Presumably most people do distinguish between kinds of negative experience, as in the common case of recognizing the long term benefits of working out despite the short term discomfort of the experience—though, how well distinguish between kinds of negative experience warrants further study.

The kinds of experiences we have also influences the ease with which we manage trust recovery. Recent work in neuroimaging shows that the amount of prior positive experience determines which cognitive process an individual's mind takes to rebuild trust after breaches occur. The ease of repairing breaches of trust depends on which brain system is activated in the recovery process (Schilke et al., 2013). The controlled social cognition system (C-system), and specifically the lateral frontal cortex, is activated when breaches of trust occur early in a relationship, prior to extensive experience (Schilke et al., 2013). Whereas the automatic social cognition system (X-system) is activated when breaches occur later in the relationship, after more experience. The C-system requires "more conscious learning, more complex planning, and increased problem solving" whereas the X-system is thought to "automatically generate the habits and impulses that guide people's daily activities," habits which then "guide decision making after breaches of trust" (Schilke et al., 2013, p. 15237). So, more experience means that rebuilding trust is easier and less experience means that the rebuilding process takes more effort (Schilke et al., 2013, p. 15239). Further research suggests that pleasure and pain are also activated depending on which system of the brain is engage (McCabe et al., 2001, p. 11834; Schilke et al., 2013, p. 15239). In addition to supporting existing research on things like the importance of protecting

[1] See also Goldenberg (2021) and my thanks to a SAS: Trust in Science Conference attendee for this example.

positive experience, this research also predicts that breaches of trust will be more difficult to overcome with limited or negative prior experience.

Finally, social context also informs the epistemic process in the decision-making. Brain studies suggest that the medial prefrontal cortex, the part of the brain responsible for complex social skills, is more active when humans know they are playing a trust game with human counterparts compared to when they know they are playing the same trust game in a probabilistic strategy with a computer counterpart (McCabe et al., 2001, p. 11834). Knowing that a game partner is human results in participant's adding higher social functions to probabilistic strategies and relying more heavily on those higher social functions to form beliefs about their human counter-part's likely behavior. The more activated levels of medial prefrontal cortex likely means that participants draw on subtle probabilities based on past social experiences with humans.

1.2 The Respectability of What We Do

Formal epistemologies try to articulate how we develop beliefs and predictions about the world around us, relying especially on updating the likelihood of future outcomes based on present and past experiences (Galavotti, 2005). There are also two adjacent conversations in epistemology of trust worth noting: the synchronic versus diachronic constraints and the agent versus dispositional accounts of trust. But first, a note on what I mean by rationally held beliefs.

I mean something quite general and non-controversial. Some beliefs are rationally held and some are not rationally held and part of what distinguishes the two sorts of beliefs has to do with whether a system of reasoning exists that both produces and justifies the belief as rational. Whether one thinks that rationality is a property of belief-formation processes and so thinks that a belief is rational in virtue of the formation process or thinks that there are additional distinctions to be made is a bit beyond the point.[2] Nor does the specific epistemology that one holds matter to this point—for example, whether one is a foundationalist or a coherentist, beliefs in either epistemology are recognizable, both to those who endorse the system and those who do not, as rational products of their given epistemology. Likewise, whether one draws a distinction between reasoning as normative and psychological and logic as deductive and non-psychological (Harman, 1988), is also a distinction that does not affect the simple, generic view of a rational belief that I have in mind. A more difficult question is what we want to say about systems of reasoning that are themselves irrational. Do we want to say that a belief produced by such a system is rational? I think not, but I bracket this further consideration by simply noting that it does raise

[2] Many thanks to Brett Sherman for comments and suggestions on this section of the paper.

additional questions about what qualifies as a rationally held belief. For this paper, I assume epistemologies that are rational.[3]

It is also important to notice a tendency we have when we hear that a belief is rationally held. The tendency is to assume that the belief should therefore be endorsed. But the distinction I am making here expressly does not imply that the belief, however rational, is therefore a moral, praiseworthy, or true belief. A rational system can birth morally reprehensible, factually untrue, or otherwise condemnable beliefs that are nonetheless rational, not irrational. Throughout this paper, talk about belief as rational is meant in a quite narrow sense such that it is entirely possible that a belief be both rational and unadvisable to hold.

Experience often functions as a primary form of evidence about the way the world is and is foundational across various epistemologies. It serves as a sort of evidence for or against a hypothesis, aids in risk assessments, and forms the foundation for various beliefs (Goldenberg, 2021). The important issue of how experience serves as evidence and the ways in which experience can mislead is complicated and bracketed for now—that it commonly does serve as evidence in both formal and informal claims to knowledge is sufficient for the purposes of this paper. Here are some examples of the role experience plays in epistemology.

In deductive epistemologies, evidence confirms past hypothesis (Weisberg, 2021). For example, if I hypothesize that a person is trustworthy, evidence for that hypothesis comes in the form of experience of that prediction being correct—thus when that person acts in a trustworthy manner, my hypothesis is confirmed and the belief established. Hypothesizing or even acting as if a person is trustworthy is different from trusting a person. Thus, belief of trustworthiness comes after some experience confirms an initial hypothesis. Even if one holds that reasoning systems are only inductive and deduction is confined to logic (Harman, 1988), the reorganization still leaves experience as foundational to the inductive process of forming beliefs about trustworthiness.

In probabilistic epistemologies, like Bayesianism, evidence makes a given hypothesis more probable (Galavotti, 2005, p. 181). Formal epistemologies do not agree on whether initial probabilities are always based on previously collected evidence, nevertheless, the prior, present, and future evidence collected is important to updating probabilities (Weisberg, 2021). While subjectivist epistemologies attribute the differences in people's initial probabilities to there being no correct initial probability, objectivist epistemologies attribute such differences to initial probabilities being incomplete (Weisberg, 2021).[4] Where belief about trustworthiness is based on predictions about the probability that, in future, a person will behave in trustworthy ways, we would expect to find that experience is foundational to probabilities and beliefs, either initially or in conformation of starting assumptions.

[3] Of course, this leaves open additional questions about what to say about conspiracy theories and such and while some of what I say in this paper is relevant to that discussion, I have largely bracketed a thorough consideration as beyond the scope of this paper (see also Bardon, 2020).

[4] Other discussions in epistemology provide additional layers of theory about the role of experience in belief formation that could be useful to analyzing how well formal accounts of belief formation map to informal processes (see Galavotti, 2005; Weisberg, 2021).

Agent versus dispositional accounts of trust may initially appear salient to this paper, however, it is important to note that the central claims of this paper do not commit to either view. On an agent view, where trust is only held between agents, then rational beliefs about trustworthiness would be, in part, based on experiences of the agent. Even though this paper largely focuses on the process of developing beliefs about trustworthiness between agents, it is consistent and possible to extend the same account to dispositional accounts of trust, such as the one proposed by Nguyen (2022). In the case of trust in objects we are still relying, in much the same way as we do with agents, on prior experience with those objects or relevantly similar ones. And we rely on experience to determine and develop beliefs about the untrustworthiness of such objects (Nguyen, 2022, p. 5). Hence, if my car breaks down three times a month, I distrust it far more than if it has not broken down in ten years of driving it.

Debates about the diachronic and synchronic constraints on rationality provide another example of how the role of experience transcends the differences of position. Where diachronic and synchronic accounts of rationality turn on the distinction between mental events at a point in time "such as events in which we form or revise our beliefs" and mental states across time, such as "statistically enduring mental states (like belief-states)" (Wedgwood, 2018, p. 19), both use experience as a form of evidence in determining degrees of correctness in belief. Including deeper diachronic requirements that seem to require trust in past beliefs and possibly past experiences (Wedgwood, 2018, p. 21). Experience is hardly the only factor in determining rationality but it is essential to things like evaluating at least some instances in the following categories: such as whether certain events obtain, degrees of correctness in beliefs, the truth values of propositions, and so on (Wedgwood, 2018). Experience also helps account for the fact that variation occurs between rational thinkers despite their having the same concepts.

> Some features of the probability function that count as the rational probability for a particular thinker at a particular time will vary between different thinkers and different times—even if these thinkers possess exactly the same concepts at these times. These features of these probability functions are in a sense empirical: they depend on contingent features of these thinkers' mental lives at these times, and not merely on the concepts that they possess or the capacities required for them to count as rational thinkers (Wedgwood, 2018, p. 16).

As the research in neuroscience shows, it is not simply that experience informs the content of our thinking, it also shapes which pathways of the brain we use when examining and weighing evidences and informs how critical we are of the new experiences and evidences. Thus, equally rational persons may arrive at different conclusions and probabilities in part because one has more negative experiences which makes her more critical of the current and future evidences and makes it physiologically more difficult for her to change her beliefs.

Likewise, rational risk assessment depends in part on experience. And the influence is two-fold. Not only does experience inform the values a person assigns to various risks and probabilities, it also determines how rational she is for exercising extra caution in her assessments of acceptable risk taking. Experience informs the degree of trust and the likelihood of trusting (Lankton et al., 2012). A person whose experience supports the belief that x is unlikely to happen and that, even if it does, nothing particularly terrible will happen, is more irrational for exercising the same amount of caution as the person without such experience. Thus, a person's judgement about whether and how much to trust or distrust will rely on prior experiences to assess how much is at stake, should her judgement be wrong (Wilholt, 2013, p. 237). Thus, when we hear a person appeal to experience as their reason for their trust and distrust, they are appealing to a *prima facia* rational justification, even if other factors subsequently provide reasons for over-riding or disregarding that experience.

2 Experiencing and Forming Beliefs About Scientific Authority

Applying the first half of the paper to the issue of scientific authority yields the obvious conclusion that one's experience of science is foundational to trusting scientific authority. More good experiences increase the likelihood of trust in scientific authority and more negative experiences decrease the likelihood of trust. We would expect that persons and communities that report initial and, on balance, more negative experiences than good of scientific authority, have more reasons to distrust scientific authority and are more likely to do so. It turns out that people do, in many cases, appeal to experience when justifying their distrust of scientific authority. But how people experience science and scientific authority is not straightforward (Goudge and Gilson, 2005).

2.1 Experiencing Science

This section considers four reasons given for distrusting scientific authority: (1) Expertise, (2) Individual versus Group Identities, (3) Shared Identity, and (4) Scientific Transparency. The list is not exhaustive but indicates some of the complexities involved in experiencing science.

1. *Expertise as Experience.* One reason people give for distrusting scientific authorities is that they do not understand the issues or the claims made. Though tempting to think that a lack of knowledge or expertise provides more reason to trust an expert, it can also be a reason for exercising extra caution. Here is why. Expertise operates as a kind of experience and, therefore, a lack of relevant expertise leaves a person with less experience for assessing the trustworthiness of an authority. A lack of expertise lends to a set of compounding problems: (1) less ability for independent evaluation and confirmation, (2) requires higher degrees of trust in authority, and (3) increases the reasonableness of risk aversion. In short, a lack of expertise increases the risks involved in trusting thereby making increased caution and distrust more reasonable.

First, less expertise means that an individual has less ability to independently evaluate and confirm the claims made by a scientific authority. Even expertise in a different field of study adds a layer of experience that can extend to the new field of research. For example, a chemist evaluating the claims made by a physicist has more experience relevant to understanding the process, the claims, and the trustworthiness of the physicist than does a lay person with no scientific expertise.

Second, having less expertise requires more trust. Expertise, as a kind of experience, replaces some of the risk involved in trust.[5] In formal epistemologies of trust, trust operates in situations where there are things like uncertainty, a loss of control, limited capacity, and vulnerability (Goldenberg, 2021). As uncertainty, lack of control and capacity, and vulnerability decrease, less trust is required between the trustee and trustor. For example, the chemist invests less trust in the authority of her fellow chemists in part because of her greater expertise, which gives her more control in accurate assessment and resources for evaluating claims and behavior. So, some persons may appear more trusting of a given authority simply because they have more experience (Lankton et al., 2012, p. 657).

Third, because a lack of expertise requires more trust, what counts as reasonable for a person to trust or distrust changes. What one risks becomes a major factor in determining whether a person engages in an activity (Rudner, 1953, p. 2; Wilholt, 2013, p. 237). If a person lacks the expertise necessary to evaluate a given scientific claim or recommendation, the fact of her ignorance increases the risk involved in trusting that authority. The risk of a negative outcome compounds with her inability to assess the likelihood that the authority is wrong.

2. *Experience and Individuals versus Groups*. Another explanation for distrust of scientific authority is the appeal to individual or community experience of some particular scientist or expert who got it wrong.[6] The first challenge requires understanding what counts as science and who counts as a scientific authority. And the answers vary by person and community. For example, if a nutritionist counts as a scientist, then a person's experience of her nutritionist likely counts in her collective experience of science and scientific authority, for better or worse. The ambiguity over what science is, who counts as a scientist, and who or what is being trusted builds into the complexity of understanding why an authority is distrusted. Nor is stipulating who counts as a scientist likely to convince everyone concerned—in part because such stipulation is an implicit appeal to authority, an authority that must first have earned the trust of those listening before such an appeal is effective.

The second is the challenge requires distinguishing individuals from groups. Often individuals within fields of science represent the whole field of science. While the

[5] Studies by Kim and Phalak (2012) show that webs of trust between content providers and content users of online information can be predicted based on the degree to which provider's area of expertise overlaps with a user's area of interest.

[6] A notable example of this in the literature on vaccine hesitancy is the use of minority groups, especially black and disabled peoples, in medical experiments, which lead to generational distrust of government and scientific authority in some communities (Goldenberg, 2021).

inference from individuals to groups is unsound, it is also a standard form of inference when and where it is impossible to examine every individual of a given group. We regularly infer that all crows are black, though we have not seen them all. Thus, individual scientists in any given field of science may stand proxy for science and all scientists. Such reductions may appear unfair and often yield incorrect conclusions.[7] But for the non-scientific individual or community, the limitations of circumstance may require that they judge the trustworthiness of a group based on what experience they have available. Increased experience allows for increased distinction between individuals and the groups to which they belong—so that as a person experiences more doctors, she is better able to distinguish the one bad doctor she encountered from doctors as a group. So, nested groups in ever broader categories are likely to receive individualized trust as a person develops enough experience to distinguish between various fields of science (Huber et al., 2019; Leshner, 2021). And indeed, research supports that people individuate between scientific authority and science so that trust in science (that is, scientific principles) may remain high even when trust in scientific authority declines (Huber et al., 2019, pp. 760–761; Leshner, 2021).

3. *Shared Identity.* A third justification for distrusting scientific authority appeals to shared identity. The first appeal is that because an expert's beliefs or cares are different from the individual or community, that expert may not take seriously the concerns and beliefs of the community in the recommendations and prescriptions she makes. The second appeal is made through sharing and adopting experiences. Recent work suggests that trust often involves a shared identity—an identity that is based on the perception of shared values between the trustee and trustor (Endreß, 2008; Kunnel & Quandt, 2016). The experience of another person's presence builds into a shared identity around future behavior and allows for trust and distrust. Some researchers argue that because "the assumption of shared identity is based on the *experience of interaction*," *any* interaction "inhabits some type of relational trust or distrust," thus leading to the conclusion that "there is no trust or distrust without interaction or perceived reciprocity" (Kunnel & Quandt, 2016, p. 35). They suggest, therefore, that experience of interaction, perceived trustworthiness, and sense of belonging, are co-present in a shared identity (Kunnel & Quandt, 2016, p. 36). They emphasize that:

> the perception of a social presence is a fundamental requisite for relational trust and distrust to develop, much as the perception of a copresence is a requisite for relational trust and distrust in face-to-face situations. If recipients or users perceive a sense of reciprocity or "international space," relational trust and distrust are not only possible but also beneficial. This sense of connectedness is also based on the feeling of reliability of the medium itself or the institutions behind it (Kunnel & Quandt, 2016, p. 43).

[7] The temptation to call such reductions irrational should be resisted since we commonly endorse just such thinking as rational caution in the face of the unknown. Since we defend probabilistic and inductive reasoning as rational in many areas of research (including our sciences), we can hardly pick and choose when we count it as rational and instead must resign ourselves to acknowledging that it is reasonable, though susceptible to yielding incorrect conclusions (not to mention being a fallacy in deductive logic). But, it is not clear that reasoning and formal logic are identical to each other (Harman, 1988).

This research helps explain why shared worldview is often important for people in determining who to trust and how much to trust. If a person is devoutly committed to the personhood of embryos, she is more likely to trust the advice of scientists who share that commitment. Where, in contrast, she may worry that the scientist who does not share this value will, possibly without malice or intent and through an innocent lack of fastidiousness, fail to report nuances in the scientific process that would in fact violate her beliefs. The question of whether her belief about embryos is correct is beside the point with respect to the rationality of her belief regarding the trustworthiness of the scientist. Given her belief, her distrust is rational, even if we think that both her beliefs—about embryos and the trustworthiness of the scientist—are wrong. Unless say that only true beliefs are rational, but we all should have numerous reasons for objecting to that idea. Other things being equal, we have the option to say she is wrong, not irrational.

The second sort of reason relies on the way in which communities develop collective experiences.[8] Where a person trusts other members of her community, she is likely to adopt their experiences as authoritative in her own evaluations of scientific authority. Such a practice is essential to offsetting to the fact of her own limited experience. For example, her experience no longer consists of her twelve trips to a doctor, but, rather, her twelve trips to the doctor plus her best friend's twenty-four trips to three different doctors, and her grandfather's six trips—and that becomes the experience on which she bases her trust in a scientific authority. Aggregate experiences highlight the importance of genuine dialogue between scientific communities and non-scientific communities to help avoid distortions of scientific evidence and to help scientific communities build positive shared experiences with non-scientific communities (Leshner, 2021, pp.; Pedersen, 2014, p. 550).

4. *Transparency in Science.* A fourth reason often given is perceived dishonesty on the part of scientific authorities and usually takes the form of, "remember when the experts told us X, and then about faced and told us the opposite—but you'll never get them to admit it." These sorts of reasons point to how the lack of transparency in science counts as negative experience. Because trust in authority is built on experience, transparency regarding the processes, risks, uncertainties, and likelihoods of predicted or desired outcomes are essential to developing long-term trust (Leshner, 2021, p. 17–18). While glossing risk and uncertainty may be effective in the short term for convincing people to behave according to scientific advice, long term effects result in a decrease of trust. The same experience can count as positive or negative, depending on how the scientific authorities handle it. If a scientific authority says that a given outcome is guaranteed when in fact it had a 95% chance of occurring, and the 5% outcome instead obtains, the community experiences, first hand, the untrustworthiness of a scientific authority (as judged by that authority's inability to correctly predict outcomes). Whereas, if the scientific authority were transparent, either outcome ends up a positive reinforcement of the scientific authority.

[8] See note 5 on well documented examples of communities sharing experiences as part of a shared identity.

One motivation for avoiding transparency is that people's degrees of trust track the degree of perceived disagreement within a field of science (Funk, 2017, p. 88).[9] Studies suggest that a lower degree of trust is not always due to a lack of trust in science as such, but rather due to the uncertainty of findings, outcomes, and views within particular fields of science (Funk, 2017, p. 86; Leshner, 2021; Millstone & van Zwanenberg, 2000). And the lower degree of trust may not mean less trust in scientific authorities but, rather, a healthy skepticism of the specific predictions about particular issues. Even where the lack of trust is about the reliability of scientific authority, some argue that absolute trust in scientists and science should not be the goal and might even be counter to good practice, were it achieved, since healthy skepticism is a hallmark of good science and is often more accurate to the empirical data available (Krause et al., 2021, see also Solomon, 2021).

Research further supports transparency between non-scientific communities and scientific ones not only for the sake of maintaining and building trust but also for the sake of furthering education. For example, informing the public about the regular disagreements between scientists and the frequency with which scientific experiments and hypotheses fail or trials end up not being reproduceable are not indications of the unreliability of science (Ophir & Jamieson, 2021; Solomon, 2021). Instead, such occurrences are the normal happenings of science—the trial and error by which scientists learn and improve and our general bodies of knowledge increase (Ophir & Jamieson, 2021; Pedersen, 2021).

So, transparency in the representation of scientific findings, risks, and predictions is important not only for transforming experiences of scientific authority into positive testaments to the reliability of that authority but is also important for educating communities about the normal processes and workings of scientific research—an education which may then act in an expertise like role of reducing the risk of trusting by increasing understanding (Millstone & van Zwanenberg, 2000; Pedersen, 2021; see also Arimoto and Sato, 2012; Barber, 1987; Sztompka, 2007; Marlow et. al., 2007).

2.2 *Trusting Science*

When breaches of trust occur, individuals and communities with fewer positive experiences will engage the C-system of their brains, requiring more effort to rebuild trust and making the recovery less likely. Because these are automatic processes based on amount of prior experience, arguments and admonishments may not be effective approaches to changing beliefs about trustworthiness. And while this study deals with interpersonal trust, if dispositional accounts of trust are correct, and it seems

[9] Cary Funk considers data on public perceptions of scientific agreement over whether the MMR Vaccine is safe, climate change is due to human activity, and GM foods are safe to eat, noting that "from the public's perspective, there is considerable scientific disagreement about all three of these issues, particularly GM foods" (2017, p. 88). See also Millstone and van Zwanenberg (2000).

fair enough to say that we do trust objects in addition to agents, then it is at least reasonable to suppose that our brains might undergo similar processes in cases that transcend interpersonal trust. Of course, additional research is needed to confirm this supposition.

Realizing that experience informs everything from conscious processes to unconscious physiological process provides added motivation to take seriously arguments for finding ways to work with people's beliefs rather than trying to directly change or argue against those beliefs (Goldenberg, 2021). Rather than assume the cause of distrust or the ease with which individuals and communities can overcome that distrust, it is necessary to examine the background conditions involved. Not only will such examinations help to rebuild trust through processes of genuine dialogue and listening, they will also suggest appropriate expectations and approaches for rebuilding trust. For example, based on things like prior experience, shared identities, expertise, and so on, we can anticipate the length of recovery, difficulty of recovery, usefulness of community activities, and whether appeals to authority will be effective. What would be required to repair such breaches of trust will no doubt depend on the particulars of the situation but it is reasonable to suppose that one requirement will be numerous positive experiences of the sort where there had been negative ones.

The perhaps surprising implication of the research reviewed in this paper is that beliefs about trustworthiness are often arrived at not through direct experiences of the person about whom we form a belief but through experiences of other persons and situations which constitutes part of our framework for belief formation. In consequence, a reputable scientific authority is limited in its ability to establish its own trustworthiness.

3 Future Work

I have argued that cases where people's distrust of reliable scientific authority is based on experience, their distrust is, *prima facie*, rational because it is based on epistemic theories that defend beliefs based on experience and unavoidable in as much as it mirrors the psychological processes by which humans do in fact form beliefs. I make no normative claims that these psychological or epistemic processes, either formal or informal, are as they ought to be—maybe even our best epistemologies are irrational at some level and our psychological processes deeply incoherent—but given that we accept beliefs produced by both our psychological and rational systems *as rational beliefs*, appeals to experience as the reason for distrust, even if the resulting distrust is unadvisable or wrong, are likewise rational.

Although I have not attempted a comprehensive list of conditions under which distrust of reputable science is rational, I have considered several examples. To the degree that experience of scientific authority ends up negative or absent, we should consider distrust in a reputable scientific authority reasonable and likely. It is not a simple matter of choosing to trust scientific authorities—complex layers of experience produce the trust. As such, where we find individuals and communities distrusting scientific authority, determining the reason for this distrust will be complicated. Moreover, because experience is essential to trust in a scientific authority, where experience does not support such trust, authoritative appeals to the trustworthiness of science are unlikely to outweigh the experiences that led to the distrust; and such appeals may in fact be tantamount to the unreasonable request that a person adopt a belief without or contra good reason.

Additional areas of work are numerous. One such area involves collecting data from individuals who distrust scientific authority regarding their prior experiences with scientific authority as well as data on the amount and kind of negative experiences required to significantly increase the probability of distrust.

A second area that requires careful consideration are biases, such as those based on race, gender, sexual orientation, among others, and the role that experience, or lack thereof, plays in the formation of biases. It is likely that some biases would qualify as rational and some would not. For example, a woman, whose experience of men has involved being abused by them and therefore believes that all men are untrustworthy, would have a rational belief, even though in the case of trustworthy men, her belief that they are untrustworthy would be incorrect. On the other hand, biases that lack experience would not count as rational (with respect to experience), in addition to also being wrong on other grounds (such as moral or factual grounds).

Another area in need of development is work on the role experience plays in the formation of shared identities of distrust—for example, in groups dedicated to scientific conspiracy theories. Such communities sometimes appeal to negative experiences of scientific authority to justify their conspiracies. The argument of this paper does not defend conspiracy theories as rational—it only defends some experience-based distrust as rational.[10] Thus, whether some conspiracy theories might be based on rational distrust without therefore being rational, returns us to the earlier, bracketed concern over whether we want to consider beliefs produced by irrational systems of reasoning, rational or irrational. It also leaves unaddressed the possibility that some conspiracy theories might be operating on more than one system of reasoning such that the distrust may be reasonable, while the further views of the conspiracy may be produced by an entirely different reasoning system, rational or irrational.

References

Arimoto, T., & Sato, Y. (2012). Rebuilding public trust in science for policy-making. *Science, 337*(6099), 1176–1177. https://doi.org/10.1126/science.1224004

Barber, B. (1987). Trust in science. *Minerva, 25*(1/2), 123–134.

Bardon, A. (2020). *The truth about denial: Bias and self-deception in science, politics, and religion.* Oxford University Press.

Bogaerts, S., Vervaeke, G., & Goethals, J. (2000). Research on predictors for sexual delinquency. *European Journal on Criminal Policy and Research, 8*, 503–504.

[10] I am grateful to attendees of the SAS: Trust in Science Conference for highlighting this topic.

Chen, Y.-H., Chien, S.-H., Wu, J.-J., & Tsai, P.-Y. (2010). Impact of signals and experience on trust and trusting behavior. *Cyberpsychology, Behavior, and Social Networking, 13*(5), 539–546. https://doi.org/10.1089/cyber.2009.0188

Fareri, D. S., Chang, L. J., & Delgado, M. R. (2012). Effects of direct social experience on trust decisions and neural reward circuitry. *Frontiers in Neuroscience, 6*,. https://doi.org/10.3389/fnins.2012.00148

Funk, C. (2017). Mixed messages about public trust in science. *Issues in Science and Technology, 34*(1), 86–88.

Galavotti, M. C. (2005). *Philosophical introduction to probability*. Center for the Study of Language and Information Lecture Series 167. Stanford California: CSLI Publications.

Goldenberg, M. J. (2021). *Vaccine hesitancy: Public trust, expertise, and the war on science*. University of Pittsburgh Press.

Goudge, J., & Gilson, L. (2005). How can trust be investigated? Drawing lessons from past experience. *Social Science & Medicine, 61*, 1439–1451.

Harman, G. (1988). *Change in view: Principles of reasoning*. Bradford Books.

Huber, B., Barnidge, M., Gil de Zúñiga, H., & Liu, J. (2019). Fostering public trust in science: The role of social media. *Public Understanding of Science, 28*(7), 759–777. https://doi.org/10.1177/0963662519869097

Kim, Y. A., & Phalak, R. (2012). A trust prediction framework in rating-based experience sharing social networks without a web of trust. *Information Sciences, 191*, 128–145.

Krause, N. M., Scheufele, D. A., Freiling, I., & Brossard, D. (2021). The trust fallacy: Scientists' search for public pathologies is unhealthy, unhelpful, and ultimately unscientific. *American Scientist, 109*(4), 0.

Kunnel, A., & Quandt, T. (2016). Relational trust and distrust: Ingredients of face-to-face and media-based communication. In B. Blöbaum (Ed.), *Trust and communication in a digitized world* (pp. 27–49). Progress in IS. Springer International Publishing. https://doi.org/10.1007/978-3-319-28059-2_2

Lankton, N. K., McKnight, D. H., & Thatcher, J. B. (2012). The moderating effects of privacy restrictiveness and experience on trusting beliefs and habit: An empirical test of intention to continue using a social networking website. *IEEE Transactions on Engineering Management, 59*(4), 654–665. https://doi.org/10.1109/TEM.2011.2179048

Leshner, A. I. (2021). Trust in science is not the problem. *Issues in Science and Technology, 37*(3), 16–18.

Marlow, L. A., Waller, J., & Wardle, J. (2007). Trust and experience as predictors of HPV vaccine acceptance. *Human Vaccines, 3*(5), 171–175. https://doi.org/10.4161/hv.3.5.4310

McCabe, K., Houser, D., Ryan, L., Smith, V., & Trouard, T. (2001). A functional imaging study of cooperation in two-person reciprocal exchange. *Proceedings of the National Academy of Sciences, 98*(20), 11832–11835. https://doi.org/10.1073/pnas.211415698

Millstone, E., & van Zwanenberg, P. (2000). A crisis of trust: For science, scientists or for institutions? *Nature Medicine, 6*(12), 1307–1308. https://doi.org/10.1038/82102

Nguyen, C. T. (2022). Trust as an unquestioning attitude. In T. S. Gendler, J. Hawthorne & J. Chung (Eds.), *Oxford studies in epistemology* (1st ed., Vol. 7, pp. 214–44). Oxford University Press. https://doi.org/10.1093/oso/9780192868978.003.0007

Ophir, Y., & Jamieson, K. H. (2021). The effects of media narratives about failures and discoveries in science on beliefs about and support for science. *Public Understanding of Science*. 096366252110126. https://doi.org/10.1177/09636625211012630

Pedersen, D. B. (2014). The political epistemology of science-based policy-making. *Society, 51*(5), 547–551. https://doi.org/10.1007/s12115-014-9820-z

Rudner, R. (1953). The scientist qua scientist makes value judgements. *Philosophy of Science, 20*, 1–6.

Schilke, O., Reimann, M., & Cook, K. S. (2013). Effect of relationship experience on trust recovery following a breach. *Proceedings of the National Academy of Sciences, 110*(38), 15236–15241. https://doi.org/10.1073/pnas.1314857110

Seto, M. C. (n.d.). *Pedophilia and sexual offending against children.* American Psychological Association.

Solomon, M. (2021). Trust: *The need for public understanding of how science works. Hastings Center Report, 51*(January), S36-39. https://doi.org/10.1002/hast.1227

Sztompka, P. (2007). Trust in science: Robert K. Merton's inspirations. *Journal of Classical Sociology, 7*(2), 211–220. https://doi.org/10.1177/1468795X07078038

Wedgwood, R. (2018). Epistemic teleology. In K. Ahlstrom-Vij & J. Dunn (Eds.), *Epistemic consequentialism* (Vol. 85). Oxford University Press.

Weisberg, J. (2021). Formal epistemology. *The Stanford Encyclopedia of Philosophy.* https://plato.stanford.edu/archives/spr2021/entries/formal-epistemology/

Wilholt, T. (2013). Epistemic trust in science. *The British Journal for the Philosophy of Science, 64*(2), 233–253. https://doi.org/10.1093/bjps/axs007

Can There Be an Epistemic Authority?

Jörn Wiengarn

Abstract Linda Zazebskis's famous account of epistemic authority has provoked a particular line of criticism that argues against the very possibility of an authority in the epistemic realm. In this paper, I would like to discuss the scope of the most common arguments in this debate and defend the conceptual possibility of epistemic authorities. To this end, I will argue for an account of epistemic authority that roughly follows Zagzebski's but adds some further elements with reference to David Enoch's theory of authority. As I will show, this account can be defended against the three most common arguments against the possibility of epistemic authorities.

1 Introduction

Discussions about the notion of authority have a long tradition in philosophy, especially in political philosophy. In this context, authorities are typically regarded as belonging to the practical realm, i.e. they are understood primarily as individuals who have the normative power to demand certain *actions* from others. More recently, however, particularly due to the work of Zagzebski (2015), a different kind of authority, namely epistemic authority, has gotten more attention. Such authorities do not rule over other people's *actions but* are essentially authorities over other's *beliefs.*[1]

Zagzebski's intriguing proposal has provoked criticism of various kinds. Among them, there has been a particularly radical strand that has even disputed the sheer possibility and reasonableness of such a concept of epistemic authority at all. According to this line of criticism, the concept of epistemic authority is in itself

[1] It is debatable whether the concept of epistemic authorities should be understood even more broadly to include for example the capacity to give laypersons *explanations* for scientific phenomenon. In any case, however, to have authority over beliefs can be regarded a necessary component of epistemic authority, and it is this point that will be under discussion in the following.

J. Wiengarn (✉)
TU Darmstadt, Darmstadt, Germany
e-mail: joern.wiengarn@tu-darmstadt.de

M. M. Resch et al. (eds.), *The Science and Art of Simulation*,
https://doi.org/10.1007/978-3-031-68058-8_4

incoherent and must therefore be jettisoned. The various arguments and alternative proposals in this context vary greatly, of course. For present purposes, I will refer to the cluster of skeptical positions as *Skepticism about Epistemic Authority* (*SEA*).

The aim of this article is to defend the possibility of authorities in an epistemic sense against SEA. For this purpose, I will first introduce an understanding of authority based on David Enoch's account. As we will see, this account is more detailed and makes some stronger commitments than what Zagzebski says on the notion of authority. Its merit is that it seems to me to be closer to the everyday understanding of authority and can also be integrated into an already prominent general theory of authority. I will then show in detail that three prominent problems that proponents of SEA see in the notion of epistemic authority do not apply to it. These problems will be called: (i) the problem of command (ii) the wrong-kind-of-reasons-problem and (iii) the problem of preemption.

The advantage of approaching things this way, instead of directly discussing only Zagzebski's account, lies in the following consideration: some skeptics of the possibility of epistemic authority may think that none of the objections of SEA applies to Zagzebkski's consideration, but that this is only due to the fact that her account is too broad. Such a critic may thus say that Zagzebski can easily sidestep the criticism, but only at the cost of advocating an understanding of authority that is too distant from how authority is normally understood. Anticipating this, it will be shown directly that the arguments for SEA fail even if we presuppose a more committed understanding of authority, which such a critic would be more likely to agree with.

This is how we will proceed in detail. We will first articulate an account of authority that, based on Enoch's theory, stresses that authorities possess a normative power to give reasons in the so-called robust sense. After that, we will go through the three main arguments for SEA and show how it can be questioned if they are applied to the proposed account of authority.

2 What is an Authority?

In the following, we will develop an account of authority that is more detailed and committed than Zagzebski's reflections on the concept. A first thing to note is that the notion of authority can be formally introduced as a three-place-concept: Thus, (i) a person A is an authority (ii) for B (iii) in the field D (cf. Constantin & Grundmann, 2018, 4115; Jäger, 2016, 170).[2] Moreover, a first more substantial observation is that it seems to be essential for an authority A that she is a source of demands on B. Authorities are by definition in a position to make demands on others.

[2] One might discuss whether groups (such as expert panels) can also have the status of epistemic authority. I do not want to rule this out. However, I concentrate here on the case of an individual authority and assume that the arguments developed in this paper can also be applied to the case of expert groups.

So far so good. However, making demands on another person as such evidently does not make one an authority. For example, if I ask my flatmate to clean my room, that alone (unfortunately) does not make me an authority for him. Something is missing in the example to attribute an authoritative status to me. But can we say what exactly that is?

A first thing to note here is that genuine authorities seem to be in position to create reasons—to do a certain action D for example—by issuing a directive to do so. This can be made plausible by considering the fact that B can refer to the authority of A when asked why he is doing D. For example, if my flatmate asks me why we actually have to clean the stairwell every week, I can respond by saying: "Our landlord has asked us to do it". In doing so, I am referring to the claim of a relevant authority (our landlord) as a justifying and motivating reason for action.

However, this naturally leads to a follow-up question: What exactly does it mean to acquire such a reason from an authority? On this, Enoch (2011, 2014) introduces a promising approach to define what the essence of an authority is. According to him, a position of authority consists in a specific way that an authority gives reasons to another person. Thus, acquiring a reason from an authority is an instance of what David Enoch calls "robust reason-giving".

On his account, robust reason-giving is a specific way of how a person can give reasons to another that can at best be understood by distinguishing it from two other ways. More specifically, robust reason-giving must first be distinguished from what Enoch calls the purely epistemic reason-giving, where someone merely points out a certain epistemic fact to another. I may for example show my flatmate the cobwebs in the corners of his room that he was not aware of. In doing so, I simply indicate to him a reason to clean his room that was there all along and that exists independently of my giving him this reason.

Secondly, robust reason-giving needs to be distinguished from what Enoch labels "merely triggering reason-giving", where to merely trigger a reason means nothing more than to create circumstances that give someone else reasons for certain actions: For example, if I secretly plant cobwebs in the corners of my flat-mate's room, I thereby "give" him a reason to clean it. Or to use Enoch's example: "By placing his foot on the road, a pedestrian can give a driver a reason to stop". According to Enoch, in this case, the driver had the conditional reason "to-stop-should-a-pedestrian-start-crossing" all along. The act of "giving" a reason thus merely consists in bringing about the circumstances of the conditional so that the reason can kick in.

In contrast, when an authority issues a directive, she does not just indicate or merely trigger a reason. Rather, she gives a reason in a more robust sense, where, according to Enoch, in a case of robust reason-giving the reason-giver's *intention* to give a reason plays a direct role in the reason-giving process. While it is not necessary for epistemic or triggering reason-giving that the reason-giver actually has the intention to give a reason, things are different in the case of robust reason-giving.

To be more precise, one needs to take into account the fact that the reason giving person intends to give a reason *in the right kind of way*, namely, by intending the addressee to immediately recognize this very intention. Let's illustrate this point with the help of one of Enoch's examples: According to Enoch, if A requests B to read a paper she wrote, under normal circumstances, she requests B to read it out of *immediate* recognition of this very intention. To see this more clearly, consider the following variation of this case: Let's imagine the addressee B reads A's paper upon her request, but only to avoid a conflict with the department chair. In that case, it is not an acknowledgment of A's intention *as such* which gives B a reason to comply with her request. Rather, A's intention is taken as a non-normative circumstance that only triggers a reason, mediated by the addressee's *actual* intention to avoid social conflicts. In the case of robust reason-giving, however, the reason-giver has "the intention to give a reason merely by the very forming of the intention to give a reason" (Enoch, 2011, 7).

All in all, Enoch ends up with the following definition of robust reason-giving.

> One person A attempts to robustly give another person B a reason to u just in case (and because):
>
> (i) A intends to give B a reason to u, and A communicates this intention to B;
> (ii) A intends B to recognize this intention;
> (iii) A intends B's given reason to u to depend in an appropriate way on B's recognition of A's communicated intention to give B a reason to u (ibid.).

This gives us so far a proposal of how reasons of authority can be classified. It should be noted, however, that there are still different *modes* of reason-giving in a robust sense: A may *ask* B to do *u*, A may *beg* B to do *u*, and A may *demand* from B to do *u* etc. All of these are cases that seem to fit into Enoch's definition of robust reason-giving. But is there something specific in the way an *authority* robustly gives her subject a reason?

On this issue, Enoch makes two points. First, he claims that it is characteristic of authorities that by issuing a directive to do *u* they create a duty to create *u*.[3] This proposal seems to have a certain plausibility, even if it can be noticed that Enoch struggles with giving a clear definition of what "duty" means in the relevant sense here (cf. ibid., 25). For present purposes, though, it seems to be sufficient to claim that authorities are in a position to create relatively *strong* normative pressure compared to asking for favours, without specifying to *what exactly* this normative pressure amounts to. Drawing a perfect line here will not be important in defending the proposed account against SEA.

At this point, in line with Zagzebski and Enoch, I will focus on Joseph Raz's claim that authorities essentially give their reasons *preemptively*. That means that authorities give reasons that *replace* other reasons. In other words, a preemptive reason is a reason "which is not to be added to all other relevant reasons when assessing what to do but should exclude and take the place of some of them" (Raz, 1988, 46).

It should be noted that I take this to be a conceptual point about what recognizing an authority essentially amounts to. As Arnon Keren pointed out, there are two sides to the preemptive thesis as introduced by Raz (Keren, 2014, 63–66). On the one side, the thesis is surely meant as a normative claim. Thus, Raz and Zagzebski argue that it

[3] Keren (2014, 66) explicitly suggests that this is also charactersitic of epistemic authorities.

is often rational to defer to an authority. I will ignore this point for present purposes, since I am primarily interested in a conceptual clarification. On the other side, the preemption thesis can be read as a *conceptual* remark on what it simply means to acquire reasons from an authority. The following example may help to bring this point to the fore: Imagine A tells B that p and that, instead of taking her words at face value and preemptively believing her that p, B starts to seriously weigh in all possible kinds of reasons which bear on whether p is true or not. In this case, it seems that to the extent that B is doing so, he does exactly *not* recognize A as an *authority*. It is this conceptual intuition that the preemption thesis promises to capture.

A similar argument was put forward by Aron Keren. He argues that if we think that someone should recognize another person as an authority, we can criticize her refusal to adopt the authorities' belief and to replace her own considerations with it: "if we point out to person *A* that she should believe that *p* on *B*'s epistemic authority, then we cannot at the same time suggest that *A* should weigh all the relevant evidence available to her" (ibid., 64). This would mean to do two "incompatible" things at once.

To sum up, the here presented account states that if A is an authority for B in a domain D, she is in a position to give reasons to B (that are relevant to D), and she does so in a, first robust, second duty-creating and third preemptive manner. Before we'll examine how the typical arguments of SEA do not apply to this approach, two more aspects need to be clarified. First, a point from above needs to be made more precise: Above, following Enoch, we distinguished robust-reason giving from *mere* triggering reason-giving. The "mere" must be emphasized as Enoch claims that robust-reason giving represents a *specific* instance of triggering reason-giving. The idea behind this thought is that in order for a mechanism of robust reason-giving to take effect there must be certain success conditions in place. More precisely, Enoch argues that an authority can only create a reason, if there is a *general reason to attribute such an authoritative status to her*. That is why in the example above, I cannot command my roommate to clean my room. There is no reasoning to regard me as an authority with regards to the sort of directive that I issue.

Similarly, to develop this point a bit, it seems to me that from the perspective of a subject, who acquires and acts on such an authority-based reason, she must assume that there really are reasons to accept the authority. Her acceptance is at least not assumed to be a completely irrational attitude. In general, there seem to be some *standards* to evaluate whether it is appropriate to recognize someone as an authority or not. And in recognizing an authority one *assumes* that these standards are being sufficiently met. This does not necessarily mean that these standards really *are* sufficiently met. One can go wrong in recognizing someone as an authority. And accordingly, someone's recognition of an authority can be criticized as not sufficiently justified, as naive or maybe as a result of manipulation. But still, from the viewpoint of the recognizing subject the standards for recognizing an authority are assumed to be sufficiently fulfilled.

A last point that needs to be stressed in anticipation of the following discussion, is that the kind of reasons that authorities give has nothing to do with threats of sanctions. Quite often, especially when we talk about political authorities, we view these as essentially having the standing to coerce others to follow, i.e., via the power to impose *sanctions* on disobedient subjects. However, according to Enoch, there are good reasons to think that getting others to follow one's commands by means of *mere* coercion constitutes a case of triggering reason-giving. To do as the political authority commands out of fear for possible sanctions is a mere way of avoiding the kind of circumstances the authority may bring about, if one does not act as she is ordered. This has nothing to do with immediately recognizing the authority's intention, thereby gaining a direct reason to follow her directive. To put the thought another way: To the extent that a political power is *not recognized* as an authority, it has to use threat and sanctions to enforce its will.[4]

Certainly, this approach can be developed in more detail and more can be said about it. For present purposes, however, I will leave it with the remarks so far. As I will show in the following, we are already in a position to show how the account is equipped to sidestep the usual objections against the idea of an epistemic authority. This being said, I can now address a first standard objection.

3 Skepticism Against the Possibility of Epistemic Authorities

3.1 The Problem of Commanding Beliefs

The problem of command has been articulated among others by Zagzebski herself. Simply put, it states that an epistemic authority is not a conceivable entity since *belief cannot be commanded.* This objection, however, can be understood in different ways. First, commanding can be understood in the sense of coercion, i.e., as a form of threat of sanctions. This is also the understanding that Zagzebski has in mind when she expresses the following worry: "how is it even possible to successfully coerce someone else's belief?" (Zagzebski, 2015, 100). Proponents of SEA may question that it is possible to believe on such a threat-based command. For how could threat-based commands possibly give one good reasons to actually believe? At best, so the skeptic may say, commands could give one reason to act *as if* one believed, but one cannot really believe *on* command.

The objection thus understood, however, poses no real threat to the conceivability of an epistemic authority, if we understand the concept of authority according to our account. That is for plain reasons: we explicitly excluded coercive power as a defining trait of an authority, in contrast to how the notion is usually understood in the political domain. Quite trivially therefore, our account circumvents the objection.

The proponent of SEA, however, may insist that even if "commanding" had nothing to do with coercive power, there is still a worry. For this purpose, she may

[4] This does not exclude that threats and sanctions can be a means to try to gain recognition as an authority.

refer to an example like the following one: Imagine a devoted and loyal servant of a king. At some point the king demands to believe that there is a herd of wild unicorns in his kingdom.[5] It now seems that no matter how loyal the servant is, he simply cannot start to believe such obvious absurdities on command.

However, our conception of epistemic authority allows a certain interpretation of this alleged counterexample that would still preserve the possibility of an epistemic authority. To see this, it needs to be recalled first that we assumed that authorities are always authorities in certain respects, i.e. in a certain domain D. Thus, we can differentiate the respects in which someone is or is not to be regarded as authority. This opens up the following reading of the given example: It may be that the servant recognizes the king as an authority in *many* respects. However, to the extent that he does not believe what he is demanded to believe, we can say that *apparently* he *simply does not* recognize the king as an epistemic authority *with regards to the kind of belief he demands*. The example thus illustrates a case where there simply *is no* relevant epistemic authority. Thus, according to this reading, we are again dealing with a *mere* "demand", but not a demand within the domain where the authority is recognized in a relevant sense. However, it is crucial that mere demands and demands from recognized authorities do not get conflated. To recall the point from above: that was the reason why I cannot demand from my flatmate to clean my room.

Thus, the alleged counterexample simply leaves open the possibility of epistemic authorities. Why exactly we shouldn't accept that a servant can believe a king on demand, as long as he is properly recognized, is still left unclear. If anything, a proponent of SEA would have to elaborate further on why it is not possible to demand a belief. At this point, a second objection can be formulated.

3.2 The Wrong-Kind-of-Reasons-Problem

The objection to the problem of command draws attention to the fact that there is a relevant distinction between mere demands and demands from a genuinely recognized authority. The wrong-kind-of-reasons-problem, however, can be regarded as going one step further. It can be read as explicitly questioning that even such a thing—gaining a belief on the basis of a demand *from a recognized authority*—is conceivable at all. I will introduce this argument in what I take to be its invariant, most general form. Afterwards I will present a more concrete version of it, formulated by David Enoch himself.[6]

Schematically, the argument works like this: The first thing to note is that the right kind of reasons for a belief that p seems to be reasons that bear on the truth of p. Following an influential account by Hieronymi (2013) right kind of reasons for an attitude are such reasons that bear on a specific question that is constitutive of

[5] Cf. the example given by Zagzebski (2015, 101).

[6] Zagzebski may also have this point in mind when she is careful to emphasise that her approach does not buy into the assumption that we can believe on command (ibid., 102).

that attitude. For example, an attitude like the belief that p can be seen as answering the question of whether p is the case in the affirmative. And right kind of reasons for this belief are such that bear positively on to the question of whether p. We will call such reasons, that bear on the question of whether something is the case, evidential reasons. Wrong kind of reasons, on the other hand, only look superficially like genuine reasons. Let's take the example of "It would be pleasant to believe p". Judging by its mere form, this looks like a reason to believe that p. After all, the fact that it would be pleasant counts in favor of believing p. However, it is not a reason for which one could *actually believe*, i.e., it is not a reason which can *constitute* a belief that p. That is because this wrong kind of reason does not respond to the question of whether p and therefore does not bear on the truth of p. Right kind of reasons for belief in contrast bear on the truth of the belief.

Against the backdrop of these remarks the skeptic may in a second step raise the question of how authoritative demands can give a reason of the right kind for beliefs. Remember that the way authorities give their subjects reasons rests on a mechanism of robust reason-giving. That means, authorities give their subjects reasons for actions or attitudes via the subject's recognition of the authorities very intention to do so. The question a skeptic of epistemic authority could now ask is: How could the mere recognition of this intention give right kind of reasons for belief? The given reason only seems to rest on the recognition of an intention, how can this in any sense speak in favor of the truth of the claimed belief? How can an authority's mere intention to want us to believe that p speak in favor of a positive answer to the question whether p is the case?

This seems to be the implicit consideration underlying the worry that David Enoch himself expresses against the possibility of epistemic authorities.[7] Enoch stresses that the question of whether one recognizes another individual's intention in the robust reason-giving sense, depends essentially on the kind of relationship one has to the other person: "Because of the centrality of the persons to robust reason-giving, it is unsurprising that personal relations are also relevant here" (Enoch, 2011, 7). For Enoch a decisive factor that determines whether an agent succeeds in giving a reason in a robust sense is the "nature of the relevant relationship" between the authority and subject. If, for example, I request a stranger to read a draft of my paper, he would probably not take my intention as a reason to do so. If I asked my best friend instead, he would probably comply. Hence, the kind of relationship seems to make a difference here with regards to the success-condition of robust reason-giving.

But how, Enoch asks, can the nature of one's relationship to a person be of *epistemic* relevance? It seems that "personal relationships are completely irrelevant when it comes to the giving of epistemic reasons" (ibid., 8). In general, therefore, according to Enoch, no epistemic reasons can be gained through robust reason-giving. And this seems to be equally plausible in the case of epistemic authorities. One may grant that in recognizing someone as an authority that person can pose a normative demand.

[7] This recalls the debate in the philosophy of testimony on how trust in a testifier as a *social* or *interpersonal* phenomenon can lead to an *epistemically* justified belief (cf. Darwall, 2006, 253; Simpson, 2018).

But how can such a demand support a *belief* for which evidential reasons are the only reasons of the right kind?

As a response to this worry, we can start by reminding ourselves that to recognize an authority means to assume that certain standards are being sufficiently met that justify recognizing her. However, it must be noted that the kinds of standards we appeal to may substantially *vary* depending on the field or domain of the authority in question: standards for an authority in domain D may be different to standards in domain D* or D**. This is a point that has already been stressed by Joseph Raz himself, who remarked that "[t]he non-excluded reasons and the grounds for challenging an authority's directives vary from case to case. They determine the conditions of legitimacy of the authority and the limits of its rightful power" (Raz, 1988, 46).

It seems to me that the wrong-kind-of-reasons-problem is silently based on the assumption that there is only one kind of standard for practical and epistemic authorities alike. However, if we allow that there can be various standards depending on the kind of authority we are talking about, a path opens up to see how recognition of an epistemic authority can be epistemically relevant. The solution may thus lie in simply accepting that the standards to recognize an *epistemic* authority are decisively *evidential* standards that *essentially bear on the trustworthiness (simply put: her competence and her honesty) of the authority and thereby indirectly on the truth of what she says.*[8] These standards could for example be: someone's reputation within a scientific community, someone's display of epistemic virtues, someone's capability to make accurate predictions, etc. We cannot start a discussion at this point about which standards should be considered most relevant for assessing the trustworthiness of epistemic authorities. What is important is the basic idea here: All of the listed standards are criteria that promise to speak in favour of the truth of what the person says and may thus serve as standards I can refer to in order to justify my recognition of her as an epistemic authority.

Nothing in our definition of authority seems to prohibit this move. If we thus just allow that such evidential standards are the kind of standards that are apt for epistemic authorities, it becomes clear the reasons to recognize the authority are at the same time reasons which evidentially support what the authority demands to believe.

3.3 *The Problem of Preemption*

We now come to what is probably the most prominent objection to the concept of epistemic authority, the problem of preemption. To recall, we postulated in agreement with Zagzebski's view that to recognize an authority *as an authority* goes along with

[8] This relates to Raz' idea that the reasons an authority's utterance gives are content-independent. To the extend that we can treat someone as an authority, we have reason to believe her, *whatever* she says (cf. Zagzebksi, 2015, 106). This links to the parallel possibility of *justifying* one's beliefs on authority independent from its content, but rather by reasoning for the authority's trustworthiness.

a preemption (cf. Raz, 1988: 42, 57–59) of reasons, i.e., to believe on authority that p *replaces* other relevant reasons for or against believing p. In other words, when one believes that p on authority, one does not consider any other reasons for or against believing p. Let's call this the Preemption Thesis.

Critics of this thesis, however, have argued that it is too strong. They say it is prone to lead to *blind trust* in authority and cannot meet even minimal standards of critical thinking (cf. Dormandy, 2018; Jäger, 2016; Lackey, 2016). To argue for this view, various counterexamples have been put forward, that prove that one can clearly be epistemically better off by not just taking an authority's words that p while screening off first-hand evidence for or against p. For present purposes, though, we cannot possibly resume the whole debate in detail, but will only sketch a broad outline of how our analysis of epistemic authority seems to remain untouched by it.

For that purpose, it should first be noted, that most of the objections against the Preemption Thesis are definitively meant as normative objections, i.e., they aim at showing that it is not advisable to just believe on authority preemptively. In this sense, however, our conceptual analysis is not challenged. Such criticism is still *logically* compatible with the above account, they only amount to making clear that *in many situations* it is not *normatively* recommendable to preemptively believe an authority. In line with this thought, Zagzebski herself hints to the fact that some objections only raise the question of *who* should be treated as an epistemic authority, but they do not threaten the concept as such (cf. Zagzebski, 2020, 284f.).

To make this point more salient, remember that our conceptual thesis only stated that *insofar as* one recognizes an authority (with regards to p), one does not weigh up reasons for or against p. To do so, we claimed above, would be a conceptually incoherent thing to do. However, this does not mean that by recognizing A as an authority, B makes a commitment to A that condemns him to recognize her *forever*.[9] Recognizing an authority does not constitute a form of obligation with only minor release conditions, as may be the case when B promises his loyalty to A. It is therefore possible to recognise an authority, while being able to seriously question her authoritative status at any time. The only point is that *to the extent that* one does question an authority, one does exactly not recognise her.

In particular, the proposed account allows for the possibility of recognising a particular authority, but at the same time staying *receptive* to reasons that could potentially question her authoritative status. To use a common formula from the philosophy of trust: to recognize an authority does not exclude a *counterfactual sensitivity* to evidence that could undermine her authoritative position. Counterfactual sensitivity here can be defined, with the help of Miranda Fricker, as a certain "critical openness", a state of remaining "alert to the plethora of signs, prompts and cues that bear on how far we should trust" (Fricker, 2007, 66). Staying counterfactually sensitive is thus one way of thinking about how one can recognize an authority without necessarily having to blindly trust her.

However, some of the criticism against the Preemption Thesis seems to entail not only a normative, but also a conceptual worry. This applies for example to a critical

[9] For a related argument, see the analysis by Bokros (2020, 12060–12064).

point that Jennifer Lackey raises. Lackey claims that there are some "alternative polices" of how to treat an authority other than accepting her belief preemptively. These are for example: "follow the advice of an authority, except when one is certain that the authority is wrong" (Lackey, 2016, 576). The interesting question to ask now is not whether the normative advice that Lackey gives is sound. The issue is rather that her argument presupposes that it is conceptually possible to regard someone as an authority while *at the same time* not following her advice preemptively, and more than that: while not following her advice at all. Lackey therefore seems to suggest that the power to preempt reasons does not represent an *essential* feature of authorities. We can regard someone as an authority and still, if there are reasons to assume that she is wrong in a particular instance, not take on her belief. Authorities and preemption are thus two completely uncoupled ideas.

How does this view square with the idea that preemption is characteristic of how authorities give their reasons? I think it is possible to hold Lackey's advice as plausible while at the same time holding that authority and preemption indeed are conceptually linked. However, they are not conceptually linked in a strict sense, i.e., in the sense that preemption constitutes a *necessary condition* of authorities. Rather, it seems to count as characteristic or *typical* for epistemic authorities to give their reasons preemptively, where typical does not mean necessary. To see more clearly what I have in mind, consider the following scenarios: First, imagine that in a particular isolated instance you question the truth of what a recognised authority says. It seems like this does not necessitate that one has to completely write her off as an authority. As long as her mistake remains an isolated exception, her authoritative status can remain unquestioned. By contrast, imagine you are dealing with an authority who repeatedly, over and over again, makes false statements. Wouldn't this be a case where you have precisely good reason to question whether this is a *genuine* epistemic authority after all? How can she still count as an authority, if even you, as a layperson can see over and over again that she is wrong? This intuition shows that there is a conceptual connection, albeit not a strict one, between authority and the power of preemptive reasons that cannot easily be denied.

4 Conclusion

The aim of this article was to present a possible way in which one can make sense of the notion of epistemic authority against some obvious objections. It thus promises to open a path to make conceivable in what sense there can be authorities over beliefs. To recapitulate in a nutshell, we saw that the problem of commanding beliefs basically conflates mere demand and demands from recognized authorities. The wrong-kind-of-reasons problem could be shown to rest on the unfounded assumption that there can be no evidential standards to evaluate an authority. And finally, the problem of preemption either makes a normative, but not a conceptual point, or as in the case of Lackey, it only shows that there is no strict conceptual relation between authorities and preemption.

References

Bokros, S. (2020). A deference model of epistemic authority. *Synthese, 198*(12), 12041–12069.

Constantin, J., & Grundmann, T. (2018). Epistemic authority: Preemption through source sensitive defeat. *Synthese, 197*(4), 1–22.

Darwall, S. (2006). *The second-person standpoint: Morality, respect, and accountability*. Harvard University Press.

Dormandy, K. (2018). Epistemic authority: Preemption or proper basing? *Erkenntnis, 83*(4), 773–791.

Enoch, D. (2011). Giving practical reasons. *The Philosopher's Imprint, 11*(4).

Enoch, D. (2014). Authority and reason-giving. *Philosophy and Phenomenological Research, 89*(2), 296–332.

Fricker, M. (2007). *Epistemic injustice: Power and the ethics of knowing*. Oxford University Press.

Hieronymi, P. (2013). The use of reasons in thought (and the use of earmarks in arguments). *Ethics, 124*(1), 114–127.

Jäger, C. (2016). Epistemic authority, preemptive reasons, and understanding. *Episteme, 13*(2), 167–185.

Keren, A. (2014). Zagzebski on authority and preemption in the domain of belief. *European Journal for Philosophy of Religion, 6*(4), 61–76.

Lackey, J. (2016). To preempt and not to preempt. *Episteme, 13*(4), 571–576.

McMyler, B. (2011). *Testimony, trust, and authority*. Oxford University Press.

Raz, J. (1988). *The morality of freedom, Paperbackedn*. Oxford University Press.

Simpson, T. W. (2018). Trust, belief, and the second-personal. *Australasian Journal of Philosophy, 96*(3), 447–459.

Zagzebski, L. T. (2015). *Epistemic authority: A theory of trust, authority, and autonomy in belief*. Oxford University Press.

Zagzebski, L. (2020). A defense of epistemic authority. In: L. Zazebski (Ed.), *Epistemic values: Collected papers in epistemology* (pp. 275–288). Oxford University Press.

Trust Science with What? Trust-Building Dialogue Between Scientists and the Public

Elizabeth Stewart

Abstract The lack of public trust in science poses a challenge for the kind of coordination required to resolve pressing social and environmental issues. The reasons for this lack of trust are many, but I argue that one such reason includes disagreement regarding the proper role of science in our shared lives. When people are asked to "trust the science", it is often unclear what that amounts to. People are rightfully wary of handing scientists a blank signed check when they don't understand how it will be cashed. In this paper, I propose a framework for trust-building negotiations aimed at identifying and resolving disagreements regarding the scope of science. I use this framework to understand current distrust in vaccine science.

Introduction

As the world faces multiple crises that require collective action to resolve, such as climate change and a global pandemic, there is an accompanying crisis that hinders the needed collective action; a crisis of trust (Goldenberg, 2021). This trust crisis is clearly exhibited in the negative reactions to the efforts to mitigate the spread of Covid-19. Throughout the pandemic, there has been widespread resistance to social distancing, mask wearing, and vaccines, despite over 4.5 million deaths worldwide as of August, 31 2021 (World Health Organization, 2021). This resistance has often been met with calls to "just trust the science", calls which, perhaps unsurprisingly, go unheeded. While trust is necessary for successful coordination, trust also leaves people vulnerable to harm (Baier, 1986; Dormandy, 2020; O'Neill, 2018). Trusting the untrustworthy can have serious consequences for trusters, especially when that trust makes a person physically or economically vulnerable. Thus, it is reasonable for the public to be cautious about trusting scientists. In crises requiring scientific expertise to resolve, such as the Covid-19 pandemic, this caution must be met with trustworthiness.

Many conversations regarding trustworthiness center around whether or not a trustee is willing and able to do what has been entrusted to them (Hardin, 2002; Jones,

E. Stewart (✉)
Department of Philosophy, University of Canterbury, Christchurch, New Zealand
e-mail: elizabeth.stewart@canterbury.ac.nz

M. M. Resch et al. (eds.), *The Science and Art of Simulation*,
https://doi.org/10.1007/978-3-031-68058-8_5

2012; O'Neill, 2018, 2020; Scheman, 2020). Disagreements regarding trustworthy science, then, amount to disagreements about whether scientists are willing and able to do science. However, conversations that focus on whether or not scientists are "doing science" properly assume that people agree on what "doing science" means. While there is certainly distrust regarding scientists' ability to "do science", another source of public distrust towards science arises from disagreement about what science is and what role it ought to play in our shared lives. In particular, there is disagreement about what role science should play in both the public sphere of policy-making and in private decision-making practices. When asked to "trust science", individuals may rightly question where the limits of the demanded trust lie. Without knowing what the limits of appropriate trust are, it is difficult for scientists and the public alike to assess the trustworthiness of science. If the public trusts science to do something that scientists don't think they should concern themselves with, and subsequently the scientists fail to meet the public's expectations, then the public feels justified in deeming science "untrustworthy", despite the scientists' pleas to the contrary.

For example, based on their knowledge of the development of previous vaccines, many members of the public expected scientists to take longer to develop and verify the safety of the Covid-19 vaccine and the negative response to its emergency authorization status emphasized the importance of this expectation. Many of the people hesitant to get the vaccine adopted the stance of "waiting to see what happens". That is, rather than trusting scientists to tell them when the vaccine was safe enough, they wanted to determine the appropriate timeframe for assessing vaccine safety themselves. Scientists, however, were less concerned with the amount of time taken than with the results of the existing trials, which were positive. Despite scientists' reassurances that the vaccine was safe, many of those who expected a longer period of testing remained distrustful of the vaccine and its safety.

Misperceptions, miscommunication and disagreements regarding the appropriate role of science in decision-making is a source of distrust that is not easily addressed through increased science education, transparency about scientific practices, etc., at least not in isolation. As Goldenberg (2021) has argued, more information does not always address the underlying value-laden disagreements. Thus, in addition to increased science education, transparency, etc., building trust requires entering into negotiations about what the public ought to expect of science and what commitments scientists should make in response. In this paper, I explore how we might build public trust in science through engaging in dialogue about the proper scope of trust in science. Through raising and addressing questions about the public's expectations of science and scientists' commitments in doing science, we can mitigate at least some sources of distrust in otherwise trustworthy science.

The paper proceeds as follows. In part one, I briefly introduce a framework for trust-negotiations. In part two, I demonstrate how this framework can be applied to current distrust toward science regarding the Covid-19 pandemic response. I will not give a comprehensive account of what I think the results of the negotiations should look like, instead I simply aim to demonstrate what these negotiations might involve. In particular, I will focus on what negotiations between scientists and the public about Covid-19 vaccinations might involve.

1 Trust Negotiations

I understand trust as a three-place relation in which a truster, A, trusts a trustee, B, with respect to a domain of trust, x (Baier, 1986; Hardin, 2001; Jones, 1996, 2015). The limits of the trust domain are fixed by the expectations that the truster places on the trustee (Stewart, 2024). Let's suppose that a student, A, trusts a professor, B, with respect to the domain of teaching philosophy. In trusting their professor, A expects that the professor knows about philosophy, is willing to share that knowledge, is capable of finding creative ways to communicate, and cares about students' well-being. Following the formalization of trust domains developed in (Stewart, 2024), we can represent A's trust domain as follows:

$$Teaching\ Philosophy|A_{trust} = \{knows\ philosophy,\ willing\ to\ teach,\ creative\ communicator,\ caring\}$$

Not all students share the same expectations of their professor. Another student, C, trusts their professor with the following trust domain:

$$Teaching\ Philosophy|C_{trust} = \{knows\ philosophy,\ willing\ to\ teach,\ responds\ to\ emails\ within\ ten\ minutes,\ easy\ grader\}$$

We can see that not only do people differ in what they trust others to do or be, they also differ in how reasonable they are in setting these expectations. People do not always place their trust well and sometimes trust things of others that they should not. Trust, then, can be more or less appropriate depending on what expectations the truster has of the trustee. Thus, trustworthiness does not simply involve meeting a truster's expectations because those expectations might be misplaced. This means that a truster with misplaced expectations may deem a trustee untrustworthy, even when there are good reasons to consider them trustworthy. We would not, for example, deem a professor that fails to reply to emails within ten minutes "untrustworthy". Furthermore, we should expect a trustworthy philosophy professor to have concrete strategies for creatively and effectively communicating with students. We can represent the professor's response in accepting their students' trust to teach them philosophy as follows:

$$Professor(Teaching\ Philosophy)_{response} = \{knowledge\ of\ philosophy,\ assign\ relevant\ homework,\ timely\ email\ response,\ care\ about\ student\ well-being,\ etc...\}$$

Trust is broken when the trustee's response does not align with the truster's expectations, that is, when the trust domain is incompatible with the response domain. Mismatches between trust and response domains are not always due to ill will or negligence. Sometimes they occur due to a simple miscommunication. Other times, they occur because there is a disagreement about what should be expected in a given situation. Thus, to foster and maintain a trust relationship, it is important for trusters and trustees to be aware of what the other expects or is willing to commit to. In many cases, this happens organically, through explicit communication or shared understanding of cultural norms. However, sometimes, aligning the trust and response domains requires negotiation regarding what is reasonable to expect, when it is sensible to change our expectations, and how best to prioritize expectations.

For example, there is a clear mismatch between how soon Student *C* expects an email response and how quickly the professor will reply. In order to foster and improve the trust relationship, it would be helpful if both asked themselves what a reasonable email response time might be. The philosophy professor might explain that they have family and responsibilities outside of teaching, and thus they cannot always be available via email, but they will do their best to respond within one business day. If Student *C* modifies their trust domain so that it no longer includes the expectation about ten-minute replies, then they are more likely to count the professor trustworthy.

I believe that this kind of negotiation process could serve as a model to guide communication processes between scientists and the public. In what follows, I propose several stages in which these negotiations could proceed. In stage one, existing expectations and commitments must be clarified. In stage two, mismatches between expectations and commitments are identified. In stage three, both parties must provide input regarding how their expectations and commitments might change in order to create greater alignment. This input must address what people, both the public and scientists, ought to expect of science (the scope of the trust domain), what expectations are most important to maintain (the ordering of the trust domain), and the extent to which those expectations ought to change in different contexts (the rigidity of the domain). The aim of these negotiations must not merely aim at changing the public's attitude. Scientists and policy-makers must also honestly assess the capabilities and limits of science. The following sections explore what these stages might look like with respect to public trust, and lack thereof, with respect to vaccination.

It is worth noting that what follows is intended to represent an idealized framework that can be applied to a variety of different trust relationships. These negotiations may occur at an individual level, between say, a patient and their doctor. Alternatively, they may occur at a broader societal level. For example, returning to the example of professor and student, the role that the syllabus plays in the relationship between student and professor has changed over time in response to social changes. A professor's syllabus used to simply provide information about what books to buy and a rough timeline of when various topics would be covered. In response to students' concerns, syllabi now provide concrete details regarding what professors are willing to commit to and what students should expect.[1] Whether occurring at an individual or social level, trust negotiations in concrete situations are likely to be far more messy than what I describe. The hope, however, is that while this framework is idealized, it may still prove useful in shaping the direction of trust-building dialogue.

[1] Many thanks to Brett Sherman, who provided this example.

2 Stage One: Clarifying Trust and Commitment Domains

Negotiations between what the public expects and what scientists are willing to commit to should begin with an investigation into what existing trust and commitment domains look like. While the expectations that bound trust domains vary across individuals and groups, there are several consistent themes of concern about vaccination. Issues of safety are primary among these concerns, but other sources of distrust include lack of representation, perceived disrespect, and financial and political conflicts of interest (Goldenberg, 2021). However, these concerns often remain under-examined. For example, many people are hesitant to get the Covid-19 vaccine because it is new and they just want to "wait and see" what happens to those who have taken it (Brown et al., 2022). It is often unclear, however, what exactly they are waiting to see or how long it will take for them to see it. For example, for those who report in surveys that they wish to wait to assess the vaccine's long-term effects, it is often left unexplored how long they wish to wait or what would count as sufficient evidence of safety (Brown et al., 2022; Latkin et al., 2021). Without a clear benchmark of what "safety" requires, trusters can continually shift the benchmark, making it impossible for trustees to meet the expectations. Having potential trusters clarify what, exactly, they expect of "trustworthy science" provides scientists with potential benchmarks to aim toward.

3 Stage Two: Identifying Mismatches

Once both parties have clarified what they expect and what they are willing to commit to doing, we can then move forward in finding where the domains align and where they diverge. For example, both scientists and the public want vaccines to be safe. However, in her book on vaccine hesitancy, Maya Goldenberg discusses how the scientific community's arguments for vaccine hesitancy are targeted at population level assessments of safety, but many people want to know if the vaccine will be safe *for them* (Goldenberg, 2021). Those who are vaccine hesitant want assurance that they will not have an adverse reaction to a vaccine. The expected benchmark for safety, is thus set impossibly high.

Assessing the safety of a vaccine for each particular individual is not something that scientists can do with one hundred percent certainty. Scientists can identify features that may make an individual more likely to have an adverse reaction. Indeed, scientists are currently investing time, money and energy into identifying the factors that make someone more likely to have an adverse reaction to the COVID vaccine, although these efforts are currently incomplete. However, even with the knowledge about risk factors, scientists can't guarantee that the vaccine will or will not cause some particular individual harm. Thus, scientists do not commit to guaranteeing this level of certainty regarding vaccine safety, nor should they. They can, however, guarantee that the vaccine will be safe for the vast majority of individuals.

Thus, while both scientists and the public are concerned with the safety of vaccines, they may diverge in how they interpret the appropriate benchmark for "safety". Many scientists conclude that, given the very small likelihood of an adverse reaction, we all should consider the vaccine "safe". However, for many individuals, population level safety is not good enough. Herein lies a mismatch between the vaccine hesitant and scientists which provides an opportunity for dialogue aimed at alignment.

4 Stage Three: Moving Toward Alignment

When asked to "just trust the science", the public are asked not only to trust that science can accurately determine what the risk of an adverse reaction is, but also whether that is a risk worth taking. However, determining what the appropriate expectation regarding vaccine safety should be, that is, determining the appropriate level of risk to take, is not something that clearly falls within the scope of science. Figuring out how safe is "safe enough" is a complex task that requires more than simply knowing the proportion of people likely to experience an adverse reaction. It also includes determining the value of that information.

Consider how knowing figures about population level safety may hold varying significance given an individual's particular social and historical context and their tolerance for risk. For example, some Black Americans were hesitant to receive the vaccine because of historical and ongoing medical discrimination (Momplaisir et al., 2021). Thus, what mattered, in part, was whether the vaccine was safe and effective not just at the population level, but for the Black community in particular. Similarly, people with compromised immune systems may have deemed the risk of getting the vaccine to be greater than the risks associated with maintaining social isolation, given their particular health context. Alternatively, some people might value "natural" approaches to maintaining health, such as certain nutritional or lifestyle choices, and may therefore find any risks of the vaccine as unnecessary (Latkin et al., 2021).

Moving toward alignment, then, might include scientists letting go of commitments to determine the appropriate benchmarks for safety. Instead, this process might include greater representation from members of the public or public representatives. To be fair to scientists, this is largely what government regulators, such as the US's FDA, are supposed to do. However, if people distrust these regulatory bodies in similar ways as they do scientists, then this simply extends the issue of broken trust beyond the realm of science to include governments as well. It may be important, however, for those who are vaccine hesitant to recognize whether their lack of trust is directed at scientists or some closely aligned, but separate, institution, such as regulators of science or public health policy-makers. As is, distrust toward governing bodies is often mistakenly aimed at scientists and vice versa.

In moving the task of determining that appropriate level of risk from the domain of science into the public sphere, the commitments of scientists change from determining how safe "safe" is, to just determining the likelihood of adverse reactions.

This likelihood is then evaluated to be "safe" or "unsafe" by the public, or representatives thereof. Thus, in order to foster trust, both parties should be open to changing their respective trust and commitment domains. Members of the public should reach some agreement on what counts as "safe" and, in response, members of the scientific community should remove this task from their commitment domain. This process may not follow the exact steps outlined above, but may involve an ongoing dialogue aimed at increasing alignment over time. As noted above, the outlined steps are simply an idealized model of how trust negotiations can unfold.

Conclusion

The hope of trust negotiations is to bring trusters' expectations and the responses of trustees to that trust into greater alignment. In the case of public trust in vaccine science this may look like changing the following existing trust and response domains:

$$\mathit{Vaccine\ Science}|\mathit{Public}_{trust} = \{\mathit{safety}\}$$
$$\mathit{Scientists}(\mathit{Vaccine\ Science})_{response} = \{\mathit{identify\ risk\ of\ adverse\ reactions},\ \mathit{determine\ safe\ if\ risk} < x\}$$

Through negotiations, the hope is that we can end up with trust and response domains that look something like the following:

$$\mathit{Vaccine\ Science}|\mathit{Public}_{trust} = \{\mathit{safe\ if\ risk} < x\}$$
$$\mathit{Scientists}(\mathit{Vaccine\ Science})_{response} = \{\mathit{identify\ whether\ risk\ of\ adverse\ reaction\ is} < x\}$$

In this example, x is the agreed upon level of risk appropriate for vaccine safety. Rather than a vague, easily altered expectation of "safety", there is a settled benchmark for what counts as safe. Notice also that the task of determining what x is, however, has been removed from the response domain of science. Thus, for science to be counted "trustworthy", all that is required of science is that it accurately identifies whether the likelihood of an adverse reaction is less than what is expected of safe vaccines.

Of course, this solution assumes that the public, or representatives for the public, can reach agreement on the appropriate level of risk, which may not be possible in practice. In the US, the FDA has deemed several Covid-19 vaccines safe, and yet many members of the public either think that the FDA's standard of safety is not high enough or distrust that the FDA did their due diligence in determining whether the vaccines met that standard. However, clarifying what tasks belong to whom can help build trust where it is possible. Thus, trust negotiations as I have described them will not always lead to perfect alignment between trusters' expectations and trustees' commitments, however, it can provide a starting place to move forward from. In this case, trust can be built around scientists' ability to determine the likelihood of events, something that science does pretty well. Once trust is built around this ability, trust may extend to other areas that science is well suited to address. Conversely, as science engages in negotiations with the public, their commitments may increasingly reflect the legitimate concerns of the public, making further space for trust-building.

References

Baier, A. (1986). Trust and antitrust. *Ethics, 96*(2), 231–260.

Brown, P., Waite, F., Larkin, M., Lambe, S., McShane, H., Pollard, A. J., & Freeman, D. (2022). "It seems impossible that it's been made so quickly": A qualitative investigation of concerns about the speed of Covid-19 vaccine development and how these may be overcome. *Human Vaccines & Immunotherapeutics, 18*(1).

Dormandy, K. (2020). Exploitative epistemic trust. In K. Dormandy (Ed.), *Trust in epistemology* (pp. 241–264). Routledge.

Goldenberg, M. J. (2021). *Vaccine hesitancy: Public trust, expertise, and the war on science.* University of Pittsburg Press.

Hardin, R. (2001). Conceptions and explanations of trust. In K. S. Cook (Ed.), *Trust in society* (pp. 3–39). Russell Sage Foundation.

Hardin, R. (2002). *Trust and trustworthiness.* Russell Sage Foundation.

Jones, K. (1996). Trust as an affective attitude. *Ethics,* 107(1).

Jones, K. (2012). Trustworthiness. *Ethics, 123*(1), 61–85.

Jones, K. (2015). Trust: Philosophical aspects In *International encyclopedia of the social & behavioral sciences* (2nd ed., p. 24).

Latkin, C.A., Dayton, L., Yi, G., Konstantopoulos, A., & Boodram, B. (2021). Trust in a COVID-19 vaccine in the U.S.: A social-ecological perspective. *Social Science & Medicine, 270.*

Momplaisir, F., Haynes, N., Nkwihoreze, H., Nelson, M., Werner, R.M. & Jemmott, J. (2021). Understanding drivers of coronavirus disease 2019 vaccine hesitancy among blacks. *Clinical Infectious Diseases,* 73(10).

O'Neill, O. (2018). Linking trust to trustworthiness. *International Journal of Philosophical Studies, 26*(2), 293–300.

O'Neill, O. (2020). Questioning trust. Chap. 1. In J. Simon (Ed.), *The Routledge handbook of trust and philosophy* (pp. 17–27). Routledge.

Scheman, N. (2020). Trust and trustworthiness. Chap. 2. In J. Simon (Ed.), *The Routledge handbook of trust and philosophy* (pp. 28–40). Routledge.

Stewart, E. (2024). Negotiating domains of trust. *Philosophical Psychology, 37*(1), 62–86.

World Health Organization. (2021). *WHO coronavirus (Covid-19) dashboard.* Retrieved August 31, 2021, from https://covid19.who.int/

Trust in Science

Scientific Experts, Epistemic Wisdom and Justified Trust

Pierluigi Barrotta and Roberto Gronda

Abstract Trust in science is a multifaceted phenomenon. The goal of our essay is to shed light on the kind of trust that citizens or laypeople should place in scientific experts as public problem-solvers. Our thesis is that such form of trust in science, which we call *epistemic-pragmatist public trust*, has unique features that prompt an ad hoc account. Traditional accounts of epistemic public trust are predicated on two conditions: citizens are justified in trusting scientific experts if the latter (a) are endowed with scientific competence and (b) are honest in rendering their testimony. We show that it is easy to find examples in which those conditions are met, and yet citizens are justified in not trusting scientific experts as public problem-solvers. We argue, therefore, that a further condition is needed to account for epistemic-pragmatist public trust: epistemic wisdom, namely, the ability to apply universal knowledge to highly specific public problems.

1 Introduction[1]

Trust in science is a multifaceted phenomenon. As an umbrella term, it encompasses as different kinds of relationships as the relations of interdependence among peers, the institutional contexts that make the interaction between scientists and policymakers possible, and the relations of unilateral dependence of citizens on scientific experts. Because of such a wide variety of forms, it should come as no surprise that the notion of trust in science presents so many different facets and dimensions.

---

[1] We would like to thank Nico Formánek and Michel Croce for their comments on an earlier draft of this paper.

This work was financially supported by the ISEED (Inclusive Science and European Democracies) project, funded from the European Union's Horizon 2020 research and innovation programme under Grant Agreement No. 960366.

The Authors contributed equally to the essay.

P. Barrotta · R. Gronda (✉)
Department of Civilizations and Forms of Knowledge, Philosophy, University of Pisa, Pisa, Italy
e-mail: roberto.gronda@unipi.it

M. M. Resch et al. (eds.), *The Science and Art of Simulation*,
https://doi.org/10.1007/978-3-031-68058-8_6

The goal of this essay is to shed light on one of those forms—namely, the trust that citizens or laypeople should place in scientific experts as public problem-solvers, in virtue of the latter's superior competence. Our thesis is that such form of trust in science, which we call *epistemic-pragmatist public trust*, has unique features that require an ad hoc account. The traditional account of epistemic public trust is made up of an epistemic condition—that is, scientific competence—plus a value-based requirement, be it moral, normative or deontological. On the contrary, we believe that a three-component account is needed to deal with epistemic-pragmatist public trust, with a third further condition to be added to the epistemic and evaluative ones. Those three conditions, taken together, provide a fine-grained analysis of the bonds of trust existing between citizens and scientific experts.

We will proceed as follows. In the first section, we single out the form of trust that we want to address, then we provide a general sketch of its main features. In the second section, we turn our attention to the most common accounts of epistemic public trust in science available on the market, and we show why they are not suited to explain what is distinctive about epistemic-pragmatist public trust. Finally, in the third section, we present our three-component account, highlight its explanatory power, and provide a conceptual clarification of the notion of epistemic wisdom.

2 Epistemic-Pragmatist Public Trust in Science

As noted above, the phrase "trust in science" is taken to mean some quite distinct phenomena. The first step is, therefore, to better clarify what kind of relationship of trust we are interested in. We pursue this goal by distinguishing the most significant forms of trust in science discussed in contemporary debates. The two main distinctions to be made are that between trust within science and epistemic public trust in science, and that between epistemic public trust in science and epistemic-pragmatist public trust in science (the latter being a subspecies of the former).

First, "trust in science" is sometimes used to point out the ubiquity of relationships of trust among scientists and researchers. A better wording would be, therefore, "trust *within* science", so as to highlight the substantial epistemic equality of the people involved in that relation, on the one hand, and the fact that all those people participate in the same of form of life (i.e., scientific activity), on the other hand. As Hardwig famously remarked, without trust in fellow scientists, scientific research would be impossible, since that would mean starting from scratch over and over again (Hardwig, 1991; Wiltholt, 2013). There is a difference between those forms of trust in which it would be possible, at least in principle, for a researcher to verify the results of an experiment or an inquiry that she has not conducted herself but falls in her area of expertise, and those in which that would be pragmatically impossible because of the truster's lack of competence. Yet, in both cases, the epistemic role of trust in scientific research cannot be eliminated or downgraded.

The importance of trust within science is also reflected at an institutional level. The scientific community is kept together by relations of trust, which are supplemented and reinforced by and through institutional organizations and practices (Longino, 1990). Much of the institutional context within which research is carried out not only depends on relationships of trust among scientists but is also designed to make them sound, accountable, and justified. So, for instance, the academic status of a researcher can also be viewed as a proxy for epistemic trustworthiness, within a well-functioning system of academic recognition.

Such a form of trust, which holds among peers or "semi-peers", should be kept separate from the bonds of trust that exist between scientists and citizens. Simply put, the relationship of trust between scientists and citizens—which we will hereinafter call *public trust*—presents a higher degree of vulnerability than that which holds among peer scientists: citizens have less resources to rely on to check the epistemic credentials of the scientists' claims (on the relationship between trust and vulnerability, see Baier, 1986). The dependence of citizens on scientists is, therefore, almost total.

Public trust in science has recently come to the fore, in particular because of the relevance acquired by issues such as vaccine hesitancy and climate change within the public sphere (Goldenberg, 2021; Oreskes, 2019). However, the inherent complexity of such phenomenon has not been disclosed and clarified well enough so far. In recent times, multiple accounts have accordingly been proposed, which focus on different aspects of the relationship of trust between scientists and citizens. Indeed, though the classical epistemic account of public trust—centred on the contribution that scientific experts can make to public decision-making and to the public debate in general in terms of knowledge—is still prevalent in the literature, and is the approach that we favour and embrace, alternative approaches are gaining momentum as well. For instance, Collins and Evans have maintained that public trust in science should be conceived of as a moral, rather than epistemic, phenomenon. The reason why science should be praised and trusted by citizens is not that it provides sound knowledge about the world—scientists ran the gamut of mistakes—but that it is a "fountainhead of values", namely, a form of life that embeds the most precious values of our civilization (Collins & Evans, 2017, 19).

Collins and Evans's proposal is as radical and provocative as it could be. However, there are many other approaches that, albeit less controversial, have highlighted non-epistemic factors of public trust as key features for understanding the nature of the relationship of trust between citizens and scientific experts. In a recent paper, Contessa has argued that the ultimate ground of public trust should be social and institutional: within a rational and effective division of cognitive labour, the best option for citizens should be to trust scientists *uncritically* (from the perspective of the individual) within a well-functioning academic institution (Contessa, 2022). Schroeder and Boulicault have insisted on the importance of scientists and citizens sharing the same values, and have emphasized that, in some very specific cases, citizens are allowed to trust scientists on the basis of such an agreement (Boulicault & Schroeder, 2021; Schroeder, 2021). Finally, Furman has argued that a complete philosophical account of trust should also include emotions (Furman, 2020).

As we said, notwithstanding such a variety of accounts, we stick to the traditional epistemic approach to the topic. In doing so, we are by no means going to deny that other aspects may be relevant to the understanding of the relationship of trust between citizens and scientific experts too; yet, we hold firm to the belief that the epistemic features of public trust should be given pride of place, since they are what accounts, both conceptually and genealogically, for its very possibility in the first place. At the same time, however, we believe the notion of epistemic public trust needs to be deepened. As we said, epistemic public trust loosely refers to the contribution that scientific experts can make to public decision-making and to the public debate, broadly conceived, in terms of knowledge.[2] Hence, it is left partially undetermined what kind of epistemic contribution such a relationship of trust is supposed to make possible.

Many people are genuinely interested in science and are happy to learn from scientists, whom they recognize as valuable and reliable sources of knowledge. Citizen science projects, for instance, show that it is far from unusual for common citizens to take part in highly theoretical scientific research, which does not have any immediate practical outcome (Denia, 2021; but see also Vohland et al., 2021; Hecker et al., 2018). Such contexts of cooperation are made possible by the bonds of trust that are built between citizens and scientists. Nonetheless, it is not difficult to show that those are relatively unproblematic relationships of trust: first of all, the citizens are voluntarily involved in the research, out of personal and unconstrained interests, and are free to withdraw as soon as they are no longer satisfied with the conditions under which the research is carried out; moreover, the citizens play an epistemic role—usually, they are instrumental to collecting the data—whose relevance is easily acknowledged by scientists and which helps cement the bonds of trust among all the members of the community of inquiry; finally, and most importantly, those who get involved in citizen science projects usually trust science—trust, therefore, does not have to be created, enhanced and fostered since it is right there from the very beginning.

Much more complicated are those cases in which citizens and scientific experts are somehow compelled to cooperate. That usually happens when public problems are at stake: while in citizen science projects it is the citizens who freely decide to join and take part in the scientists' form of life, with public problems it is quite the other way around, and science is often perceived of as an external authority threatening the interests and wellbeing of the people, who nonetheless have to endure the decisions that policymakers make, based on knowledge made available and certified by scientific experts. The literature is replete with examples of contexts of inquiry in which scientific experts failed to gain the citizens' trust, also spectacularly failing to solve the problems they were asked to tackle (Barrotta & Montuschi, 2018; Wynne, 1996). It is this very specific kind of relationship of trust between citizens and scientific experts—which we call *Epistemic-Pragmatist Public Trust in Science*—that we are interested in investigating.

Before moving on to a discussion of the different ways in which those relationships of trust can be framed, however, a terminological remark is in order. So far, we have not yet formulated—though we have rigorously adhered to—a distinction, though it is of the utmost importance for our investigation. Contrary to what seems to be

[2] The phrase "Epistemic Public Trust in Science" comes from Irzik and Kurtulmus (2019).

the standard use, we draw a clear-cut distinction between scientists and scientific experts: we believe that the two terms are by no means synonyms.[3] An exhaustive and full-fledged defence of that distinction would mean anticipating the conclusion of the argument put forth in the following sections: indeed, the conceptual distinction between scientists and scientific experts goes hand in hand with, and depends on, the difference between epistemic-pragmatist public trust in science and the more general phenomenon of epistemic public trust in science, of which the former is a species. Accordingly, we won't provide any further justification here for our decision to distinguish between scientists and scientific experts: we will use "scientific expert" to refer *solely* to those scientists who (and when they) undertake an inquiry aimed at solving a public problem; we will use "scientist" in any other situation. We hope that our terminological decision will look less arbitrary at the end of our argumentation. For now, we ask our readers to keep our unconventional use of these expressions in mind and to patiently wait for a conceptual clarification.

3 The Two-Component Account of Epistemic Public Trust in Science

Now that we have singled out the particular form of relationship of trust which will be the topic of our analysis, the next step is to find a model that may account for its distinctive features. In the present section, we introduce the standard two-component account of epistemic public trust and assess its overall validity. We will show that the latter is not rich enough to faithfully represent epistemic-pragmatist public trust in science.

As we said, citizens are expected to trust scientists and scientific experts, and scientists and scientific experts are expected to be trustworthy.[4] The question is: on what grounds can such a relationship of trust be rational?[5] Or, more precisely, what are the features, virtues or qualities of scientists and scientific experts that make it

---

[3] The rationale behind that distinction can be rephrased as follows: there is a difference between the kind of activities that scientists perform when they carry out their scientific inquiries and those that scientific experts are expected to perform when they carry out public inquiries. We have argued in favour of that distinction in Barrotta and Gronda (2022) and in Barrotta and Gronda (2019)—we refer to those texts for arguments in support of that distinction. See also Gundersen (2018) and Grundmann (2017) for similar approaches.

[4] Since the two-component account does not possess—or, at least, that is our thesis—the expressive resources needed to make the distinction between scientists and scientific experts that we are trying to carve out, we assume that a supporter of that account would be likely to look for a unified account of what we have called the epistemic public trust in science. For this reason, we will use the laborious wording "scientists and scientific experts" to stress that the two-component account is supposed to provide an explanation for both forms of trust.

[5] A remark is needed here. We are not too interested in the epistemological question of how citizens can find out whether the scientific experts are trustworthy. Our concern is, rather, with the features that the scientific experts must have to be trustworthy. In a way, ours is more of an ontological than an epistemological analysis.

rational for citizens to trust them?[6] The two-component account of epistemic public trust purports to provide an elegant answer to that question.

Following Keren, we take those two components to be (a) the competence of the scientists and scientific experts in the relevant domain and (b) their honesty in rendering their testimony (Keren, 2014).[7] It is worth stressing that the two-component account of epistemic public trust is predicated on two basic assumptions: (a) it is a doxastic account in that it assumes that A's (the citizens) epistemic trust in B (the scientists and scientific experts) involves A's having beliefs about B; (b) it is a non-reductionist account in that the two components are, by principle, not reducible to one another. The two components are, indeed, jointly needed to substantiate the content of the beliefs that ground A's trust in B (Rolin, 2020).

The choice of those two components seems straightforward. For a scientist or a scientific expert to be perceived as trustworthy by citizens, the former needs to have and display sound competence in the field of research for which she is claiming recognition by the public. Unless we are trying to get rid of the epistemic element of public trust, this seems quite a straightforward assumption: at the end of the day, it seems reasonable to say that the ultimate epistemic reason why citizens should defer to scientists and scientific experts is because the latter are supposed to know more.

The idea that scientists and scientific experts have more reliable knowledge than citizens can be formulated in different ways. It can be said, for instance, that they have actual knowledge of true general laws that help explain and predict highly complex patterns of behaviour; alternatively, it can be argued that their superior knowledge amounts to their ability to use refined models to shed light on selected aspects of natural and social phenomena; or, finally, one may opt for a dispositional approach, according to which the knowledge that scientists and scientific experts have (and citizens lack) consists in a set of practical skills and theoretical understanding that enable them to set up and carry out a satisfactory inquiry. We are not interested in settling the issue here, so we leave it open for further discussion: any position on the nature of scientific knowledge is perfectly compatible with the point we are trying to make.

The competence condition is meant to warrant that scientists and scientific experts can be taken as reliable sources of information by the citizens, thus justifying—though only incompletely—the latter's trust in the former. However, since it is usually

[6] Here, we do not want to address the issue of how to conceive of those features, virtues or qualities. We believe that virtue-based epistemology is likely the best theoretical framework for understanding them, but we won't discuss the question any longer since any explanatory model is equivalent for our present purposes. For a promising line of investigation, see Croce (2019).

[7] See also Irzik and Kurtulmus (2019, 1151) for a more refined and yet substantially equivalent account: "Pulling these together, we offer the following analysis. M has warranted epistemic trust in S as a provider of P only if (1) S believes that P and honestly (i.e. truthfully, accurately and wholly) communicates it to M either directly or indirectly, (2) M takes the fact that S believes and has communicated that P to be a (strong but defeasible) reason to believe that P, (3) P is the output of reliable scientific research carried out by S, and (4) M relies on S because she has good reasons to believe that P is the output of such research and that S has communicated P honestly. While (1) and (3) spell out the necessary conditions for S to be trustworthy, (2) and (4) specify the necessary conditions for M to trust S and do so with good reasons".

acknowledged that trust cannot boil down to mere reliance, a further condition needs to be added, which spells out the moral or evaluative dimension that is constitutive of that notion. The idea of honestly rendering the testimony is wide enough and broad enough to meet such a requirement.

It is apparent that there must be some kind of commitment on the side of scientists and scientific experts to help citizens get reliable knowledge, otherwise the latter would not have any clue about what propositions to accept and what views to endorse. Their epistemic vulnerability requires a leap of faith which, to be justified, must be grounded on the scientists or experts' moral dispositions. Indeed, it is not difficult to imagine a group of highly competent scientists whose efforts are directed to concealing the results of their inquiries from the public—say, because they fear that the uncontrolled use of those results may have dreadful consequences on a societal level. That shows that the scientists and scientific experts' competence is not enough to guarantee and justify the trust that citizens may happen to place in them.

It is worth noting, once again, that we do not specify what the ultimate reasons that lead scientists and scientific experts to honestly render their testimony are. Ours is a sketch of an account that is compatible with many different explanations on the nature and reasons for such an altruistic behaviour. It may be that scientists and scientific experts act out of good will; or it may be that they act out of self-interest, along the lines famously described by Hardin (2002); or it may even be that scientists and scientific experts act trustworthily just because they feel constrained by social norms and expectations, though they would be happy to behave differently. Any of these accounts would be fine with us—pick whichever you like as a potential explanation of the honesty factor. What matters is that a moral or evaluative component should be preserved in the account.

We concede that the two-component account satisfactorily portrays the main features of epistemic public trust. In other words, we believe that it provides a satisfactory account of the relationship of trust between citizens and scientists. We hold, however, that it does not provide a satisfactory account of what we have called the epistemic-pragmatist public trust in science, namely those relationships of trust built in contexts in which public problems are at stake that may have deep consequences on citizens. Why so? What does it lack?

Our point is that it is easy to envision a situation in which it would be rational for citizens *not* to trust scientific experts even though the latter turn out to be highly competent in their domain and willing to honestly render their testimony. The most straightforward way to put this idea is in terms of the epistemic irreducibility of individual cases to knowledge of general laws or patterns.

The application of universal knowledge to particular cases is often taken to be utterly unproblematic. According to this view, theories—let's assume, for the sake of simplicity, that knowledge of general laws comes encoded in theories—are like instruction books: they provide a list of instructions that can be easily and successfully carried out when and as needed. Cartwright has notoriously labelled such a conception of scientific theories as the "vending machine" view: "[t]he theory is a vending machine: you feed it input in certain prescribed forms for the desired output;

it gurgitates for a while; then it drops out the sought-for representation, plonk, on the tray, fully formed, as Athena from the brain of Zeus" (Cartwright, 1999, 184).

When it comes down to tackling public problems, things are much more complicated because scientific experts are expected, first, to understand the problematic situation in its entire complexity and then to figure out reasonable lines of conduct that could eventually solve the problems that called for the public inquiry. In both cases—and this is the most relevant aspect—the conditions that scientific experts are facing are completely new and unprecedented. The experts who were in charge of dealing with the consequences of the radioactive fallout from the Chernobyl accident did not know anything about the peculiarity of the Cumbrian soil on which sheep grazed (Wynne, 1996); the experts who were commissioned by the Italian government to build the Vajont dam did not know anything about the stability of Mount Toc (Barrotta & Montuschi, 2018). In both cases, the experts were highly competent in their field of expertise and well respected by the international scientific community; nonetheless, their unquestioned and unquestionable competence did not prevent them from making serious epistemic mistakes that led to an unsatisfactory conclusion to the inquiry. Those mistakes also led to the citizens' deep and well-grounded mistrust of them, which in turn contributed to the scientists' failure as problem-solvers.

Those case-studies show, therefore, that the epistemic-pragmatist relationship of trust between citizens and scientific experts cannot be explained in the light of the two-component account. This, we think, is a welcome result: it accounts for the scientific experts' poor performance—which obviously provide a very good reason for not trusting them—without assuming either that they lacked competence in their fields of research (which is false most of the time) or that they were morally vicious (which would undermine any possible bonds of trust). The possibility is thus left open that more reasonable causes can be found for the citizens' mistrust in scientific experts, thus paving the way to a constructive mechanism for rebuilding sound relationship of trust between them.

4 The Three-Component Account of Epistemic-Pragmatist Public Trust in Science

We have repeatedly argued that epistemic-pragmatist public trust in science calls for (at least) a further condition in addition to the standard two-component account of epistemic public trust. In the last section, we have better qualified such a claim by pointing out that the missing condition has to do with the application of universal knowledge to highly specific public problems, on the assumption that application is—at least in these cases—a highly creative epistemic act.

Before moving on to a closer discussion of this point, a remark is worth making about some relevant features of our approach, and its theoretical advantages over its competitor. It should be noted, indeed, that our account is compliant with the traditional conception of trust as a three-place relation. According to this view, the fundamental form of trust has the following shape: A trusts B to perform C. Our insistence on the context of application, as well as on the distinctive traits of the problematic situation which scientific experts are asked to tackle, easily accommodates such insight: citizens trust scientific experts to solve a certain public problem. On the contrary, the standard two-component account is more apt to portray trust as a two-place relation: citizens trust scientists and scientific experts, full stop. The boundaries and reasons of their trust are thus left completely unaccounted for.

With this remark in mind, let's go back to our quest for the missing component. Keep in mind that our focus is on what makes scientific experts trustworthy for citizens—more precisely, what makes scientific experts trustworthy for citizens with regard to a specific epistemic performance, i.e., the activity of solving public problems. Accordingly, the question that we should try to answer is the following: what is the quality, feature, or disposition that scientific experts must have in order for citizens to be justified in trusting them as public problem solvers?

As we said, citizens are justified in trusting scientific experts if they are led to believe that, in addition to scientific competence and moral integrity, the latter are also capable of perceiving and understanding the distinctive traits of the problematic situation, and of addressing it properly.[8] "Epistemic wisdom" is the label we have chosen for such ability. Accordingly, epistemic wisdom is our missing component.

It is worth stressing that the kind of wisdom that we have in mind is distinctively epistemic. Wisdom is usually associated with the domain of practical philosophy, in particular with the Aristotelian conception of *phronesis*. Our use of that notion is remarkably different, though, in that we are exclusively concerned with the epistemic demands that must be met by an agent to apply the knowledge of universal laws or patterns of behaviour to specific existential conditions.[9] In clearer terms, epistemic wisdom can be properly described as the scientific experts' ability to put the body of knowledge that they possess at work to solve public problems. Epistemic wisdom is, therefore, what makes the application of universal knowledge to specific circumstances an epistemically creative activity rather than an act of sheer repetition.

[8] The notion of problematic situation is drawn from Dewey's logic of inquiry: it refers to the broad context encompassing both the investigating subject and the investigated object (Dewey, 2008, 72). The notion of situation is, therefore, a reminder that every inquiry is both temporally and spatially situated, and entirely depends on the epistemic resources available to the agent.

[9] This does not amount to saying, however, that there are no structural similarities between the Aristotelian wisdom and the epistemic wisdom that we advocate. Take, for instance, the following quotations from McDowell: "[o]ccasion by occasion, one knows what to do, if one does, not by applying universal principles but by being a certain kind of person: one who sees situations in a certain distinctive way"; "the virtuous person's reliably right judgements as to what he should do, occasion by occasion, can be explained in terms of interaction between this universal knowledge and some appropriate piece of particular knowledge about the situation at hand" (McDowell, 1998, 73 and 57). The difference between our conception of wisdom and the Aristotelian tradition can be summarized by noting that, by being pragmatist through and through, our analytical framework rejects the distinction between theoretical and practical knowledge that underlies the Aristotelian accounts.

The notion of application—and, consequently, the notion of epistemic wisdom—is far from being self-evident. We are thus left with the task to answer some questions about the forms of application that scientific experts are asked to perform, and, accordingly, the shapes that public epistemic-pragmatist trust can take. Usually, the relationship of epistemic trust between citizens and scientific experts is conceived of in cognitive terms: scientific experts supply citizens with reliable knowledge. More precisely, scientific experts are most often expected to provide useful knowledge to reach a certain goal. This is undoubtedly correct; scientific experts must be epistemically wise to perform that function in a proper manner: figuring out what means are needed to satisfactorily reach a goal is an activity that requires dealing with the specificity of the situation and perceiving the distinctive features of the problem at stake. For instance, it might well be that relevant information is encoded in local knowledge; the scientific experts then have to find a way to translate that information into their scientific language (see, for instance, Epstein, 1996). If that body of knowledge is not paid attention to, it is highly likely that the public problem will not be solved.

However, the process of inquiry aimed at solving a public problem is much more complex than a mere search for means to reach a goal. As pragmatists, we give pride of place to the notion of situation; following Dewey, we argue that the most important and decisive part of any inquiry lies in the processes that lead to the statement of the problem—in Deweyan terms, to the transformation of an indeterminate situation into a clearly defined problem (Dewey, 2008, 108). It is only if the problem is well and clearly defined that a course of inquiry can be envisioned.

Our thesis is that a relevant epistemic interaction between citizens and scientific experts should also take place *before* the public problem is defined and dealt with and *for the purpose of* defining the problem that will be addressed by the public inquiry. Scientific experts must be epistemically wise to understand the distinctive features of the public problem that they are facing and provide a satisfactory framework for conceiving of and directing the activities that aim at solving it. Every genuine public problem is as inherently complex and articulated as it is highly specific: solving it involves the ability to imagine the potential, yet unintended and unexplored consequences of certain decisions.

So, for instance, deciding that the public problem of, say, providing a community with electricity is to be narrowly defined as a purely economic and/or engineering problem puts some very significant constraints on what the best solution will look like. By fixing the parameters and criteria by which the alternative solutions will be judged, the decision about how to define the problem lays out the conditions for success: for instance, those solutions which promise to produce energy at a lower price will be favoured over those that turn out to be more expensive because they are more sensitive to the local environment, to sustainability or to potential repercussions on the job and real estate markets, and so on and so forth. Deciding that building a nuclear power plant is a better option than building a hydroelectric power plant may be justified in the light of the parameters set forth by the public inquiry, but it may well be a very bad option when measured in different terms. No public problem is satisfactorily addressed—we believe this is a rather uncontroversial assumption—if the implemented solution would lead to a host of undesired and harmful consequences.

From that perspective, it should also be clear why epistemic wisdom cannot be reduced to scientific competence—which implies, among other things, that our three-component account is not spurious, since all three components are necessary and relatively independent of one another.[10] The skills and abilities required to be a good scientific expert are different from those needed to be a good scientist. Scientific experts must be capable of engaging in inter- and transdisciplinary activities; they must be able to translate concepts, points of view and values from other disciplines into the language of their own discipline; they must be able to create new trading zones and develop new conceptual tools that allow them to properly define the public problem and envision reasonable lines of inquiry; they must be able to address the citizens' concerns and to evaluate whether and how their local knowledge should be accepted as a source of relevant information.

The list is by no means exhaustive: many other abilities can be singled out. What is worth remarking is that nothing similar is expected of scientists: indeed, none of these features is relevant to the possession of scientific competence. When we listen to a scientist illustrate aspects of her work, we as citizens trust her because we think she knows more and she has no reason not to be honest in her testimony. We are not interested in ascertaining whether she is endowed with epistemic wisdom because the latter does not affect her epistemic performance *as a scientist*. Epistemic wisdom plays its role as a ground for the citizens' trust in scientific experts only when public problems are at stake.

5 Conclusion

The goal of this article was to argue for the thesis that epistemic-pragmatist public trust in science has a unique and distinctive structure, which cannot be explained by the traditional accounts of epistemic public trust. We have shown that epistemic

[10] Independence is qualified as relative rather than absolute, because we believe that there can be no epistemic wisdom without scientific competence—or, put otherwise, that one cannot be a scientific expert without being competent—at a professional level—in one or more sciences. We acknowledge that this is quite strong an assumption since it seems to rule out the possibility that interactive experts can properly count as scientific experts (for the notion of interactive experts, see Collins & Evans, 2002). This is an important point that we cannot address here. We will simply provide a couple of suggestions, which we hope may be useful for properly framing the question. First, our assumption is not as restrictive as it may seem: in fact, it by no means purports to underestimate the importance of interactional expertise; it only excludes the citizens from the scope of who may be considered a scientific expert. This seems to be compliant with the shared insight that citizens are not scientific experts, even though they can contribute, in their own ways, to public inquiries. Secondly, within a broader and less idealized analytical framework, which takes into consideration the political and institutional factors that make the activity of public problem-solving possible, a much broader spectrum of activities can be acknowledged, which can be performed by citizens who are also interactive experts.

wisdom is the component that should be added in order to account for the possibility of justified bonds of trust between citizens and scientific experts.

We believe that our approach has some important advantages. It highlights the multi-layered character of the relationship of trust; it provides quite accurate recommendations on where to act to enhance the citizens' trust in scientific experts; it prompts a contextual shift in the study of public trust towards a performance-based epistemology; it provides an overall framework for understanding the intrinsic relation between epistemic wisdom and public problem-solving. We believe that all these are promising lines of investigation.

What is yet to be done is to find a way to de-idealize some powerful theoretical assumptions—for instance, the lack of external (political and institutional) pressure on scientific experts or the absence of time constraints on public inquiries—that underlie our three-component account of epistemic-pragmatist public trust in science. The next step should be, therefore, backing our normative and idealized account with empirical evidence: we are confident that its application will yield quite a few interesting and original insights.

References

Baier, A. (1986). Trust and antitrust. *Ethics, 96*(2), 231–260. https://doi.org/10.1086/292745

Barrotta, P., & Gronda, R. (2019). Scientific experts and citizens' trust: Where the third wave of social studies of science goes wrong. *Teoria, XXXIX*(1), 9–27. https://doi.org/10.4454/teoria.v39i1.54

Barrotta, P., & Gronda, R. (2022). Intelligence and scientific expertise. *Synthese, 200*, 142. https://doi.org/10.1007/s11229-022-03513-4

Barrotta, P., & Montuschi, E. (2018a). The dam project: Who are the experts? A philosophical lesson from the Vajont Disaster. In P. Barrotta & G. Scarafile (Eds.), *Science and democracy. Controversies and conflicts* (pp. 17–33). John Benjamins Publishing Company.

Boulicault, M., & Schroeder, S. A. (2021). Public trust in science: Exploring the idiosyncrasy-free ideal. In K. Vallier & M. Weber (Eds.), *Social trust* (pp. 102–121). Routledge.

Cartwright, N. (1999). *The dappled world. A study of the boundaries of science.* Cambridge University Press.

Collins, H., & Evans, R. (2002). The third wave of science studies: Studies of expertise and experience. *Social Studies of Science, 32*(2), 235–296. https://doi.org/10.1177/030631270203200 2003

Collins, H., & Evans, R. (2017). *Why democracies need science.* Polity Press.

Contessa, G. (2022). It takes a village to trust science: Towards a (thoroughly) social approach to public trust in science. *Erkenntnis.* https://doi.org/10.1007/s10670-021-00485-8

Croce, M. (2019). On what it takes to be an expert. *The Philosophical Quarterly, 69*(274), 1–21. https://doi.org/10.1093/pq/pqy044

Denia, E. (2021). *Report on citizen science—A conceptual overview for ISEED team.*

Dewey, J. (2008). *Logic: The theory of inquiry.* In Id., *The Later Works*, 1925–1953 (Vol. 12). Southern Illinois University Press.

Epstein, S. (1996). *Impure science. AIDS, activism, and the politics of knowledge.* University of California Press.

Furman, K. (2020). Emotions and distrust in science. *International Journal of Philosophical Studies, 28*(5), 713–730. https://doi.org/10.1080/09672559.2020.1846281

Goldenberg, M. (2021). *Vaccine hesitancy: Public trust, expertise, and the war on science.* University of Pittsburgh Press.
Grundmann, R. (2017). The problem of expertise in knowledge societies. *Minerva, 55*, 25–48.
Gundersen, T. (2018). Scientists as experts: A distinct role? *Studies in History and Philosophy of Science Part A, 69*, 52–59. https://doi.org/10.1016/j.shpsa.2018.02.006
Hardin, R. (2002). *Trust and trustworthiness.* Sage.
Hardwig, J. (1991). The role of trust in knowledge. *Journal of Philosophy, 88*(12), 693–708.
Hecker, S., et al. (2018). *Citizen science innovation in open science, society and policy.* UCL Press.
Irzik, G., & Kurtulmus, F. (2019). What is epistemic public trust in science? *The British Journal for the Philosophy of Science, 70*(4), 1145–1166. https://doi.org/10.1093/bjps/axy007
Keren, A. (2014). Trust and belief: A preemptive reasons account. *Synthese, 191*(12), 2593–2615. https://doi.org/10.1007/s11229-014-0416-3
Longino, H. (1990). *Science as social knowledge.* Princeton University Press.
McDowell, J. (1998). Virtue and reason. In J. McDowell (Ed.), *Mind, value, and reality* (pp. 50–73). Harvard University Press.
Oreskes, N. (2019). *Why trust science.* Princeton University Press.
Rolin, K. (2020). Trust in science. In J. Simon (Ed.), *The Routledge handbook of trust and philosophy* (pp. 354–366). Routledge.
Schroeder, S. A. (2021). Democratic values: A better foundation for public trust in science. *The British Journal for the Philosophy of Science, 72*(2), 546–562. https://doi.org/10.1093/bjps/axz023
Vohland, K., et al. (2021). *The science of citizen science.* Springer.
Wilholt, T. (2013). Epistemic trust in science. *The British Journal for the Philosophy of Science, 64*(2), 233–253. https://doi.org/10.1093/bjps/axs007
Wynne, B. (1996). May the sheep safely graze? A reflexive view of the expert-lay knowledge divide. In S. Lash, B. Szerszynski, & B. Wynne (Eds.), *Risk, environment & modernity* (pp. 44–83). Sage.

Confidence: Calibrating Trust in Science

Lara Huber

Abstract Systematic reviews call for improved strategies of evaluating and communicating the degree of certainty and quality of a given body of scientific information. Drawing on environmental assessment projects, the paper argues, that building, promoting and maintaining trust in scientific methodology and outcomes, first of all, is a necessary prerequisite of scientific practice itself. Especially, as the paper shall illustrate, when the systematic review process expands across multiple disciplines, and brings authors from a broad range of research traditions to the table. In 2010 the *Intergovernmental Panel on Climate Change* (IPCC), the international body for assessing the science related to climate change, introduced a revised process of evaluating the degree of certainty in key findings. Drawing on the growing attention towards the reporting of *confidence levels*, the paper explores recent frameworks in large-scale assessment projects and their merit in *calibrating* trust in the sciences.

1 Introduction

Systematic reviews call for improved strategies of evaluating and communicating the degree of certainty and quality of a given body of scientific information. Since the early 2000s the *Intergovernmental Panel on Climate Change* (IPCC), the international body for assessing the science related to climate change, has engaged in addressing and evaluating the degree of (un)certainty in key findings (cf. IPCC, 2005, 2010a; Mach et al., 2017).

Created in 1988 by the *World Meteorological Organization* (WMO) and the *United Nations Environment Programme* (UNEP), the IPCC assesses impacts, risks and possible response options to climate change to adequately inform and guide policy makers on topics such as assessing and mitigating climate change worldwide. Quite notably, the IPCC performs no original research. All IPCC-reports since the very

L. Huber (✉)
Institute of Philosophy, Christian-Albrechts-Universität zu Kiel, Leibnizstr. 4, 24118 Kiel, Germany
e-mail: huber@philsem.uni-kiel.de

M. M. Resch et al. (eds.), *The Science and Art of Simulation*,
https://doi.org/10.1007/978-3-031-68058-8_7

first one in 1990 synthesise and assess information from a broad range of published sources, including so-called grey literature (i.e., white papers).

Three working groups collect and assess information on relevant topics, namely on the physical sciences of climate change, impacts of climate change, and mitigation methods. The guidance notes of the IPCC are intended to assist lead authors within these groups, who take responsibility for the production of chapter sections of the report within a given assessment cycle.

In the eve of the Fourth Assessment cycle (AR4) the IPCC introduced a new regime of evaluating the degree of certainty in key finding, including the reporting of so-called *confidence levels*. The regime has undergone slight revisions over the years and is now said to represent the degree of a synthesised judgment about the validity of findings based on the "type, amount, quality and consistency of evidence" and the "degree of agreement" within the peer group, respectively (IPCC, 2010a, 1). Due to its reputation in the field of environmental assessment, the IPCC's turn towards confidence grading hardly went unnoticed. In taking the case of the *Intergovernmental Science-Policy Platform on Biodiversity and Ecosystem Services* (IPBES), the question is raised, how similar institutions responded to this development.

Drawing on these new regimes the paper illustrates how "confidence" as an *epistemic* concept evolves in reference to a peer community and given standards of scientific excellence in this field (peer review in assessment projects). In particular, the paper elaborates on the question if confidence grading could be regarded as an asset in large scale assessment projects, given the overall aim of building, promoting and maintaining trust in scientific methodology and outcomes on the one hand and of informing and guiding policy makers on key topics on the other.

2 Confidence as an *Epistemic* Concept

Confidence is commonly understood as a subject's *strength of belief*. The term depicts the feeling that a person can trust someone or something to be good, work well, or produce good results. Being confident might ground the subjective perception that one's trust—for example in the sciences—will not be violated, given the certainty of a proposition or assertion or the sureness with regard to a fact. As regards to well-earned trust in the sciences, we have to presuppose the integrity of the scientific system as a necessary but not sufficient prerequisite (cf. Grasswick, 2014, 2017; Wilholt, 2013). The latter enables a person to *rely on* scientific methods and outcomes respectively. Much has already been speculated about the question if lay persons are justified in trusting scientists, extending to the question if scientists themselves are justified in doing so—given that they are dependent on building on epistemic trust in everyday research (Ridder, 2022). The paper refrains from adding to the debate on subject-based accounts of trust and trustworthiness. In acknowledging the system-relatedness of trust formation in the sciences, the paper introduces confidence as an *intersubjective* concept, as *strength of shared belief*. In doing so, the paper presupposes (a) that a peer community has been formed or could be identified as such by others; and (b) that this peer community, as part of community building itself, has established a procedure that, at least in principle, allows for arriving at a "shared view" of a given question. Any such procedure might be described as a more or less formalised practice of how to address problems, discuss conjectures, critically reflect underlying assumptions, and last but not least, assess if a consensus on a given matter is reachable. For the sake of scientific scrutiny, the ability to depict the amount and nature of remaining dissent, might also be key. Building, promoting and maintaining trust in scientific methodology and outcomes, is regarded as a precondition of scientific practice itself: Engaging in a scientific enquiry commits scholars to community standards which are instrumental for evaluating and communicating outcomes. A special subset of community standards orients the very methods of evaluating and communicating themselves. For example, in the case of consensus formation practices. Drawing on the existence and integrity of such practices, scientists are enabled to come to an agreement, a "scientific consensus", concerning the value of findings or the credibility of data for a scientific purpose at hand.

In proposing an *intersubjective* account of "confidence" the paper addresses two questions: First off, on what grounds the sharing of scientific assumptions is enabled. The paper sheds light to the very steps that are key to support and empower decision-making, especially in cases when it is mandatory to invite scholars from a broad range of scientific fields. Of specific import, secondly, is the question, how the validity and strength of an agreement is addressed and evaluated, concurrently. In taking the case of intergovernmental bodies and large-scale environmental assessment projects the paper illustrates, how the reporting of confidence levels, given their suitability for any such aim, might pave the way for assessing the validity of consensus formation practices, especially in complex and contested topics such as climate change and biodiversity.

3 On *Grading* Confidence

Given the scale and import of climate change assessments, it may come as no surprise, that scientists involved in any such project aim for adequate frameworks to allow for valid expert judgments. Especially, when we take into consideration the pivotal role of the IPCC to inform and guide policy makers on topics such as assessing and mitigating climate change worldwide. The paper sets forth a number of implications that comes with this very endeavour. The first part is occupied with the question why the IPCC turned to the topic of "confidence" at all. That is to say, on what grounds the grading of confidence started out and developed until today, respectively. In the second part, taking the case of the IPBES, the paper sets out to investigate, why confidence grading is adopted by similar institutions and for what reasons. Both case studies are preparatory to the discussion in the last part of the paper (Sect. 4), which aims to infer if frameworks that include confidence grading come with the merit of providing elaborate accounts of "calibrated trust" in scientific sources, data and methods, for example by furthering the comparability of expert judgment within a given peer community.

3.1 *The Case of the IPCC*

In the eve of the Fourth Assessment cycle (AR4) the IPCC introduced a new regime of evaluating the degree of certainty in key findings, including the reporting of so-called *confidence levels*. As already mentioned, all IPCC reports are based on the assessment of published literature, including so-called grey literature (i.e., white papers). Three interdisciplinary working groups collect and assess information on relevant topics, namely on the physical science of climate change (WGI) and on research into the impacts, adaptions and vulnerabilities of climate change (WGII). The latter includes for example, the vulnerability of natural and socio-economic systems. Additionally, the objective is to assess climate change mitigation methods (WGIII), for example, by evaluating scientific strategies for reducing greenhouse gas emission.[1] Given the scale and interdisciplinary context across the IPCC's working groups, fundamental differences due to methodology and research questions are regarded as main challenges of consistently reporting in general and dealing with risk and uncertainty in particular (cf. Swart et al., 2009).

The guidance notes of the IPCC in general, are intended to assist lead authors within these groups, who take responsibility for the production of chapter sections of the report within a given assessment cycle.

The IPCC introduced a first regimen of assessing confidence in 2005: The *Guidance Note for Lead Authors of the IPCC Fourth Assessment Report on Addressing Uncertainties* addressed confidence as a *qualitative* criterion for scientific assessment. Additionally, a roadmap for calibrating confidence *quantitatively* was introduced with the intention "to characterize uncertainty based on expert judgment regarding the correctness of a model, analysis or statement".[2] Accordingly, the degree of confidence was said to correspond to the probability of being correct. To support this assessment, the guidance note differentiated between the *amount of evidence* available in support of findings on the one hand, and the *degree of consensus* among experts on its interpretation (IPCC, 2005, 3). Quite noticeably, both terms were intended to be used in a relative sense to summarise judgments of the scientific understanding relevant to an issue, or to express uncertainty in a finding where there was no basis for making more quantitative statements.

[1] Further information about the history of the IPCC, the assessment reports and the review processes within the working groups (WGI-WGIII) could be found online: https://www.ipcc.ch. Accessed, 16 Jan 2023.

[2] IPCC (2005, 3). Originally, the concept of "confidence levels" was introduced by a guidance paper in the eve of the Third Assessment Report (IPCC TAR: Moss & Schneider, 2000). The paper not only provided the climate science community with a set of criteria and uncertainty terms, but also influenced graphical approaches to communicating uncertainty within the field of environmental assessment altogether.

In 2010 the IPCC changed its policy regarding the question how to address confidence adequately. The *Guidance Note for Lead Authors of the IPCC Fifth Assessment Report on Consistent Treatment of Uncertainties* (IPCC, 2010a) retains the term, but no longer defines it quantitatively. Instead, levels of confidence are now intended "to synthesize author teams' judgments about the validity of findings as determined through their evaluation of evidence and agreement, and to communicate their relative level of confidence qualitatively" (Ibid., IPCC, 2010b, 1). The guidance note itself is the result of an *IPCC Cross-Working Group Meeting* answering to major challenges of reporting and assessing information on climate change consistently. Topics were, for example, the integration of observations and simulations or the evaluation of given scenarios and knowledge gaps in published literature on relevant topics. Its overall aim could be described as follows: to *align* the very process of assessing, synthesising and evaluating relevant information and to *harmonise* the reporting of uncertainties across all three working groups.

At the heart of the IPCC's endeavour to frame and align expert judgment in the field of climate reporting and assessment lies its *nine-boxed model* of evaluating confidence based on evidence and agreement (Fig. 1).

According to the guidance note, the figure depicts "evidence and agreement statements and their relationship to confidence" (Ibid., 3). In a nutshell, confidence as a qualitative metric is since defined as the degree of a synthesised judgment about the validity of findings based on both, the "type, amount, quality, and consistency of evidence" and the "degree of agreement" within the peer group, respectively (IPCC, 2010a, 1). According to IPCC's "confidence scale" defined non-probabilistically, confidence is said to increase "towards the top-right corner as suggested by the increasing strength of shading" (Ibid., 3). To measure the level of confidence,

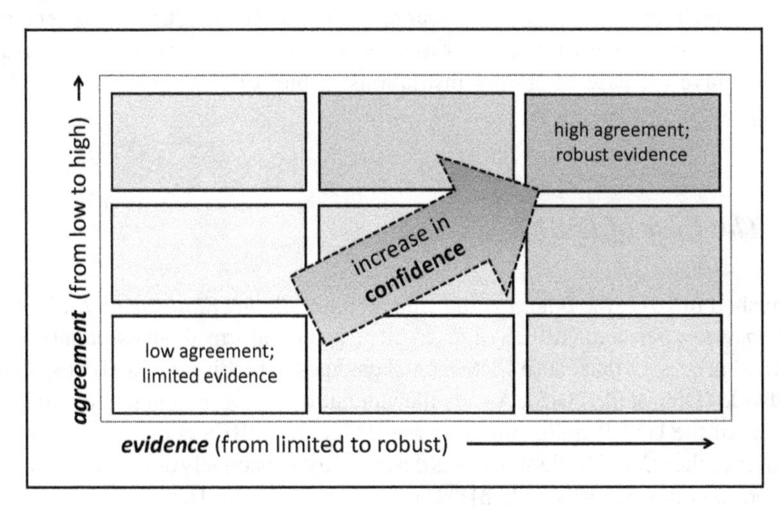

Fig. 1 Simplified and annotated version of the "nine-boxed model of evaluating confidence based on evidence and agreement" (IPCC, 2010a, 3)

authors are provided with "five qualifiers" (very low, low, medium, high, very high confidence).

Closing in on the Sixth Assessment cycle (AR6), the assessment reports were due in 2022, the added value of confidence grading was again under review. A draft version of the Assessment Report of the first Working Group (WGI), published in September 2021, disclosed that the reports in AR6 will follow the overall approach established in 2010. With regard to the assessment of confidence in detail the report reads as follows:

> For a given evidence and agreement statement, different confidence levels can be assigned depending on the context, but increasing levels of evidence and degrees of agreement correlate with increasing confidence. When confidence in a finding is assessed to be *low*, this does not necessarily mean that confidence in its opposite is *high*, and vice versa. Similarly, *low confidence* does not imply distrust in the finding; instead, it means that the statement is the best conclusion based on currently available knowledge (IPCC, 2021, 31).

Quite noteworthy, confidence grading is an additional metric to inform about the degree of a (synthesised) judgment about the validity of findings. Since the Fifth Assessment cycle (AR5) lead authors should begin the progress of assessment by evaluating *evidence* and *agreement*. In cases where sufficient evidence and agreement exist, authors address the issue of *confidence* in the aforementioned way. In cases where uncertainties could be quantified probabilistically, the authors subsequently inform about *likelihood*, that is to say the outcome probability: Whereas evidence and agreement are said to "underpin confidence assignments", confidence also "supports likelihood assignments"—if additional quantitative or probabilistic evidence is available (Mach et al., 2017, 3). Given that this "idealized step-by-step process" was newly adopted within the Sixth Assessment cycle (AR6) and became official with the above mentioned draft version of WGI, it might be assumed that confidence grading is no longer a singular endeavour in the history of the IPCC, but is becoming a common feature in environmental assessments altogether. This becomes apparent when we take the case of similar institutions in the following second part of the analysis.

3.2 The Case of IPBES

Established in 2012, the *Intergovernmental Science-Policy Platform on Biodiversity and Ecosystem Services* (IPBES) aims to identify and inform about essential components and processes that cause "detrimental changes in biodiversity and ecosystems" worldwide (Díaz et al., 2015). As an independent intergovernmental body under the auspices of the UNEP, quite similar to the IPCC, the IPBES is open to all member countries of the United Nations and was designed to "proactively develop assessments matched to policy needs" (Ibid., 3).[3] The assessments focus from supranational and

[3] Further information about the history of the IPBES, its conceptual framework, policies and procedure could be found online: https://www.ipbes.net. Accessed 16 Jan 2023.

regional to global scales and encompass thematic as well as methodological assessments. The first deals with topics such as pollination or marine wild fisheries, the second is concerned with scenarios and issues of valuation. Also, comprehensive assessments of biodiversity and ecosystem services are conducted. Given the need to support and frame assessments, the IPBES has produced a number of roadmaps. In its *Guide on the Production and Integration of Assessments across Scales*, the body for the first time referred to "confidence" as a criterion in assessment (IPBES, 2014). The guide informs about two different approaches of confidence grading in the field of environment assessments, including the IPCC's revised account of the "nine-boxed model" of 2010, and a "four-boxed model" that was introduced within the realm of a technical report by *UK Nation Ecosystem Assessment* (UK NEA) in 2011.[4] According to the IPBES' roadmap both frameworks shall be used in future assessments reports—depending on the question, if the authors aim for a "more detailed qualification of agreement and evidence", that is to say a "finer grained perspective", or if a "simplified version is desired" (IPBES, 2014, 77). In the first case, authors should refer to the *nine-boxed* model, in the second, the *four-boxed* model will suffice (Ibid.). By 2016 a trend towards the later model could be observed.[5] Two years later *The IPBES Guide on the production of assessments* introduced the "four-box model" with a slightly revised terminology as unique strategy for the purpose at hand (Fig. 2).

According to the qualitative assessment of confidence favoured by the IPBES, authors should apply one of four confidence terms—depending on the judgment level of evidence and agreement: "inconclusive" in cases with no or limited evidence, "unresolved" in cases when findings from independent studies come to different conclusions, "established but incomplete" if a general agreement could be reported but a comprehensive synthesis of findings is still underway, or "well established" if sufficient evidence does exists, for example if multiple independent studies come to a similar conclusion (IPBES, 2018, 30). As regards IPBES's "certainty scale", colour coded from red to green, "confidence increases towards the top-right corner as suggested by the increases by the increasing strength of shading" (Ibid.).[6]

[4] UK NEA, being part of the United Nations *Environmental Programme World Conservation Monitoring Centre* (UNEP WCMC), refers to both the IPCC as much as the Millennium Ecosystem Assessment (MA), as important sources for their own approach (UK NEA, 2011, 61). The latter was established in 2001 by a number of international bodies, such as the UNEP, the World Bank and others. Quite similar to IPCC and IPBES the MA primarily synthesises and assesses relevant findings of already published data. Further information about the history of the MA, the conceptual framework of the body as much as all assessment reports could be found online: https://www.millenniumassessment.org/en/index.html. Accessed 16 Jan 2023.

[5] Namely, within the realm of the thematic assessment of *Pollinators, Pollination and Food Production* (IPBES, 2017, xiv). The later presents the "four-box model for the qualitative communication of confidence" as a modified depiction of "supplemental qualitative uncertainty terms" introduced in the eve of IPCC TAR (Moss & Schneider, 2000, 45).

[6] Contrary to the IPCC the display practice of the IPBES changed continuously since 2014. Most obviously, this includes the explicit use of false colours. Also, inconsistencies concerning coding regimes (i.e., single colour scale; two colour scale) could be reported. Given the attention towards standardisation of display practices in the sciences, and pitfalls of colour coding in communicating scientific data especially, this topic, clearly, deserves a paper on its own.

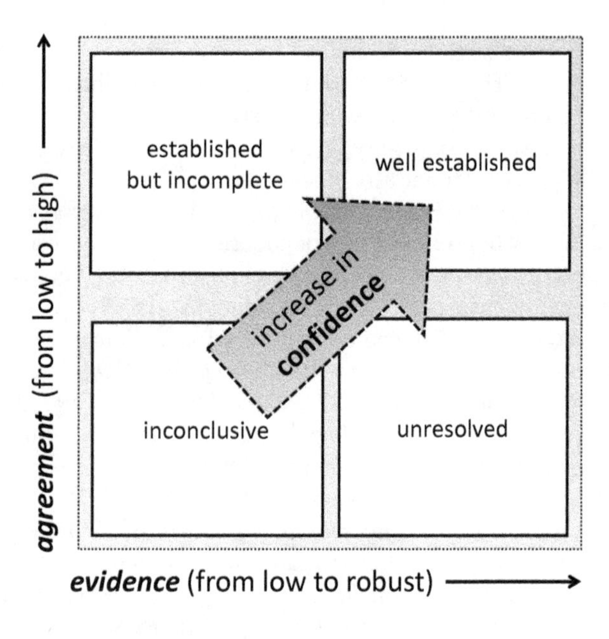

Fig. 2 Simplified and annotated version of the "four-box model for the qualitative communication of confidence" (IPBES, 2018, 30)

Despite obvious differences, both frameworks of confidence grading aim at increasing the comparability of expert judgments across key topics in environmental assessments. The internal consistency of "multiple-metric" frameworks has been criticised by scholars for a number of reasons and most famously with regard to the IPCC's model.[7] Further improvement of assessment frameworks might decide upon their potential of becoming instrumental for similar assessment projects. In the following part, the paper elaborates on the question how confidence grading as a key asset in assessment frameworks relates to the overall aim of building, promoting and maintaining trust in scientific methodology and outcomes.

4 Calibrating Trust in Science

The IPCC and the IPBES are two intergovernmental bodies that participate in an ongoing endeavour to develop and implement refined and expanded frameworks that support and empower consensus formation practices within the realm of environmental assessments. By framing and aligning expert judgments, both case studies

[7] The internal inconsistency of the framework has been criticised for a number of reasons: (1) unclear relationship between the qualitative metrics of "evidence/agreement" and "confidence" (cf. Janzwood 2020); (2) unclear conceptual distinction between the qualitative metrics of "confidence" and the quantitative metrics of "likelihood" (cf. Helgeson et al., 2018). Besides, differences in interpretations of the relationship between metrics and inconsistent use of criteria within the working groups have been reported by scholars involved in the revision process and, hence, are said to leave room for further improvement (cf. Mastrandrea & Mach, 2011; Mach et al., 2017).

offer valuable insight into the micro processes of evaluating and assessing information from a plethora of sources. As a matter of course, the validity of consensus formation practices themselves are under review, especially in complex and contested topics. Given the scientific ends of environmental assessments, and their pivotal role for informing policy makers, it seems noticeable to reflect on the query how existing standards of assessment in science are currently re-evaluated and revised. With regard to the field of environmental assessment, the necessity of "an iterative improvement" of assessment frameworks has already been addressed (Mach et al., 2017, 10). Drawing on said frameworks, key criteria deserve further scrutiny. Especially, when we take into consideration that "evidence" on the one side and "agreement" on the other are said to allow for addressing and grading "confidence" in the first place.

4.1 Evidence and Credibility of Information

Today, what exactly counts as "evidence" more often than not is a matter of methodological debate: Take, for instance, the hierarchical system of classifying evidence in evidence-based medicine (EBM). According to evidence grading systems, a systematic review of controlled randomised clinical trials (RCTs) provides the highest level of evidence in evaluating treatment benefits and ensures the highest strength of recommendation respectively.[8]

Within the realm of environmental assessments, peer-reviewed literature is regarded as the most favourable source of information. Given that assessments are not supposed to generate new "primary" findings, but synthesise relevant information from a plethora of sources, *credibility* of information is of major concern. To begin with, relevant information has to be identified and accessed when needed. Besides, contributing scientists are encouraged to "update grey literature to peer review literature", for example, by publishing a synthesis of key issues addressed in relevant workshop proceedings, white papers and similar reports from a broad range of scientific, political and economic institutions, including governmental sources and industry journals (IPCC, 2019, 4). Incorporating information from grey literature is vital to map useful information which could not be accessed otherwise, namely by consulting peer-reviewed data exclusively. Nevertheless, they are also regarded as main sources of error, take for instance the prominent case of the wrongful prediction of the disappearance of Himalayan glaciers by 2035 in the IPCC's *Fourth Assessment Report* (AR4) in 2007.[9] Assessment projects are confronted with a number of challenges with regard to the evaluation of different sources and types of evidential support. This includes the ability of "tracing" information, that is to say, correctly

[8] Take, for instance, the one developed by the *Oxford Centre for Evidence-Based Medicine* (OCEBM, 2011). For an introduction into the rationale of EBM and its pitfalls from a philosophy of science's perspective see Worrall (2010).

[9] For a detailed account of the events and a critical assessment of how they could have been prevented, see Kosolosky (2015).

attributing and referencing sources, and ensuring comparability and consistency of evidence. As a matter of course, the need to address *uncertainty* in data is getting more and more attention in recent years. In the IPCC's guidance note on *Addressing Uncertainties* in 2005, the lead authors are provided with a table informing about important, but simplified types of uncertainty, such as "unpredictability" of scenarios, "structural uncertainty" due to inadequate climate models, and "value uncertainty" in the course of missing or non-representative data.[10] A key aim of the IPCC (and similar bodies) was to raise awareness of different kinds and sources of uncertainties, and to provide scholars with a toolbox of reporting and describing findings, adequately and consistently—whatever their source might be.

Concurrently, the very aim of aligning and empowering reporting has to be considered against the background of the plethora of scientific disciplines and research traditions that are vital for environmental assessment projects (i.e., disciplinary bias, disagreement as concerns methods and outcomes, expert-judgment bias, overconfidence due to unbalanced group dynamics).[11]

4.2 *Agreement and Diversity of Interests*

Given the overall aim to provide a well-informed synthesis of available knowledge in the field on the one side and to inform and guide policy-makers on key topics on the other, a "clear consensus" in the underlying assessment report is said to be a *prerequisite*.[12] Still, the emphasis on consensus is also regarded as a troublesome limitation in environmental assessments. Especially, when it comes to acknowledging differing views on key topics that are supported by relevant and sufficient evidence. In general, consensus formation practices are challenged by a number of caveats. Namely, to allow for framing and aligning communication within a peer community or for responding to known disciplinary bias and methodological disagreement. Inter- and transdisciplinary research assessment comes with the additional need to integrate different disciplinary viewpoints. Systematic challenges of inviting and empowering participation are of specific concern within the realm of environmental assessments.

In its brochure *How to participate in the IPCC*, published in 2019, the body highlights that the selection of authors (lead authors, contributing authors, chapter scientists and review editors) aims to "encompass as wide a range of views, expertise and geographical representation as possible" (IPCC, 2019, 11). Acknowledging that *agreement* on a given topic is regarded as "measure of the consensus achieved across

[10] IPCC (2005, 1). About the sources, types and levels of uncertainty in climate modelling in general, see Petersen (2006).

[11] As regards to the question of how the IPCC's frameworks align expert judgment in climate change assessment but also is faced with consistent differences due to methodology and research questions across its working groups, see, for example, Swart et al. (2009).

[12] Ideally based on "multiple lines of independent, high-level evidence" (Mastrandrea et al., 2011, 682).

the scientific community [...] not just across the author team",[13] the review process is regarded as key in moderating known challenges of consensus formation practices such as bias in methodology and misrepresentation of interests.

Then again, the latter does not compensate for misrepresentation of interests as such. Given that not only the scope and evidential strength of information but also the diversity of authorship affect structure and content of assessment reports, the need has been addressed, "to expand the scope of content by diversifying authorship" (Ford et al., 2012, 203). Whereas the former (*scope* of information/content) might be regarded as mere epistemic challenge, namely, to allow for adequate and sufficient informed reporting, the latter comes with an explicitly social and political dimension. The call for *diversifying* authorship addresses two aims, primarily, the need to include a broad range of expertise, that is to say, individuals with a unique access to a certain body of knowledge (expert knowledge, practitioner's knowledge etc.). Secondly, it impacts on the question if specific group interests (local population, indigenous population) are represented in decision-making, and what is more, how to ensure, that they are represented *adequately*. To allow for the fact that any such demands are addressed within an assessment process, "at-risk" populations have to be identified beforehand. Take, for instance, the need to empower communities that are specifically vulnerable to climate change.[14]

Some critics scrutinise necessary revisions regarding the recruitment of authors in environmental assessments. To include scholars with "first-hand expertise of working with (and publishing on) Indigenous populations" is regarded an important step towards identifying specific vulnerabilities and disproportionate burdens (Ford et al., 2012, 209). But, this does not necessarily compensate for the traditionally low prominence given to non-scientific knowledge in general and local populations' perspectives.

Increasing awareness towards diversifying authorship has as of yet not led the IPCC and similar bodies to meet these demands sufficiently. Nevertheless, one might argue, that the overall aim has been acknowledged, for example, within the realm of the United Nations Climate Change Conferences, that is to say, the *Conferences of the Parties* (COP), where the parties that have signed the *United Nations Framework Convention on Climate Change* (UNFCCC) come together annually to assess progress in dealing with climate change.[15] To empower local and indigenous groups expressively, the *Local Communities and Indigenous Peoples Platform Web Portal* (LCIPP) under the auspices of the United Nations secretariat for climate change (UN Climate Change) was launched in 2017.[16] This platform is of

[13] For example, Mastrandrae et al. (2011, 678).

[14] For instance, local populations that engage in the "People of the Pacific Islands Development Forum" as mentioned in the *Nadi Bay Declaration on the Climate Change Crisis in the Pacific* published in July 2019, see: https://cop23.com.fj/nadi-bay-declaration-on-the-climate-change-crisis-in-the-pacific/. Accessed 16 Jan 2023.

[15] Further information on the *United Nations Framework* (UNFCCC) could be found online: https://unfccc.int. Accessed 16 Jan 2023.

[16] The LCIPP comes with three essential functions: (i) to "promote exchange of experiences and good practices for addressing climate change in a holistic way"; (ii) "build capacity for engagement";

specific import, given that indigenous populations have experienced marginalisation, discrimination and exploitation in the past—by individual scientists or in the name of science. To acknowledge mistreatment and inadequate representation of populations in the history of science might be key in building trust in general and to diversify participation in scientific assessments in particular (cf. Grasswick, 2017).

Taken together, consensus forming practices within the field of environmental assessments are faced with the query of how to fruitfully connect the plethora of scientific disciplines and research traditions that are vital for environmental assessment projects. Adequately framing and aligning information that result from different "knowledge systems" in the sciences (e.g., disciplinary ways of thinking and investigating) and beyond, including, "Traditional Ecological Knowledge" (TEK), could be regarded as an additional challenge in large-scale assessment projects.[17]

4.3 *Confidence and Trust in the Sciences*

Both frameworks analysed in this paper implement confidence grading as an additional metric to inform the degree of a (synthesised) judgment on the validity of findings. The validity of a finding is said to reflect both, the consistency of a body of evidence on the one side and the degree of agreement within a peer group on the other (IPCC, 2010a; IPBES, 2018). Whereas the IPCC relies on a detailed approach ("nine-boxed-model"), the IPBES restricts its framework to a simplified ratio ("four-box-model"). Taken together, within the field of environmental assessment a move towards frameworks that implement confidence grading might be reported. As of yet it is undecided, if these frameworks will mature to essential building blocks of generating and maintaining "calibrated trust" in the sciences, especially when it comes to fruitfully connecting different scientific cultures of evaluating information and prioritising evidential support.

In regards to the question of which framework might be better suited for being adopted by other bodies or scientific institutions within and outside environmental assessment, the paper does not claim to provide an answer. Suffice to say, any approach that offers rigorous expert judgment without implementing an over complex regime might be at an advantage. Quite noteworthy, a trend within the IPCC's own community favouring a simplified model to "advance the accessibility" of said framework could be reported (Mach et al., 2017, 10).

---

and (iii) "bring together diverse ways of knowing for designing and implementing climate policies and actions". See: https://lcipp.unfccc.int. Accessed 16 Jan 2023.

[17] The conceptual framework of IPBES, for instance, explicitly addresses the need to acknowledge "different knowledge systems (western science, indigenous, local and practitioner's knowledge)" in environmental assessments (Díaz et al., 2015, 3). A draft version of the most recent "methodological guidance for recognizing and working with indigenous and local knowledge" (April 2021) could be found online: https://ipbes.net/modules-assessment-guide. Accessed 16 Jan 2023.

Having analysed confidence grading as an additional qualitative metric the question remains if any such metric comes with an added value of furthering the comparability of expert judgments on key topics. At best, confidence grading informs about the *distribution* and *strength of shared belief* across a given peer community. In other words, confidence grading provides us with an elaborate account of "calibrated trust" within this very community. In measurement *calibration* addresses the need to ensure the reliability of measuring devices, for example, by properly tuning a generic device according to technical standards to allow for its functionality as an instrument. Calibration, therefore, could be regarded as a methodological means of reducing or circumventing (known) error to a certain degree.[18] In the case of environmental assessment, regimes that provide a standardised terminology ("calibrated language") seemingly respond to a similar endeavour, namely to "minimize possible misinterpretation and ambiguity" in assessing the value of data in reference to a set of established criteria such as evidence.[19] In measurement as well as in assessment, calibration not only comes with the methodological merit of circumventing (known) error and reducing conceptual ambiguity. Its specific value could be addressed as contributing to the credibility of scientific practices within and beyond the sciences (i.e., reliability of methods; validity of outcomes).

Taken together, confidence grading could be regarded as an (additional) asset in responding to major caveats in assessment and addressing these respectively. For instance, concerning the question how scientific and (non-scientific) experts grapple with threats to sharing judgment in assessment: robust evidence and high agreement might not guarantee correctness of expert judgments, but surely, it increases the confidence with which findings can be communicated within the peer community. Then again, given the specific needs of policy-makers ("usability" of information), consistency of evidence and confidence might be regarded as necessary but not sufficient prerequisites of environmental assessments (cf. Gramelsberger & Feichter, 2010; Moss, 2016). Still, by addressing and grading confidence explicitly within their frameworks, the IPCC and the IPBES further awareness towards the *complexity* of integrative judgments across scientific disciplines in large-scale assessment projects. Especially this might be the case, when it comes to match policy needs in complex and contested research topics such as climate change and biodiversity. In acknowledging these caveats, the reporting of confidence levels might prove to be especially suitable to inform about and assess the peer community's "trust" in key building blocks of large-scale assessment projects.

[18] For a definition of calibration as a basic concept of metrology see *International Vocabulary of Metrology* (JCGM, 2012).

[19] From the beginning said frameworks (cf. Section 3) dealt with topics such as the "appropriate level of precision" in reporting and describing findings, and how to use "calibrated language", that is to say language, "that minimizes possible misinterpretation and ambiguity" in communicating assessment results (IPCC, 2005, 2f.; IPBES, 2014). Then again, the implementation of a common lingo ("uncertainty language") for the purpose of framing and aligning communication has been a challenging issue from the start, given, for example, the heterogenous use of said criteria across the IPCC's working groups and the question how closely the frameworks were adopted. For an introduction see Mastrandrea and Mach (2011).

Drawing on this development in environmental assessments, the paper has elaborated on "confidence" as an epistemic concept: By introducing an *intersubjective* account, confidence is conceptualised as *strength of shared belief*—in reference to a given peer group. From an epistemological point of view, this understanding allows to scrutinise three prerequisites of "calibrated trust" in science as follows:

A peer community is able

(1) to *assess* the scientific potential of a given body of information *adequately* (*reliability* of methods; *validity* of outcomes; *consistency* of evidence),

(2) to *reach* a consensus about how this body of information has to be evaluated in state-of-the-art assessment projects, acknowledging both inherent challenges (i.e., persistent uncertainty in modelling) and the given purpose at hand (*adequacy* and *sufficiency* of information as regards policy-making), and, subsequently

(3) is able to *rely* on assessment outcomes for the purposes at hand (*credibility* of assessment conclusions; *integrity* of scientific peer review).

The latter presupposes that potent strategies are in place that come to pass in all those cases where either data is non-representative, evidence is insufficient or an agreement in assessing a body of information could not be achieved. This not only includes a nuanced regimen on how to address the uncertainty of projections, the fragility of evidential support, or the fallibility of assessment conclusions, but also about how to adequately document heterogenous interpretations of methods and data or to identify relevant disparate views—given they are supported by sufficient evidence.

Last but not least, the very process of *scientific community building* itself is put under a microscope: Given that rigorous scientific judgment demands a "well-evolved" peer community, the question how it evolves within the realm of science while how their evolvement is driven and framed on the basis of large-scale assessment projects, might be of specific interest to further studies on this topic.

5 Conclusion

Taking the cases of IPCC and IPBES the paper presented two frameworks of confidence grading that was introduced with the purpose of furthering rigorous expert judgment in large-scale environmental assessments. The paper illustrated why the IPCC turned to the topic of "confidence", that is to say, on what grounds the grading of confidence started out and developed until today. In the case of the IPBES, the question was raised, how similar institutions responded to this development. Drawing on an analysis of key criteria within these frameworks, the paper addressed "confidence" as an epistemic concept: By introducing an *intersubjective* account, confidence was conceptualised as *strength of shared belief*—in reference to a given peer group. It was argued that confidence grading comes with an elaborate account of "calibrated trust" within this very community. The latter informs about a peer community's reliance on scientific methods, projections as well as outcomes of a given field of research. Also, it invites reflection and critical assessment of community building within large-scale assessment projects. Hence, the paper concludes that epistemic challenges, commonly regarded as internal to scientific analysis and evaluation, are deeply connected to the discourse on trustworthiness of scientific practices and trust in the sciences, respectively.

References

Díaz, S., et al. (2015). The IPBES conceptual framework—Connecting nature and people. *Current Opinion in Environmental Sustainability, 14*, 1–16.

de Ridder, J. (2022). How to trust a scientist. *Studies in History and Philosophy of Science, 93*, 11–20.

Ford, J. D., Vanderbilt, W., & Berrang-Ford, L. (2012). Authorship in IPCC AR 5 and its implications for content: Climate change and Indigenous populations in WGII. *Climatic Change, 113*, 201–213.

Gramelsberger, G., & Feichter, J. (Eds.). (2010). *Climate change and policy: The calculability of climate change and the challenge of uncertainty*. Springer.

Grasswick, H. (2014). Climate change and responsible trust: A situated approach. *Hypatia, 29*(3), 541–557.

Grasswick, H. (2017). Epistemic injustice in science. In I. J. Kidd, J. Medina & G. Pohlhaus (Eds.), *The Routledge handbook of epistemic injustice* (pp. 313–321). Routledge.

Helgeson, C., Bradley, R., & Hill, B. (2018). Combining probability with qualitative degree-of-certainty metrics in assessment. *Climatic Change, 149*, 517–525.

Intergovernmental Science-Policy Platform on Biodiversity and Ecosystem Services. (2014). Guide on the production and integration of assessments from and across all scales (deliverable 2 (a)). IPBES/3/INF/4. IPBES. Retrieved January 16, 2023, from https://ipbes.net/sites/default/files/downloads/IPBES_3_INF_4.pdf

Intergovernmental Science-Policy Platform on Biodiversity and Ecosystem Services. (2017). The assessment report on Pollinators, Pollination and Food Production of the Intergovernmental Science-Policy Platform on Biodiversity and Ecosystem services. Full Report. IPBES. Retrieved January 16, 2023, from https://ipbes.net/assessment-reports/pollinators

Intergovernmental Science-Policy Platform on Biodiversity and Ecosystem Services. (2018). Guide on the production of assessments. Core Version. IPBES. Retrieved January 16, 2023, from https://ipbes.net/guide-production-assessments

Intergovernmental Panel on Climate Change. (2005). Guidance Notes for Lead Authors of the IPCC Fourth Assessment Report on Addressing Uncertainties. IPCC. Retrieved January 16, 2023, from https://www.ipcc.ch/report/ar4/wg1/

Intergovernmental Panel on Climate Change. (2010a). Guidance Note for Lead Authors of the IPCC Fifth Assessment Report on Consistent Treatment of Uncertainties. IPCC. Retrieved January 16, 2023, from https://www.ipcc.ch/working-group/wg1/

Intergovernmental Panel on Climate Change. (2010b). Annex A to Guidance note for Lead Authors of the IPCC Fifth Assessment Report on Consistent Treatment of Uncertainties. Comparison of AR4 and AR5 Approaches. IPCC. Retrieved January 16, 2023, from https://www.ipcc.ch/working-group/wg1/

Intergovernmental Panel on Climate Change. (2019). How to participate in the IPCC. IPCC. Retrieved January 16, 2023, from http://www.ipcc.ch/how-to-participate-in-the-ipcc/

Intergovernmental Panel on Climate Change. (2021). Climate Change 2021. The Physical Science Basis. AR 6. WRI. Full Report. Chapter 1. Accepted version. IPCC. Retrieved January 16, 2023, from https://www.ipcc.ch/report/ar6/wg1/#FullReport

Janzwood, S. (2020). Confident, likely, or both? The implementation of the uncertainty language framework in IPCC special reports. *Climatic Change, 162*, 1655–1675.
Joint Committee for Guides in Metrology (JCGM). (2012). International Vocabulary of Metrology—Basic and General Concepts and Associated Terms (VIM), Third Edition, 2008 version with minor corrections 2012. Retrieved January 16, 2023, from https://www.bipm.org/en/committees/jc/jcgm/publications
Kosolosky, L. (2015). "Peer Review is Melting Our Glaciers": What Led the Intergovernmental Panel on Climate Change (IPCC) to go astray? *Journal for General Philosophy of Science, 46*(2), 351–366.
Mach, K. J., Mastrandrea, M. D., Freeman, P. T., & Field, C. B. (2017). Unleashing expert judgment in assessment. *Global Environmental Change, 44*, 1–14.
Mastrandrea, M. D., Mach, K. J., Plattner, G.-K., Edenhofer, O., Stocker, T. F., Field, C. B., Ebi, K. L., & Matschoss, P. R. (2011). The IPCC AR5 guidance note on consistent treatment of uncertainties: A common approach across the working groups. *Climatic Change, 108*, 675–691.
Mastrandrea, M. D., & Katharine, J. (2011). Treatment of uncertainties in the IPCC Assessment Reports: Past approaches and considerations for the Fifth Assessment Report. *Climatic Change, 108*, 659–673.
Moss, R. H. (2016). Assessing decision support systems and levels of confidence to narrow the climate information "usability gap." *Climatic Change, 135*, 143–155.
Moss, R. H., & Schneider, S. H. (2000). Uncertainties in the IPCC TAR: Recommendations to lead authors for more consistent assessment and reporting. In R. Pachauri, T. Taniguchi & K. Tanaka (Eds.), *Guidance papers on the cross cutting Issues of the Third Assessment Report of the IPCC* (pp. 33–51). World Meteorological Organization.
OCEBM Levels of Evidence Working Group. (2011). The Oxford Levels of Evidence 2. Oxford Centre of Evidence-Based Medicine. Retrieved January 16, 2023, from https://www.cebm.ox.ac.uk/resources/levels-of-evidence/ocebm-levels-of-evidence
Petersen, A. C. (2006). Simulating nature. In *A philosophical study of computer-simulation uncertainties and their role in climate science and policy advice*. Het Spinhuis.
Swart, R., Bernstein, L., Ha-Duong, M., & Petersen, A. (2009). Agreeing to disagree: Uncertainty management in assessing climate change, impacts and responses by the IPCC. *Climatic Change, 92*, 1–29.
UK National Ecosystem Assessment. (2011). *Technical Report. Chapter 3: The Drivers of Change in UK ecosystems and ecosystem services*. UK NEA. Retrieved January 16, 2023, from http://uknea.unep-wcmc.org/Resources/tabid/82/Default.aspx
Wilholt, T. (2013). Epistemic trust in science. *The British Journal for the Philosophy of Science, 64*(2), 233–253.
Worrall, J. (2010). Evidence: Philosophy of science meets medicine. *Journal of Evaluation in Clinical Practice, 16*, 356–362.

$#*! Scientists Say: Monitoring Trust with Content Analysis

Petar Nurkić

Abstract In addition to existing practices and norms within the institutions to which they belong, scientists form an epistemic community. The flow of information, the processes of belief formation, and the assumptions scientists are making determine the structure of an epistemic community. Members of such communities can be categorized as epistemic experts or epistemic agents. In exceptional circumstances, such as the crisis caused by the global coronavirus pandemic, the existing organization, conventions, and rules within the epistemic community are disrupted. In times of crisis, epistemic agents' trust in experts becomes as crucial as scientific knowledge. Trust epistemic agents have in experts may decrease in crisis situations and affect experts' roles within the epistemic community. This puts epistemic experts in the position of needing to reestablish their trustworthiness through various rhetorical attempts. Here, we examine some of those attempts from the existing framework of rhetorical strategies used to reestablish or defend one's own expert role, but on novel data. To that end, we use examples of epistemic experts' justifications when confronted with public questioning about their behavior amidst the coronavirus crisis. We analyze experts' rhetorical approaches via qualitative content analysis and replicate four rhetorical strategies experts use when defending their epistemic positions that are previously established in analyses of similar cases in other crisis situations.

1 Introduction

This article focuses on the behavior of epistemic experts in crisis situations and a way to monitor their attempts to reestablish the trust of other epistemic agents, that is, the public. When we speak of epistemic experts, we usually mean epistemic agents who possess expert knowledge or skills (Goldberg, 2009). More broadly, where the context determines the need for a skill relevant to that context, epistemic expertise usually refers to scientists, bankers, physicians, and policymakers—that is, a group of people with the knowledge relevant or crucial to the problem at hand.

P. Nurkić (✉)
Faculty of Philosophy, Institute for Philosophy, University of Belgrade, Belgrade, Serbia
e-mail: petar.nurkic@f.bg.ac.rs

M. M. Resch et al. (eds.), *The Science and Art of Simulation*,
https://doi.org/10.1007/978-3-031-68058-8_8

Here, we will focus on physicians and scientists since the situation of crisis examined pertains to the coronavirus pandemic. Focusing on physicians and scientists as epistemic experts in the coronavirus pandemic will provide us with the following epistemic network: "epistemic experts" refer to mentioned scientists and physicians as undeniable experts for the coronavirus, "epistemic agents" represent the general public understood as all other people without the expertise relevant to the context of the coronavirus pandemic, and "epistemic community" pertains to both epistemic experts within their institutions and agents as a whole. Figure 1 shows an epistemic network in which we will operate.

Our goal here is twofold. First, we are interested in examining the relationship between experts and other epistemic agents, understood as the general public (i.e., non-experts), in the context of trust. Given that experts' positions within the epistemic community are especially vulnerable in times of crises (Kruglanski et al., 2005)—when experts are questioned by other epistemic agents in the community—trust other epistemic agents have in experts can deflate. Crisis situations, such as the coronavirus pandemic, hence seem as a suitable context in which we can examine the relationship of trust between scientists as epistemic experts and the public as epistemic agents. Since the relationship of trust between scientists and the public entails a two-way perspective (a perspective of the trustee and the trustor) and is too complex for the scope of this paper, we will focus on a more specific problem of interest: the epistemic experts' attitude towards other epistemic agents in circumstances in which their epistemic positions are threatened. That is, we are interested in the way epistemic

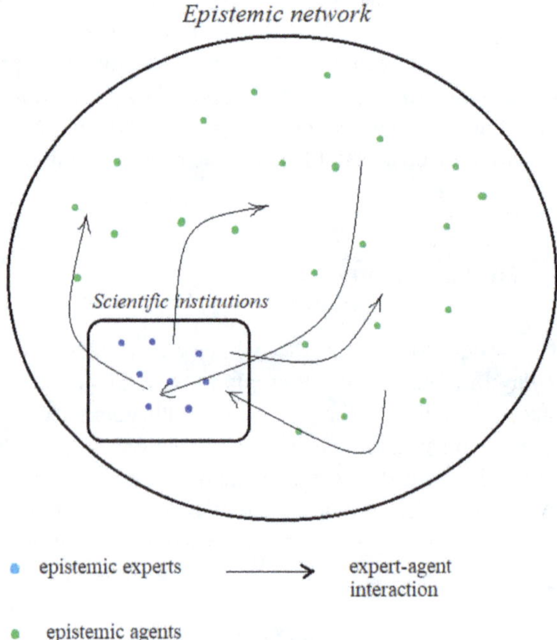

Fig. 1 Interaction between epistemic experts (e.g., scientists) and epistemic agents (e.g., the public) within an epistemic community

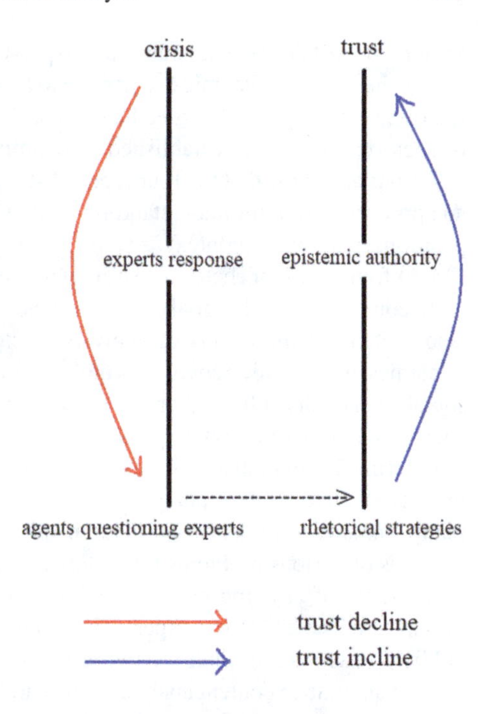

Fig. 2 Relation between trust and epistemic experts' responses in crisis situations

experts use their language to manage public perceptions of their trustworthiness. To further clarify our conceptual apparatus, in this paper *epistemic communities* consist of both epistemic experts and epistemic agents. Unlike *epistemic experts* who are here defined as scientists and physicians because of the skills relevant to the coronavirus pandemic, *epistemic agents* should be understood as the general public, that is, non-experts. In the usual circumstances, epistemic communities are more confined to people who are directly relevant to said community (Zollman, 2013). For example, the scientific or medical epistemic community usually refers only to various epistemic experts for that community (e.g., scientists, physicians, medical personnel, epidemiologists). In situations of crisis, however, epistemic communities are widened and include other epistemic agents (i.e., non-experts, the public) since the context of crisis creates crucial problems and epistemic needs that become relevant to agents other than immediate, usual members of said community. In other words, in crisis situations such as ours, the medico-scientific epistemic community also includes the general public since the epistemic needs (e.g., knowledge about the virus) of the previously localized community also become the primary needs of the other agents, that is, the public. Furthermore, epistemic experts work within the boundaries of *medico-scientific institutions* of public importance that can influence and shape the opinions and actions of agents in crisis situations by gaining their trust. The way epistemic experts gain trust is closely related to establishing and strengthening their epistemic authority. Authority is a direct result of how epistemic experts linguistically defend their positions within the community during times of crisis. Drawing on research from social psychology (Kruglanski et al., 2009) and the ancient roots of effective language use to persuade and gain the trust of others (Garver, 1994), the "tools" epistemic experts use to (re)establish trust we are interested in four rhetorical strategies.[1] We specifically use Riaz and colleagues (2016) framework of rhetorical strategies established in examining experts' language in other crises situations and show that the four types of strategies that emerged in their research are an appropriate tool for understanding experts' linguistical attempts to stabilize their positions in our crisis context as well. That is, we will show that Riaz and colleagues (2016) framework of rhetorical strategies is applicable to novel data.

Second, our broader goal that transcends the previous problem of interest is to also sketch out the means of analysis epistemologists can utilize to research the dynamics of epistemic networks, which are traditionally used in other, more empirical social science disciplines (Forman & Damschroder, 2007; Hsieh & Shannon, 2005) such as social psychology (e.g., Kruglanski et al., 2009). The suitable approach for monitoring the relationship of trust between scientists and the public we are interested in is qualitative content analysis. Even though this is a standard methodology in many humanities, philosophers seldom endeavor to incorporate it in their conceptual analyses of various problems that interest them. For that reason, we wish to present an example of utilizing content analysis to examine epistemological questions—in our case, the epistemic experts' language use for the purposes of managing the public's perception of their trustworthiness. As an example of how epistemologists can use qualitative content analysis to research their questions of interest, we analyze illustrative statements from eight epistemic experts which arose in response to the public's questioning of those experts' behavior. These statements become our units of analysis from which we identify different rhetorical strategies scientists employ for managing public trust and perception, using Riaz and colleagues (2016) framework.

2 Theoretical Background

To demonstrate how the method of qualitative analysis can serve to monitor trust, we must first establish the theoretical assumptions for the suitability of such a methodology to our question of interest: the way epistemic experts use their language to manage public perceptions of their trustworthiness in situations where this trust is questioned. Therefore, we will first describe what epistemic communities and epistemic experts are. We will then explain the relationship between epistemic authority and trust within the epistemic community. Finally, we will present the central part of the investigation, namely the experts' use of rhetorical strategies to reinforce their epistemic authority in order to gain the general public's trust.

---

[1] On the types of rhetorical strategies, see more in Jarzabkowski and Sillince (2007), Riaz et al. (2016), and Radenović and Nurkić (2021, 2022).

2.1 *Epistemic Communities and Epistemic Experts*

Epistemic networks are an appropriate framework for understanding the relationships between epistemic agents. Epistemic communities consist of epistemic agents who exchange information, form beliefs, provide justifications for the truth of their beliefs, and form a network of information semantics (Zollman, 2013). In what follows, we will refer to the epistemic community as an epistemic network, precisely because of the informational nature of the relationships between the agents within that community. The relationships within an epistemic network are determined by skills, trustworthiness, values, and rules given by a particular context (Zollman, 2007). Based on the communication between the aforementioned epistemic experts and the public as other epistemic agents, we can identify patterns and epistemic nodes (themes). The rhetorical claims we will discuss later constitute one such pattern.

Particularly important for our analysis is how the experts' answers to the public's questions turn out and how they match the sources of information. This correspondence determines the success of a particular rhetorical strategy in gaining the trust of others. Our interest, therefore, focuses on the answers that epistemic experts give in public appearances and the application of qualitative analysis to those answers.

2.2 *Epistemic Authority and Trust*

There is extensive research in social psychology dealing with epistemic authority (Kruglanski et al., 2005, 2009). Although the primary goal of this paper is to illustrate how scientists, through their public statements, manage and shape the trust that other epistemic agents (i.e., the public) have in them, epistemic authority is the first step in achieving this goal. Epistemic authority plays a unique role in unusual circumstances such as pandemics, financial and social crises, and environmental disasters. Trust is an important social factor that enables certain members of the epistemic community to interpret the circumstances that led to the crisis and to suggest possible solutions and strategies to overcome the crisis context. The concept of trust that we have in mind is multifaceted. On the one hand, it has the theoretical character of epistemic trust, in which we seek a justified true belief (McCraw, 2015). Trust thus serves as an adequate substitute in situations where we accept the justification of an epistemic expert because time and opportunities do not allow us to seek it independently. On the other hand, it also has a practical side, such as trust in high-reliability organizations (Cox et al., 2006). In such organizations, trust is the most critical factor because research in real-time would be infeasible if the actors in the organization had to monitor and check each other constantly. In general, the trust epistemic agents have in epistemic experts is represented through experts' trustworthiness. Trustworthiness is the expert's epistemic virtue and can be expressed as the conviction of other agents about the reliability of information spreading from the expert. As we mentioned earlier, epistemic authority is based on how other agents perceive the expertise of experts. That means that authority, in return, provides the trust that other epistemic community members, such as the public, instill in their skills and knowledge. The expert's high authority enables more pronounced trust in her from other epistemic agents. That allows the expert the epistemic privilege to act as a source of information in future circumstances (Kruglanski et al., 2005) and influence how other epistemic agents in the network will behave.

Our focus is on the epistemic authority of experts such as pulmonary specialists, virologists, and epidemiologists and their management of trust other epistemic agents, namely the public, have in them during crises such as global pandemics. What makes an epistemic agent an expert are exceptional circumstances that require a certain amount of skills, abilities, and knowledge to prevent a catastrophe. In a pandemic, physicians and scientists are given the status of epistemic experts, although their expertise would not matter to economists in financial crisis circumstances. Just as economic experts are mere epistemic agents in a situation where medical decisions must be made, pulmonary physicians and immunologists are only the members of an epistemic community who instill their trust in a financial expert during a recession. Here we make a distinction between whom we consider an epistemic expert and who an epistemic agent in the coronavirus pandemic. As we said, epistemic experts of interest are physicians and scientists (e.g., immunologists, epidemiologists, virologists), while epistemic agents are all other non-experts in the context of the pandemic. Since experts' attempts to reestablish their trustworthiness in situations of global crises are mostly represented through public statements, epistemic agents in this situation are best understood as the general public in relation to which experts manage trust.

2.3 Rhetorical Strategy and Internal Organization of Epistemic Networks

In cases of broader domain crisis, the structure of epistemic networks becomes fluid and unstable. Previously established practices may become untenable and unprofitable for epistemic experts to defend them (Hoffman, 1999). In the lack of respirators, it is not possible to adhere to previously established practices; new measures must be taken to maintain the sustainability and stability of the medical healthcare system. On the other hand, reconfiguring the epistemic network due to new circumstances poses a threat to the preservation of experts and epistemic authority (Hoffman, 1999). Experts are set before strict standards in a crisis. Due to structural inertia to preserve epistemic authority, experts resort to rhetorical strategies. These linguistic maneuvers ensure the preservation of their role in the epistemic network and the position that allows the expert to control the outcomes of the crisis circumstances.

The first thing that can be noticed in the expert's attempts to maintain the position of epistemic authority is the extent of self-attribution of expertise, i.e., frequent statements about the amount of knowledge and abilities that the self-praising expert possesses (Riaz et al., 2016). The goal of such a rhetorical strategy is to get the other epistemic agents in the network to accept the expert's interpretation of the crisis circumstances (Lesfrud & Meyer, 2012). Our research examples focus on situations in which experts answered questions from other agents using rhetorical strategies to justify their epistemic authority. The experts' rhetorical utterances thus constitute our unit of analysis.

As we mentioned earlier, rhetoric has deep historical roots. For example, Aristotle describes how language can be used to persuade others (Garver, 1994). Thus, rhetoric is the ability to persuade others using arguments that aim to advance the actors' interests (Scott, 1967). Specifically, rhetorical strategies are an instrument for determining how other epistemic agents will perceive various social factors (Warnick, 2000). The most apparent use of rhetorical strategies boils down to presenting specific actions as morally acceptable, necessary, mutually beneficial, and rational (Vaara et al., 2006). The same can be done in the opposite rhetorical direction by presenting the actions of other epistemic experts as unacceptable and irrational. Through rhetorical strategies, experts gain support for their specific interpretation of the crisis context and the trust that others place in their attempts to shape the outcomes of the crisis. Previous research has shown that these strategies are highly cost-effective in epistemic terms and, by and large, allow experts to retain their authority and gain the trust of others (Creed et al., 2002; Hardy et al., 2000).

Our examples aim to reveal how experts can maintain or regain their epistemic authority by, for example, imposing epistemic obligations on other agents, attributing responsibility, or convincing others of their epistemic legitimacy and an authentic interpretation of circumstances. Also, we want to show that content analysis can be used by philosophers as a standard means to understand interactions and relationships within epistemic networks, that is, to analyze the dynamics of epistemic communities. Before moving on to the methodological part of our paper, we present a figure illustrating the assumptions of our theoretical framework. Here, we can see that in situations of crises, where epistemic experts' trustworthiness is questioned, rhetorical strategies are used as a way to mitigate the potential decline of trust other epistemic agents have in said experts. Our analysis will further illustrate strategies for reestablishing trustworthiness expert use when their epistemic authority is questioned by the public (Fig. 2).

3 Methodology

First, we must emphasize that here provided examples of content analysis are illustratively oriented. We present examples of limited novel data, with the intention of (1) analyzing the relationship of trust between scientists and the public, with a specific focus on language strategies epistemic experts use to manage the trust of the public; and (2) providing an illustration of viable and standardized methodology, a qualitative content analysis, epistemologists can utilize in their research. To that end, we wish to show that content analysis can serve in further understanding and mapping the processes of information flow, belief formation, and trust through ecologically valid, real-life interactions between epistemic agents (experts and other members). Even though qualitative content analysis is a standard empirical tool widely used in many other disciplines of humanities (Hsieh & Shannon, 2005), it is not a common practice in philosophy. Hence, we want to demonstrate content analysis methodology on examples of interesting novel data as a way to illustrate the benefits of empirical, besides theoretical, research of epistemic properties.

Content analysis is a research method traditionally used for analyzing textual data and is usually divided into quantitative and qualitative approaches to texts or problems of interest (Krippendorff, 2004). Quantitative content analysis is used to code textual data (or, in some cases visual or audio data) into meaningful and previously theoretically established categories. Codes that are used here are numerical in nature and thus lead to further statistical analyses that test different research hypotheses, given the problem of interest (Rourke & Anderson, 2004). Qualitative content analysis is more "meaning oriented" in that it is focused on various language uses, narratives, communication, and general contextual meanings of units of analysis (Hsieh & Shannon, 2005). Those units of analysis, depending on the problem of interest, can be whole texts, paragraphs, sentences, or words and the researcher approaches them from a specific theoretical framework of interpreting their meaning (Hsieh & Shannon, 2005; Krippendorff, 2004). Unlike the process of quantitative content analysis, the qualitative approach does not consist of numerical coding, but the researcher approaches units of analysis by observing the patterns and contextual (both explicit and implicit) meanings in them, which are then systematically categorized into different thematical groups by researcher's "close reading" of the text (Forman & Damschroder, 2007). Those themes are then also interpreted through a greater theoretical framework of the research in order to answer specific research questions, and provide knowledge and understanding of the problem (Forman & Damschroder, 2007; Hsieh & Shannon, 2005; Krippendorff, 2004). The crucial characteristic of qualitative content analysis that differs from quantitative one is that the "coding" of the text into categories and meaningful themes is done inductively, from the data, and not before that by using "predetermined categories" generated from other sources, as in qualitative content research (Forman & Damschroder, 2007). This does not mean that qualitative content analysis does not utilize an existing theoretical framework for making sense of and understanding the significance of particular meanings for the research problem; it just means that themes are determined depending on the meaning of the texts (i.e., the data), while the theoretical background is utilized for a comprehensive understanding of the meaning of those themes and their connection to the research question. In our case, Riaz and colleagues (2016) framework of rhetorical strategies—which explains the broader goals of particular language usage in relation to reestablishing epistemic authority and classifies it into four rhetorical strategies—is our theoretical background for understanding and meaningfully reading the experts' statements. Because our goal is to better understand whether and how epistemic experts use their language to manage the public's trust in them, we only qualitatively analyze their statements without numerical coding and further statistical analysis. We begin with experts' statements as units of analysis and through contextual reading of texts subjectively interpret the meaning and goals of experts' statements at hand. Then, we generate themes of meaning (i.e., rhetorical strategies) using the existing framework that defines those strategies (e.g., Riaz et al., 2016).

3.1 Context and Units of Analysis

Units of our analysis are whole statements of epistemic experts that arose as their responses to public questioning of their previous statements about coronavirus, mistakes, and behavior regarding the coronavirus. As mentioned earlier, we have placed a particular focus on the rhetorical strategies contained in these statements as experts' means to manage the fading trust of other epistemic agents (i.e., the public) in them. We were guided by existing research and similar analysis approaches (Brown et al., 2012; Erkama & Vaara, 2010; Nurkić, 2022; Riaz et al., 2016; Radenović and Nurkić, 2021). Crisis circumstances, such as the global pandemic, provide a unique opportunity to examine the dynamics of epistemic authority. These circumstances put scientists in a position of epistemic experts that guide governmental decisions and are subsequently responsible for both successful and unsuccessful outcomes. Furthermore, other epistemic agents have the unique opportunity to question those epistemic experts. This gives us valuable data for understanding the means experts thus use to justify and establish their authority, as well as the trust of epistemic agents, through content analysis of experts' linguistical attempts to straighten their authority when said authority is declining. Such situations present valuable opportunities to analyze ecologically valid data, that is, real-life behavior in the context of epistemic networks.

Since our investigative goal was to monitor the language attempts of scientists to (re)establish the public's trust in them, we chose the context of the crisis caused by the SARS-CoV-2 virus and situations of public questioning of scientists' behavior and epistemic authority amidst the crisis. This provides us with "background context" for further close reading of statements of interest—all examples of scientists' statements are made from the position of defending their epistemic authority when met with questioning from the public. As this is only an illustrative analysis, we focused on eight central actors whose decisions directly influenced the shaping of public opinion of their epistemic communities from March 2020 to November 2021 and who were in the position of defending their epistemic authority. We selected eight events from various countries that focus on different epistemic experts. Table 1 provides an overview of selected epistemic experts, the institutions to which they belong, and the events that constitute their behavior which was questioned by the public. The roles and titles of these epistemic experts have provided them with a position of authority from which they influence the perception of the public and can, subsequently, try to preserve their trust through rhetorical strategies.

Table 1 Epistemic experts, their areas of expertise, affiliations, and crisis events in which they participated

Epistemic experts	Area of expertise	Affiliation	Crisis events
Neil Ferguson	Epidemiology, immunology, mathematical modeling	Professor of Mathematical Biology & Head of Department of Infectious Disease Epidemiology at Imperial College London; Adviser at Scientific Advisory Group for Emergencies (SAGE)	Event 1
Catherine Calderwood	Obstetrics and gynecology	Former Chief Medical Officer of Scotland	Event 2
Darija Kisić Tepavčević	Epidemiology	Adviser at Serbia's Crisis Staff for the Suppression of Infectious Diseases COVID 19	Event 3
Anthony Fauci	Epidemiology, immunology	Director of the USA National Institute of Allergy and Infectious Diseases	Event 4
Michael Levitt	Biophysics	Professor of Structural Biology at Stanford University	Event 5
Roman Prymula	Epidemiology	Professor of Epidemiology at Charles University; former Minister of Health of the Czech Republic and the head of the government's Central Crisis Board	Event 6
Zlatibor Lončar	General surgery	Minister of Health of the Republic of Serbia	Event 7
Gordan Lauc	Biochemistry, molecular biology	Professor of Biochemistry and Molecular Biology at the University of Zagreb; member of Croatia's Government Scientific Council	Event 8

3.2 *Key Crisis Events and Defining a Relevant Data Sampling Framework*

In order to define a set of relevant statements for content analysis, we first identified eight events that represent controversial moments in which the expertise and behavior of our epistemic experts are questioned. Then, the statements made by epistemic experts' during such examinations were observed and analyzed to classify them into appropriate rhetorical strategies that were used to strengthen their epistemic authority.

Event 1 On March 23, 2020, U.K. Prime Minister announced the first lockdown for all four U.K. nations as a protective measure against the spread of coronavirus throughout the country, and on March 25, the Coronavirus Act 2020 officially received Royal Assent. This Act officially established the coronavirus lockdown and imposed it on the public. A month later, on May 5, Dr. Neil Ferguson, the lead epidemiologist and advisor to the U.K. government who had helped shape the coronavirus lockdown strategy, resigned from his advisory post after reporting that he had violated the lockdown measures during March and April.

Event 2 Under the same U.K. Coronavirus Act 2020, Scotland entered coronavirus lockdown on March 25, 2020. On April 5, 2020, Scotland's Chief Medical Officer,

Catherine Calderwood, resigned from her government post after it was reported that she had violated the lockdown rules she had advocated in March.

Event 3 On October 13, 2020, journalists from the Balkan Investigative Reporting Network (BIRN) reported that Serbia's Crisis Staff provided the official numbers of deaths and cases of infection in Serbia for the Suppression of Infectious Diseases COVID-19 (i.e., Crisis Staff) were false. BIRN's reports explicitly referred to the statements and data provided by Dr. Darija Kisić Tepavčević, a member of the Crisis Staff, and Serbian Prime Minister Ana Brnabić. In response to these reports and the government's relaxation of protective epidemic measures during the pre-election period, 350 physicians published an open letter to the government of Serbia demanding the resignation of all members of the Crisis Staff and the appointment of a new Crisis Staff. At the press conference of the acting Crisis Staff, Dr. Darija Kisić Tepavčević responded to the open letter by questioning the expertise of the signatories.

Event 4 At a hearing of the Senate Health, Education, Labor, and Pensions Committee on November 4, 2021, Dr. Anthony Fauci, director of the U.S. National Institute of Allergy and Infectious Diseases, refuted the claims that the National Institute of Health funded the acquisition of operational research at the Institute of Virology in Wuhan prior to the pandemic COVID-19. U.S. Senators Ted Cruz and Rand Paul subsequently accused Dr. Fauci of lying to Congress and urged the U.S. Attorney General to appoint a special investigation into Dr. Fauci's statements. In a November 29 interview with CNN News, Dr. Fauci responded to the senators' allegations.

Event 5 Dr. Michael Levitt, a professor of structural biology at Stanford and a Nobel laureate, spoke out publicly during the coronavirus outbreak, making public his erroneous predictions about the virus' spread patterns and outcomes of the pandemic in several countries. On July 25, 2020, he published his prediction that the coronavirus outbreak in the United States would be over by the end of August of that year and that the total number of deaths would be less than 170000. He later commented on several of his flawed coronavirus predictions. Many epidemiologists and fellow academics have publicly criticized Prof. Levitt's actions as irresponsible and misleading to the public and not in accordance with the academic community.

Event 6 On October 21, 2020, the Czech newspaper Blesk published the reports of Dr. Prymula, epidemiologist and Acting Minister of Health of the Czech Republic, on coronavirus violations. Dr. Prymula was subsequently relieved of his duties as Minister of Health but remained in an advisory position with the Czech Prime Minister. Four months later, on February 19, 2021, Dr. Prymula was photographed attending a soccer game despite regulations on social gatherings due to the coronavirus outbreak. Following public criticism, Dr. Prymula was also relieved of his advisory post to the prime minister. In an interview with Czech television, Dr. Prymula stated that it was his own decision to leave the position of government advisor.

Event 7 In July 2020, a French international news agency (franc. Agence France-Presse, AFP) published a hospital report on the death of a patient at the Zemun Clinical Center, Serbia, due to a lack of space in the hospital's respiratory center. At the July 23 government Crisis Staff press conference, the Serbian prime minister and health minister answered reporters' questions about the lack of respirators and the subsequent death published in the AFP report.

Event 8 On February 22, 2021, Gordan Lauc, Professor of Biochemistry and Molecular Biology at the University of Zagreb, announced the preliminary results of the first phase of his study, "Assessment of the serological response of the population of the city of Zagreb to contact with SARS-CoV-2" for the Croatian evening newspaper. He stated that the peak of the coronavirus pandemic in Croatia is over, as more than a third of the Croatian population has recovered from the virus. In March 2021, Ivan Štagljar, professor of molecular genetics and biochemistry at the University of Toronto, whose team has developed a new antigen test for coronavirus, commented on Lauc's study for Nova Croatian News T.V. Prof. Štagljar stated that Lauc's study contained incorrect information and did not comply with scientific methodology. On the same day, Prof. Lauc responded to these comments.

3.3 Data Sampling Procedure

Epistemic experts' public statements regarding described events were collected from electronic media sources. These experts' statements were made in response to public questioning of their behavior and can be seen in Table 2 We collected eight expert statements from electronic media using a *fake news tracker*, browser *news.Google*, and *factcheck.org*. The eight events and corresponding experts' statements were collected from the following media sources: *The Guardian*, *BBC*, *Independent*, *021*, *UnHerd*, *IndexHR*, *Newsbeezer,* and *Radio Slobodna Evropa*. These eight statements were chosen as they represent illustrative examples for the content analysis of rhetorical strategies.

3.4 Results

Previous research on different crisis circumstances distinguishes internal and external rhetorical strategies that experts use to straighten their authority (Brown et al., 2012; Erkama & Vaara, 2010; Riaz et al., 2016). In accordance with Riaz and colleagues (2016), we use four rhetorical strategies they distinguish as our framework for naming, defining, and interpreting meanings of experts' statements in this example data. These strategies are also semantically comparable with the strategies other authors delineate (e.g., Brown et al., 2012; Erkama & Vaara, 2010), but we use Riaz and colleagues (2016) taxonomy as the generation of themes from our data fits the narrative themes of Riaz's strategies. Table 2 shows four representative examples of our experts' statements categorized according to each of the rhetorical strategies distinguished by Riaz et al. (2016), while the remaining four statements were used in the text below to further describe the contextual meaning of the statements that represent each strategy.

Table 2 Rhetorical strategies and their representative examples

Rhetorical strategies	Representative examples
Internally directed strategy 1: Rationalization	"I accept I made an error of judgment and took the wrong course of action. I have therefore stepped back from my involvement in Sage. I acted in the belief that I was immune, having tested positive for coronavirus, and completely isolated myself for almost two weeks after developing symptoms. I deeply regret any undermining of the clear messages around the continued need for social distancing to control this devastating epidemic. The government guidance is unequivocal and is there to protect all of us." Neil Ferguson, epidemiologist, UK government adviser at Scientific Advisory Group for Emergencies (Sage) (Stewart, 2020)
Internally directed strategy 2: Emphasizing normative responsibilities	"The First Minister and I have had a further conversation this evening and we have agreed that the justifiable focus on my behavior risks becoming a distraction from the hugely important job that government and the medical profession has to do in getting the country through this coronavirus pandemic. Having worked so hard on the government's response, that is the last thing I want. The most important thing to me now and over the next few very difficult months is that people across Scotland know what they need to do to reduce the spread of this virus and that means they must have complete trust in those who give them advice. It is with a heavy heart that I resign as Chief Medical Officer." Catherine Calderwood, consultant obstetrician and gynecologist, Scotland's Chief Medical Officer ("Scotland's chief medical officer resigns over lockdown trips", 2020)
Externally directed strategy 1: Disproving the expertise of other epistemic experts	"What should be known is that there are 33,000 doctors in the Republic of Serbia. Colleagues who signed a certain statement, I assume that they understand and have competencies when it comes to infectious diseases, although there are no immunologists, epidemiologists, or infectiologists among them. I assume that they have competencies when they gave themselves the right to criticize," Darija Kisić Tepavčević, epidemiologist, Serbia's adviser at Crisis Staff for the Suppression of Infectious Diseases COVID 19 ("Kisić Tepavčević dovela u pitanje kompetentnost 350 lekara", 2020)

(continued)

Table 2 (continued)

Rhetorical strategies	Representative examples
Externally directed strategy 2: Doubting the motives of other epistemic experts	"I'm just going to do my job and I'm going to be saving lives and they're going to be lying. Anybody who's looking at this carefully realizes that there's a distinct anti-science flavor to this. If they get up and criticize science, nobody's going to know what they're talking about. But if they get up and really aim their bullets at Tony Fauci, people could recognize there's a person there... it is easy to criticize, but they are really criticizing science because I represent science. That's dangerous. To me, that's more dangerous than the slings and the arrows that get thrown at me. And if you damage science, you are doing something very detrimental to society." Anthony Fauci, epidemiologist, director of the US National Institute of Allergy and Infectious Diseases (Sarkar, 2021)

Rationalization of provided guarantees

Rationalization represents an internally oriented rhetorical strategy in which epistemic experts justify their behavior through rationalization of their mistakes (Riaz et al., 2016). The contextual reading of the following statement indicates rationalizations that allow us to categorize this statement into the strategy of rationalization:

> The prediction has turned out less well than I had hoped... There are about 55,000 deaths in the U.S. every week, and right now there are about 5,000 more. So I think the prediction turned out less well than I had hoped, but it served as a milestone and clarified what we mean by 'over'. It made it clear how important it is to look at excess deaths and the prediction when it's over. My mistake was that I should have given a range instead of a number. Michael Levitt, a structural biologist and winner of the 2013 Nobel Prize in Chemistry (Sayers, 2020)

In this way, Prof. Levitt strengthened his epistemic authority and reinforced his role in the network that provided him with the trust of others in the further privilege of interpreting crisis circumstances and shaping crisis outcomes. By emphasizing their knowledge and skills, experts have the opportunity to explain the means used to achieve their goals and to justify why such a link between goals and means was rational (Brown et al., 2012). By rationalizing one's errors without clearly categorizing them as such, experts minimize the effects of those errors. Levitt's case shows that emphasizing the justification of one's errors, rather than confronting the error itself, contributes to strengthening the authority of the epistemic expert.

Emphasizing normative responsibilities

This is the second type of internally oriented strategy where experts emphasize their responsibilities towards other network experts and other epistemic agents (Riaz et al., 2016). In the following statement, the expert highlights his normative responsibilities to justify his mistakes and further establish his epistemic position in the epistemic community:

> I wanted to promote professional opinions, but I became something of a symbol for both parties and I do not want to polarize this society further, so I decided to leave. Roman Prymula, epidemiologist. ("Prymula: There was a reason to attend soccer", 2021)

Experts' responsibilities are normative, meaning that they express concern for other epistemic agents by referring to existing norms and rules that they have adhered to and thus contributed as much as possible to the interests of others (Riaz et al., 2016). When Prymula emphasizes his normative responsibility to unite the society (e.g. not polarizing the society), he expresses a personal and emotional level of responsibility that obligates him to act toward such a goal. In this way, he emphasized the normative responsibility of the epistemic experts who belong to his institution. In this way, expressed responsibility to adhere to some norm, rather than recognizing one's errors, serves as a defense and a justification of one's behavior without explaining said behavior. Through normative responsibility references that do not serve the recognition and explanation of the errors at hand, experts can rhetorically avoid the responsibility for those errors and subsequently strengthen their authority and trustworthiness.

Challenging the expertise of others

In contrast to the previous two rhetorical strategies, which internally aim to reinforce authority and conversely gain trust by making statements about one's abilities, knowledge, or norms, epistemic experts also use external rhetorical strategies. These external forms of rhetoric allow epistemic experts to consolidate their epistemic authority by challenging other experts. This challenge of expertise has two aspects: it can be aimed at challenging (1) the level of skills and knowledge of others (usually the ones that question said expert) or (2) the motives of other experts (Riaz et al., 2016).

In the rhetoric that utilizes challenging the expertise of others, experts focus on the failures or the expertise levels of other experts, instead of their own errors, thereby influencing the level of trust that other epistemic agents place in other experts and, consequently, them (Warnick, 2000). Here, the expert is defending his epistemic authority by questioning the expertise of other epistemic agents (i.e., journalists) that challenge him and thereby tries to render the "challengers" inadequate to even question the expert:

> You should get a degree in medicine to talk about this. The point is that the man was on a noninvasive ventilator, he had a saturation of 96, and he had been for 4 days – it would be a crime to intubate a patient with a saturation like that. Why would you not put someone on a ventilator when you have 34 unused at this point? Just in case there is any doubt, there is monitoring, there is always something that controls the first thing that is done - concede to us that we know a little bit more than you do. You do not understand the documents, but you are not in the medical field, I understand that, but you would have to go to medical school. Zlatibor Lončar, physician, Minister of Health of Serbia ("Niko nije umro zato što je čekao na respirator", 2020).

By highlighting the non-medical background of the journalist who questions him, dr Lončar places his own position above the position of the other epistemic agent, rather than answering the question at hand. With this strategy, epistemic experts allow themselves not to answer questions posed by epistemic agents who do not possess the same level of expertise.

Doubts about the motives of others

This is the second type of externally oriented rhetorical strategy where epistemic experts challenge the authority of other experts or epistemic agents by doubting the motivation behind any questioning the agents' may have (Riaz et al., 2016). We can take the following example:

> When someone excuses falsehoods and insults others for no reason on television, one has to ask why they do it. The first possibility to consider is a hidden interest. In this case, it is probably patent application 63/121,689 for a serological test for coronaviruses, in which a certain I.S. is listed as an inventor. When someone on a government salary and with the help of public funds makes a discovery that he personally files as an inventor, it is not hard to understand that he would not like the pandemic to end. Of course, it would be much better for this hypothetical person if we were all locked in basements and the pandemic continued for years, with a portion of the proceeds from each test sold going to his personal account ... to call for the introduction of a completely unnecessary austerity measures and unreasonable panic in a situation where it was clear that the pandemic was coming to an end ... Such a person certainly cannot understand that there are people who have accomplished enough to afford to work for the common good in a crisis situation without any hidden motives or interests ... That's probably completely inconceivable to him. Gordan Lauc, professor of biochemistry and molecular biology at the College of Zagreb, ("Lauc odgovorio Štagljaru", 2021)

In this way, the expert who disputes the benevolence of others' common-sense motives reinforces the trust that other epistemic agents have in him (Radenović & Nurkić, 2021). In this way, dr Lauc questions the motives of another expert that publicly criticized him without providing a direct answer to the criticism itself. Here, dr Lauc is implicitly emphasizing others' immorality in an attempt to defend his own and, more importantly, to evade publicly addressing the criticism the general public was interested in. In this way, he aims to straighten his epistemic authority and public trust that was threatened by other experts' criticism.

What is distinctive about these externally-oriented rhetorical strategies is the lack of a denial of responsibility that can be noticed in internally-oriented strategies. In addition to tacitly acknowledging responsibility for controversial actions, epistemic experts simultaneously emphasize the responsibility of other experts for their failures and focus critically on others to reinforce their authority and maintain the trust placed in them by other agents in the epistemic network.

4 Discussion

In this paper, we have considered the issue of trust in science and the attitude of scientists and medical practitioners toward the public using the content analysis method. To examine the concept of trust, we first had to establish a link between trust and epistemic authority. A qualitative approach to communication between scientists, that is, epistemic experts, and other agents has proven extremely useful in monitoring trust in crises, especially in the analysis of the crisis caused by the SARS-CoV-2 virus, where the existing rules and norms in the epistemic network of epidemiologists, pulmonologists and virologists become unsustainable and require reconfiguration.

We applied the conceptual framework of rhetorical strategies to the sampled statements for the media given by scientists and physicians from March 2020 to November 2021. Epistemic authority is widely considered in the field of social psychology, while rhetorical strategies as a conceptual framework were used in the qualitative analysis of financial and institutional crises (Brown et al., 2012; Erkama & Vaara, 2010), but also concerning global pandemics (Nurkić, 2022; Radenović and Nurkić, 2021). Additionally, qualitative content analysis is widely used in many humanities as a standard research methodology (Forman & Damschroder, 2007). However, this combination of methodology and conceptual framework has not been exclusively applied to epistemic networks in crisis situations, especially as a way for epistemologists to empirically enrich their research and further the understanding of epistemic network dynamics. With the goal of both understanding the linguistical means epistemic experts use to manage the public perception of their trustworthiness and authority, and emboldening other epistemologists to utilize content analysis in their research of epistemic networks, we provided illustrative examples of experts' statements amidst the coronavirus pandemic and analyzed their meanings in the context of rhetorical strategies. We can note that linguistical self-emphasis and self-attribution of abilities and skills are intrinsic features of rhetorical strategies. In internally directed strategies one either rationalizes their own behavior and errors or emphasizes the normative dimensions of the responsibility they take on within the epistemic network, rather than directly accounting for their errors, to keep the authority and reestablish the trust of the public. In externally oriented strategies, we see that experts are focused on other experts or agents rather than on themselves as a means of defending themselves from criticism, thus avoiding but implicitly confirming their accountability. They either challenge the epistemic authority and expertise of others or their motivation.

When unable to invoke previously established rules and norms within an epistemic network, experts can resort to rhetorical strategies to preserve their position and authority. Thus, crisis situations, because they entail the unsustainability of previously established practices, present a good basis for examining the dynamic within an epistemic community. Specifically, the relationship between experts and other epistemic agents. Qualitative content analysis can be utilized for these purposes. The usage of language in establishing one's position and consolidating the trust of others within an epistemic community is a suitable subject for the content analysis method that philosophers can use to further their understanding of the epistemic properties of such communities. Figure 3 presents a diagram containing Riaz and colleagues (2016) division of rhetorical strategies that can be found in cases of defending one's own epistemic authority.

We applied content analysis to a limited number of representative examples of experts' statements in situations in which they were challenged by the public.

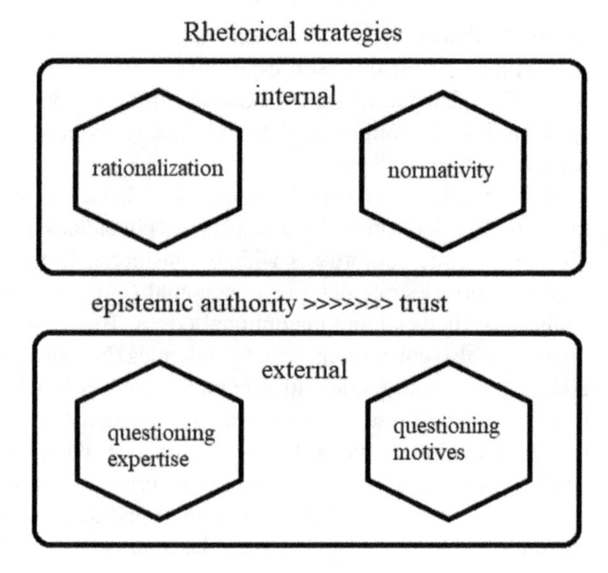

Fig. 3 Internal and external rhetorical strategies

However, we believe that these situations represent suitable contexts for further investigation of epistemic expert-agent dynamics that can shed light on a variety of rhetorical means experts use to re-establish their positions within an epistemic community.

References

Brnabić: Niko nije umro zato što je čekao na respirator. (2020). Radio Slobodna Evropa. https://www.slobodnaevropa.org/a/30743472.html

Brown, A. D., Ainsworth, S., & Grant, D. (2012). The rhetoric of institutional change. *Organization Studies, 33*(3), 297–321. https://doi.org/10.1177/0170840611435598

Coronavirus: Scotland's chief medical officer resigns over lockdown trips. (2020). BBC. https://www.bbc.com/news/uk-scotland-52177171

Cox, S., Jones, B., & Collinson, D. (2006). Trust relations in high-reliability organizations. *Risk Analysis, 26*(5), 1123–1138. https://doi.org/10.1111/j.1539-6924.2006.00820.x

Creed, W. E. D., Scully, M. A., & Austin, J. R. (2002). Clothes make the person? The tailoring of legitimating accounts and the social construction of identity. *Organization Science, 13*(5), 475–496. https://doi.org/10.1287/orsc.13.5.475.7814

Erkama, N., & Vaara, E. (2010). Struggles over legitimacy in global organizational restructuring: A rhetorical perspective on legitimation strategies and dynamics in a shutdown case. *Organization Studies, 31*(7), 813–839. https://doi.org/10.1177/0170840609346924

Forman, J., & Damschroder, L. (2007). Qualitative content analysis. In *Empirical methods for bioethics: A primer*. Emerald Group Publishing Limited. https://doi.org/10.1016/S1479-3709(07)11003-7

Garver, E. (1994). *Aristotle's rhetoric: An art of character*. University of Chicago Press.

Goldberg, S. (2009). Experts, semantic and epistemic. *Noûs, 43*(4), 581–598. https://doi.org/10.1111/j.1468-0068.2009.00720.x

Hardy, C., Palmer, I., & Phillips, N. (2000). Discourse as a strategic resource. *Human Relations, 53*(9), 1227–1248. https://doi.org/10.1177/0018726700539006

Hoffman, A. J. (1999). Institutional evolution and change: Environmentalism and the US chemical industry. *Academy of Management Journal, 42*(4), 351–371. https://doi.org/10.5465/257008

Hsieh, H.-F., & Shannon, S. E. (2005). Three approaches to qualitative content analysis. *Qualitative Health Research, 15*(9), 1277–1288. https://doi.org/10.1177/1049732305276687

Jarzabkowski, P., & Sillince, J. (2007). A rhetoric-in-context approach to building commitment to multiple strategic goals. *Organization Studies, 28*(11), 1639–1665. https://doi.org/10.1177/0170840607075266

Kisić Tepavčević dovela u pitanje kompetentnost 350 lekara - potpisnika otvorenog pisma. (2020). https://www.021.rs/story/Info/Srbija/248823/Kisic-Tepavcevic-dovela-u-pitanje-kompetentnost-350-lekara-potpisnika-otvorenog-pisma.html

Krippendorff, K. (2004). *Content analysis: An introduction to its methodology*, 2nd (ed.). Sage.

Kruglanski, A. W., Raviv, A., Bar-Tal, D., Raviv, A., Sharvit, K., Ellis, S., & Mannetti, L. (2005). Says who? Epistemic authority effects in social judgment. *Advances in Experimental Social Psychology, 37*, 345–392.

Kruglanski, A. W., Dechesne, M., Orehek, E., & Pierro, A. (2009). Three decades of lay epistemics: The why, how, and who of knowledge formation. *European Review of Social Psychology, 20*(1), 146–191. https://doi.org/10.1080/10463280902860037

Lauc odgovorio Štagljaru: Ne odgovara mu da pandemija ide kraju, ima skriveni interes. (2021). IndexHR. https://www.index.hr/vijesti/clanak/lauc-odgovorio-stagljaru-njemu-je-nepojmljivo-da-netko-radi-za-opce-dobro/2263127.aspx

Lefsrud, L. M., & Meyer, R. E. (2012). Science or science fiction? Professionals' discursive construction of climate change. *Organization Studies, 33*(11), 1477–1506. https://doi.org/10.1177/0170840612463317

McCraw, B. W. (2015). The nature of epistemic trust. *Social Epistemology, 29*(4), 413–430. https://doi.org/10.1080/02691728.2014.971907

Nurkić, P. (2022). Retorika državne nestabilnosti. *Međunarodne studije, 22*(1), 97–113. https://doi.org/10.46672/ms.22.1.5

Prymula. (2021). There was a reason to attend soccer. I wanted to show that it works. 2021. Newsbeezer. https://newsbeezer.com/czechrepubliceng/prymula-there-was-a-reason-to-attend-soccer-i-wanted-to-show-that-it-works/

Radenović, L., & Nurkić, P. (2021). Epistemic authority and rhetorical strategies in crisis circumstances. In N. Cekić (Ed.), *Етика и истина у доба кризе* (pp. 153–180). Faculty of Philosophy. https://nauka.f.bg.ac.rs/wp-content/uploads/2022/02/Etika-i-istina-u-doba-krize-NBS.pdf

Riaz, S., Buchanan, S., & Ruebottom, T. (2016). Rhetoric of epistemic authority: Defending field positions during the financial crisis. *Human Relations, 69*(7), 1533–1561. https://doi.org/10.1177/0018726715614385

Rourke, L., & Anderson, T. (2004). Validity in quantitative content analysis. *Educational Technology Research and Development, 52*(1), 5–18. https://doi.org/10.1007/BF02504769

Scott, R. L. (1967). On viewing rhetoric as epistemic. *Communication Studies, 18*(1), 9–17. https://doi.org/10.1080/10510976709362856

Stewart, H. (2020). Neil Ferguson: UK coronavirus adviser resigns after breaking lockdown rules. The Guardian. https://www.theguardian.com/uk-news/2020/may/05/uk-coronavirus-adviser-prof-neil-ferguson-resigns-after-breaking-lockdown-rules

Sayers, F. (2020). Prof Michael Levitt: Here's what I got wrong. UnHerd. https://unherd.com/thepost/prof-michael-levitt-heres-what-i-got-wrong/

Sarkar, A. R. (2021). Fauci hits back at Ted Cruz saying he should be prosecuted: 'What happened on January 6, senator?'. Independent. https://www.independent.co.uk/news/world/americas/us-politics/fauci-ted-cruz-capitol-riot-b1965808.html

Vaara, E., Tienari, J., & Laurila, J. (2006). Pulp and paper fiction: On the discursive legitimation of global industrial restructuring. *Organization Studies, 27*(6), 789–813. https://doi.org/10.1177/0170840606061071

Warnick, B. (2000). Two systems of inventions: The topics in the Rhetoric and The New Rhetoric. In *Rereading Aristotles Rhetoric* (pp. 107–129).
Zollman, K. J. S. (2007). The communication structure of epistemic communities. *Philosophy of Science, 74*(5), 574–587. https://doi.org/10.1086/525605
Zollman, K. J. S. (2013). Network epistemology: Communication in epistemic communities. *Philosophy Compass, 8*(1), 15–27. https://doi.org/10.1111/j.1747-9991.2012.00534.x

Trust and Policy

Trust in Science During Global Challenges: The Pandemic and Trustworthy AI

Vlasta Sikimić

Abstract Through analyzing examples from the COVID-19 pandemic and legislative strategies for the responsible use of AI, we argue that trustworthy science is an intrinsic value with which every good research practice should align. This means that science has to be conducted in a responsible way. Still, this is a necessary but not sufficient condition for building long-lasting trust in science. The social component plays a significant role in this process. Education is important because it increases the scientific literacy of laypeople and the responsible applications of technology by experts. Moreover, when applying scientific research to a larger population one has to keep in mind the individual, cultural and national context of the measures in question. In order to achieve cross-national trust in science, social and natural sciences have to act in synergy with educational institutions over a long period of time. While the normative questions have to be answered by philosophers and policymakers. Especially in the case of AI, science has to be designed for the benefit of humans and for their use, constituting the so-called human-centered application of AI.

1 Introduction

We expect scientists to provide solutions that will save our time, and increase productivity and overall wealth. When big changes in the world happen, such as the COVID-19 pandemic, people turn to researchers expecting cures and solutions. Even though science has brought significant benefits to humanity it also has its dark side mirrored in, for example, the construction of weapons of mass destruction. Since humans witness the side effects of scientific results, their skepticism towards it increases. In particular, when it comes to drug development, philosophers have raised concerns because the financial incentives and interests of private pharmaceutical companies might clash with the medical benefits for individuals (Sismondo, 2021). In this sense,

V. Sikimić (✉)
Philosophy and Ethics Group, Department of Industrial Engineering and Innovation Sciences, Eindhoven University of Technology, Eindhoven, The Netherlands
e-mail: v.sikimic@tue.nl

M. M. Resch et al. (eds.), *The Science and Art of Simulation*,
https://doi.org/10.1007/978-3-031-68058-8_9

we can differentiate between science that deserves and that does not deserve our trust. From a more general perspective, it was often argued that science mainly benefited privileged groups. Consequently, underprivileged groups believe less in the benefits of science for society (Funk et al., 2019). These valid reasons for taking a critical stance towards science can be mitigated by the inclusion of underprivileged groups in science (cf. Vučković & Sikimić, 2022). Doing responsible and inclusive science contributes to its trustworthiness. More specifically, from the perspective of virtue epistemology, it is reasonable both to trust in intellectual authority and to be aware of the limitations of the scientific method. One has to find a balance between uncritically accepting scientific claims—a stance that leads to scientism, and science skepticism. Finally, in order to be worthy of trust, science has to follow ethical principles.

When it comes to trust in advanced digital solutions, philosophers have discussed computational reliabilism in relation to trust (e.g., Durán & Formanek, 2018; Resch & Kaminski, 2019), while the concept of trustworthy science and technology has been introduced in legal guidelines (e.g., European Commission, 2019). Legislators are aware that building trust is important when it comes to motivating people to use the technological advancements that AI offers. Moreover, building trust in science is a long-term project that, apart from the legal framework, requires raising general scientific literacy and consistent policies with an emphasis on non-discriminatory practices (Sikimić, 2022).

Apart from educational measures, an interdisciplinary approach that combines knowledge from natural and social sciences with policy measures is beneficial. One illustrative example of why a collaborative and interdisciplinary approach together with adequate policy choices is important for addressing vaccine skepticism comes from the Ebola outbreak in Congo (Maxmen, 2019). A vaccinated person was killed by his frightened neighbors because the people believed that the vaccine made him infectious. To make people willing to take the vaccine feel safe from assaults by their community, the policies were changed. Instead of vaccinating people close to their homes, they were directed to vaccination sites in other cities. This solution required an understanding of local beliefs and showed how valuable insights from social sciences are when we are responding to the global challenges in front of us. Still, science itself can only give answers to *how-to* questions, predictions, and descriptions, but not to normative concerns. Normative questions are given to policymakers, but in order to address them responsibly philosophical perspective is beneficial.

Even though we aim for value-free science when it comes to the socio-political views of researchers, it is often argued that different epistemic and non-epistemic values play a role in scientific decisions (Douglas, 2009). For example, researchers evaluate inductive risks, ethical consequences, etc. (Douglas, 2000). In particular, researchers working in the domain of applied science need to consider the applications of their findings. The very first thing is that research needs to be ethical. The second layer is the epistemic evaluation of the research. Trustworthy applied science is science that is properly constructed, inclusive, and honestly communicated with the public. As such, trustworthiness is an intrinsic value of science and all research should aim at achieving it.

The case of the COVID-19 pandemic is particularly illustrative when it comes to understanding how to create trustworthy science since it was a recent global event that revealed several layers and origins of science skepticism. Applied science that has the goal of directly impacting human lives needs to follow ethical guidelines and has to measure epistemic trade-offs. Moreover, we will argue that trustworthy science is a transparent and inclusive endeavor that considers individual, cultural, and national circumstances in which it will be applied. Finally, we will shortly illustrate how these insights also hold for the use of AI as one instantiation of applied science with global impact.

2 Global Challenges

To grasp the future global challenges to which science could contribute, let us first reflect upon the significant scientific and social advancements of the past. First, let us consider the past practices that we find shocking nowadays. For instance, one might be terrified by the fact that slavery existed, that females were not allowed to go to universities, or how they were forced to dress. During these times, science typically aligned with these values, which is one reason to be critical when evaluating the scientific practices of today. To identify what are the global challenges of the future, we can think about what the next generations will find most distressing about our lives and practices. They might be shocked by the inequalities, such as the unequal wealth distribution in the world, including the lack of access to medical care and food. Furthermore, the way how we treat non-human animals, and how we exploit and consume them, has already shifted significantly and will probably continue to change. The way we ignore global challenges, such as climate change, and thereby increase the problems for the following generations, will likely be considered unjust. Finally, future generations will most likely find the COVID-19 pandemic intimidating and might think of our responses to it as unsatisfactory.

Some of the important measures for increasing trust in science on a global level are education and solutions adapted for a specific cultural context or balanced use of new technologies that are evaluated on a case-by-case basis. In order to identify these contexts, an interdisciplinary approach is beneficial. For instance, in the communication about vaccines, apart from medical facts, there is a significant social component. Moreover, we argue that such a strategy is helpful for responding to at least two global challenges: increased use of AI and taming the COVID-19 pandemic. By outlining some lessons from the COVID-19 pandemic we will illustrate what this extreme and accelerated global challenge taught us about trusting science.

Science plays an important role both in understanding the listed global challenges and in responding to them. Social science can, for example, study how to motivate people to get vaccinated, or how to gain support for environmental protection. Furthermore, insights from social sciences are invaluable for designing the legal framework for using AI. A relevant component of responding to future challenges is a globally coordinated action that is only possible with understanding and reasonably trusting scientific results.

3 Trust in Science and the COVID-19 Pandemic

During the COVID-19 pandemic, we noticed the need for coordinated global action as a response to it. The virus was spreading quickly across borders and it was mutating. The plan was a massive vaccination of the global population in a short period of time. This was challenging both for scientists who had to create vaccines, for companies who had to produce them, and for policymakers who needed to secure the doses and administer them to the population. Communication about vaccines played an important role here. Moreover, the inclusion of all countries and minority populations is something that increases general trust in science, and in the case of the COVID-19 pandemic, there was even a strong pragmatic argument for it since it is hard to keep respiratory viruses within borders.

An example of suboptimal communication and changing knowledge are recommendations concerning the use of masks and the spread of COVID-19. When the pandemic started the World Health Organization (WHO) and many western health authorities discouraged the use of medical masks, citing the lack of evidence for their benefit. This statement was frequently overinterpreted as evidence of their inefficiency (Tso & Cowling, 2020), showing that education about the scientific method is necessary to properly respond to health advice. COVID-19 was thought to spread through respiratory droplets and surfaces. However, the knowledge about COVID-19 increased, and aerosols were established as a main driver of infections (Wang et al., 2021), resulting in the recommendation to wear masks. Such changed guidelines due to new knowledge is not necessarily a problem for the educated population, as long as this increase in knowledge is properly communicated and erroneous early assumptions are honestly acknowledged and corrected. However, the mask-wearing recommendations were also influenced by concerns over the availability for health care professionals (Chiang et al., 2020). If such additional interests are untransparent and are influencing the guidelines, they can decrease trust in science and governmental institutions.

The first vaccines available were the RNA vaccines developed by BioNTech/Pfizer and Moderna. Even though their technology enabled the fast deployment of large amounts of highly effective vaccines, the technology was new. COVID-19 was the first time most people would even hear about the approach, hence some people felt insecure about the long-term consequences. In particular, the rare side effects associated with the vaccines let some people believe that they were deployed before it was thoroughly tested. Similar fears were raised when rare side effects caused by the vaccine developed by the University of Oxford and AstraZeneca resulted in the discontinuation of vaccination of certain age groups in some countries (Mahase, 2021). The extensive media coverage of any potential side effects, in combination with contradicting recommendations regarding the target populations by the authorities of different countries, and changing national policies, certainly increased the skepticism towards COVID-19 vaccines. Finally, the general approach in contemporary media is that conflicting viewpoints are what attract people's attention. Such populistic and commercially driven journalism can result in giving too much attention to pseudo-scientific claims.

3.1 Vaccine Skepticism

Vaccine skepticism represents an important challenge in front of modern science that goes beyond the COVID-19 pandemic. On the one hand, vaccines are among the most important breakthroughs in medicine. They are the only kind of treatment that has the potential to eradicate a disease. For example, smallpox which used to kill millions every year was fully eradicated in 1977. While previous national eradication programs frequently failed, this worldwide effort, led by the World Health Organization and supported by vaccine doses from the Soviet Union and the United State, was declared completed in 1980 (Belongia & Naleway, 2003).

On the other hand, the general vaccine success is built on the treatment of healthy individuals who can face rare side effects. From the perspective of the individual, lowering the risk of contracting a disease by vaccinating everybody else, can be the best option, however, this approach does not work from the perspective of the community (King, 1999). Therefore, the perception of the individual risk of contracting the disease and the side effects of the potential vaccine play important roles in the decisions of the individual. There are three contexts in which it is important to address vaccine skepticism: individual, cultural, and national.

The individual context is concerned with personal medical history such as connected diseases but also trust in medical institutions based on previous experience. If this experience was negative, people will tend to question new medication more. Also, research in psychology has shown that there are correlations between certain values of an individual such as religiosity and vaccine hesitancy (Rutjens et al., 2018). For example, conservative political orientation, religiosity, conspiracy thinking, and lower level of education correlate negatively with COVID-19 vaccine uptake (Haakonsen & Furnham, 2022).

During the COVID-19 pandemic, we noticed that the vaccination campaigns were not equally successful in different parts of the world or different populations within the countries, and there are good reasons for that. In order to address science skepticism, we should try to understand the reasons behind it.

Minorities in certain countries might be more skeptical of the dominant medical approaches or the mainstream media because of previous bad experiences. For example, people of color in the United States are less likely to get vaccinated (Nguyen et al., 2022). Untrustworthy science has contributed to this skepticism, e.g. the infamous Tuskegee syphilis study is one of the reasons for lower vaccine uptake (Momplaisir et al., 2021). The observatory syphilis study had the aim to study the course of untreated syphilis. It started in 1932 and terminated only in 1972. It enrolled about 400 African-Americans with syphilis under the pretext that they would receive free medical care. However, the participants were never informed about their diagnosis and were not treated, even though penicillium was established as an effective treatment and was widely available in the 1940s. As a consequence, more than 100 participants died and many infected their wives further affecting their children (Gamble, 1997). Such experiences led to widespread mistrust towards the medical community, facilitating vaccine hesitancy (Hsu et al., 2022). To address these concerns, the work with trusted messengers can facilitate vaccine uptake (Privor-Dumm & King, 2020).

Other reasons that cause vaccine skepticism are related to private funding of science. The trust in public research might also decrease because of its links with the private sector (e.g. Holman & Elliott, 2018). People have concerns about the motivation behind different types of research and the elitist nature of contemporary science. The exclusion and the elitist approach that gives no access to underprivileged groups lacks both the virtues of honesty and epistemic justice.

When it comes to global efforts, we notice significant national differences. Different vaccines are more welcomed and trusted in different countries (Steinert et al., 2022). For example, in Denmark and Germany, the deployment of the Oxford/AstraZeneca COVID-19 vaccine was limited to small parts of the population (e.g. people over 60) due to rare side effects, while it remained one of the main vaccines in other countries such as the UK (Mahase, 2021). Conflicting medical recommendations from different health authorities can undermine the trust in the underlying science on the global level, emphasizing the importance of international bodies such as WHO.

3.2 *Lessons on Increasing Trust in Science*

Several measures can be implemented to increase trust in vaccinations. The obvious one is to increase scientific literacy. Cross-national studies have shown a context-dependency of science skepticism. For example, climate change is strongly politicized in the US (Rutjens et al., 2021). Moreover, misinformation bubbles do not only affect lay people. A recent review comparing studies on vaccine hesitancy among medical professionals in different countries reported hesitance rates between 4% in China and 72% in Congo (Biswas et al., 2021). Because local doctors are among the most trusted sources, improving the education of medical professionals and teaching them about the safety and efficiency of vaccines appears to be one of the most important long-term strategies for increasing vaccine uptake. Empirical studies suggest that academics trust scientific results more than the general population (e.g. Sikimić et al., 2021). Moreover, their trust in science increases with seniority. Such findings are encouraging because they are pointing in the direction that education plays an important role when it comes to trusting science.

Even though the situation is context-dependent, certain criteria constitute responsible and trustworthy science, such as transparency about the methods and results and openness to their revision. The general goal of applied research should be driven to help humans and accommodate their needs, i.e., it should be human-centered. These criteria for responsible applied science are often explicated in the theoretical and legal considerations regarding AI for which it is required to be transparent and human-centered.

The following measures were useful in addressing vaccine hesitancy during the COVID-19 pandemic. First, ensuring easy access to vaccines was important. Secondly, it was helpful when people had the choice between different manufacturers. Giving people the option to choose can be an important factor in promoting vaccine uptake, as people might trust one but not the other producer (Steinert et al., 2022). Third, hesitant populations have to be directly addressed by trustworthy sources. Many individuals unwilling to become vaccinated suffer from access to reliable information.

When it comes to communication, misinformation can spread easily in small groups, as people tend to trust the information they receive from their peers more than from official sources. In other words, we trust people that are our friends because we trust their intentions and do not assume other hidden interests. However, what is neglected is the question of their actual epistemic authority when it comes to scientific questions which is often lacking. Moreover, laypeople can be manipulated by the intentional spread of misinformation. The intentions of a source are not sufficient for making trustworthy claims; actual knowledge and expertise on the matter are necessary. The general population needs to have access to experts willing to take the time to explain possible side effects and address the individual concerns of the people. However, medical practices frequently do not include in-depth discussions and many people do not have a local doctor they would trust.

Building trust in science is not something that can be done overnight, it is a process that requires long-lasting efforts. In the best case, this starts with good school education, explaining how medical studies are performed and evaluated. Ideally, it could include personal communication with researchers at universities or pharmaceutical companies. Moreover, experts should be familiar with the ethical and methodological limitations of their work. They would also benefit from learning how to communicate their findings and basic philosophical notions of intellectual virtues.

To sum up, for increasing trust in science it is important to understand the local circumstances. The general observation we can make is that trust in science is dependent on important factors: national circumstances, cultural context, and individual experience. These three factors are often the basis of epistemically justified mistrust. For overcoming the mistrust we should use an interdisciplinary approach that takes into account both social factors and the scientific results that people should understand.

3.3 Freedom Versus Safety

When it comes to responding to global challenges, we often face the dilemma between freedom and safety. For instance, some lockdown measures such as curfews were rather drastic. On the one hand, they had the goal of saving human lives. On the other hand, they were restricting the freedom of movement and indirectly also affected both mental and physical health, e.g., orthopedic problems that require movement. Policies have to balance the benefits of increased safety and decreased freedom. In general, any kind of lockdown measure should be evaluated for its potential to improve the health of the population and to decrease the freedom of the individual. The vastly different responses in countries facing similar challenges show how different this assessment can be. Addressing global warming as another important challenge is politically even more complex. The response of every single country affects the freedom of choice of the current voters, however, the benefits of providing a safer place for future generations, are not necessarily their biggest concern. Moreover, past decisions of one country affect the present state of the others.

While in some circumstances it is easy to assess the outcomes of certain measures, in others this is far from trivial. In particular, when it comes to the use of AI, especially algorithms requiring large data sets, we are faced with the dilemma of trading-off privacy for either safety or less important concepts such as efficiency that still have an immediate appeal. On the one hand, data gathering can include unequal representation of underprivileged groups, thus, violating the principle of epistemic justice and inclusion. On the other hand, such data can be abused for tracing the members of vulnerable groups. When it comes to data collection, curation, and availability. All these three steps need to pass strict ethical approvals. In different national contexts, people's fear of data privacy violations varies based on the historical experiences of tracing and discriminating against underprivileged groups. There is a good reason for that. We can set to evaluate epistemic trade-offs only once all morally unacceptable uses are excluded. These trade-offs are context dependent.

4 Trustworthy AI

There are many dangers of the use of AI, such as violation of data privacy, biases in algorithms that lead to unfair decisions, dehumanized decision-making, etc. On the other hand, AI provides significant epistemic benefits, since it can detect regularities in large data sets in an efficient way. These benefits reach science. For example, in mathematics, AI can be used for finding new results and it has even been argued that AI can lead human intuition about the field (Davies et al., 2021). Thus, finding the right measure of trust in AI is an important task. In this context, policymakers introduced the notion of trustworthy AI (European Commission, 2019). AI has to pass the test of reasonable trust. The lessons from the pandemic that can also be applied in the case of AI are the benefits of the interdisciplinary approach, strengthening the scientific literacy of laypeople and broadening the expert knowledge, as well as understanding specific circumstances when a technology is being applied.

There is a solid theoretical ground for assuming that specific ethical virtues are behind the principles for the responsible use of AI (Hagendorff, 2022). For example, our sense of justice is behind the principle of algorithmic fairness and non-discrimination, while the virtue of honesty is important when it comes to the principle of transparency of AI (Hagendorff, 2022). Epistemic principles, such as epistemic inclusion, epistemic justice, but also epistemic, i.e., rational authority play a role in securing trustworthy research and its responsible application.

4.1 Trade-Offs in the Application of AI

Assessing both epistemic and practical trade-offs is significant when it comes to the use of AI. On the positive side, AI can be of great use for the improvement of public health. For example, AI can identify genes of interest in large databases to identify the ones contributing to cancer risks. With this data, pharmaceutical companies can develop targeted therapies. Furthermore, AI can help in the early diagnosis and prognosis of individual cases. It is important to keep in mind that the benefits of such applications depend on the coverage of the training data. If the algorithms are mainly trained with data from people of European origin, the developed treatments will most likely mainly benefit these populations. Initiatives, such as Human Heredity and Health in Africa (Wonkam, 2021) are, therefore, of uttermost importance to ensure that the promises of AI in healthcare are benefiting everybody. The Human Heredity and Health in Africa initiative aims to make sequence data from Africa available to encourage the development of drugs that benefit everyone. Other uses of AI in medicine include the design of drug candidates (Elemento et al., 2021). Based on previous data on beneficial drug properties and information on the toxicity of similar compounds, AI can help researchers to narrow down a list of drug candidates.

From a negative side, we are noticing that the development of AI is very fast and opens many possibilities for its misuse. Moreover, its long-lasting effects on human behavior cannot be fully estimated. Thus, a dynamic approach to AI that allows for revising the views and normative guidelines is beneficial. Moreover, the irresponsible use of AI might lead to stronger mistrust.

Lockdown measures increased the use of digital technologies both for work and education which led to both positive and negative outcomes. For example, the UK government replaced exams with computer-generated scores which disadvantaged pupils from poorer neighborhoods because the algorithm considered not only their individual performance but also the ranks of their schools (Satariano, 2020). Only after an outcry and after students lost their spots at their favorite universities, the government reconsidered. Therefore, the responsible use of machine learning in education requires ethical considerations regarding data collection, parameter choices, and the development of algorithms. The trade-offs that are present in the outcomes of different AI applications emphasize the need for careful assessment and balanced use.[1]

4.2 Measures for Ensuring the Responsible and Balanced Use of AI

In order to create trustworthy AI, several principles were brought forward which include fairness of the results, transparency of the procedures, traceability of the person responsible for the results, data protection, motivation, and a design that is benefiting humans, etc. (European Commission, 2019). Moreover, standardization of its use allows for prescribing in which domains applications of AI are riskier. For instance, in Germany, the use of machine translation in court is assessed as riskier than its use for medical purposes, while its use for instructions and technical manuals carries even fewer risks, though it is still not risk-free (DIN & DKE, 2020).

As in the context of vaccine hesitancy, education plays a key role when it comes to the integration of AI into society. Computational literacy should enable humans to make informed decisions regarding AI. Moreover, it is important to educate both experts and laypeople about ethically sensitive aspects of AI. When constructing and applying new technologies, experts should consider their ethical consequences. However, this is not always the case. Similar to medical professionals doubting vaccine safety (Biswas et al., 2021), computer scientists are sometimes unaware of the data protection principles. In a recent survey almost one-fifth of the experts working in the field of AI feel free to use pictures that they find online for facial recognition software training, while only 40% of them consider the need for informed consent (Van Noorden, 2020). The education of experts should focus on raising awareness of ethical norms, the reasons behind them, and the prevention of unintentional violations.

Also in the case of AI, an interdisciplinary approach is beneficial for dealing with a global challenge. Insights coming both from social and natural sciences help maximize the benefits of AI in society and minimise its negative effects. Moreover, new technology influences many areas of our lives, and some of them belong to the social realm. Straightforwardly, the use of AI in education requires the collaboration of educational researchers with computer scientists, while policymakers are responsible for its use on children.

Finally, since the development of AI is currently happening and we cannot fully anticipate and assess all of its possible consequences, it is important to have a flexible view of it and be ready to update rules and beliefs as changes happen. This is similar to the approach that scientists had to take when dealing with the new virus about which there was little information in the beginning and that was mutating over time.

[1] The idea of finding the middle ground when deciding about AI is in line with the approach of virtue ethics (e.g. Hobbs, 2021).

5 Conclusions

Future challenges require coordinated global action. From the experience of the COVID-19 pandemic, we learned that once broken trust in science cannot be restored overnight. On the contrary, building trust in science is a long-lasting process that requires both open-mindedness and solidarity of different parties. National, cultural, and individual contexts vary and have specific needs. In order to understand these needs, an interdisciplinary approach is beneficial because it allows us to assess the social components necessary for building responsible trust in science. This conclusion can be implemented on a broader scale, but it is relevant for the application of trustworthy AI.

Nowadays, it is becoming increasingly hard to assess different scientific findings, and one has to rely on the testimony of others. For the general population, it is challenging to interpret scientific findings. Moreover, even from the perspective of experts, it is difficult to evaluate the reliability of scientific results. The fact that medical professionals doubt vaccines is one example (Biswas et al., 2021). Thus, education as a long-lasting measure for building appropriate relationships with science should encompass both laypeople and experts. Such measures might seem costly, but they prepare society for future challenges.

References

Belongia, E. A., & Naleway, A. L. (2003). Smallpox vaccine: The good, the bad, and the ugly. *Clinical Medicine & Research, 1*(2), 87–92.

Biswas, N., Mustapha, T., Khubchandani, J., & Price, J. H. (2021). The nature and extent of COVID-19 vaccination hesitancy in healthcare workers. *Journal of Community Health, 46*(6), 1244–1251.

Chiang, C.-H., Chiang, C.-H., & Chiang, C.-H. (2020). Maintaining mask stockpiles in the COVID-19 pandemic: Taiwan as a learning model. *Infection Control & Hospital Epidemiology, 42*(2), 244–245.

Davies, A., Veličković, P., Buesing, L., Blackwell, S., Zheng, D., Tomašev, N., Tanburn, R., et al. (2021). Advancing mathematics by guiding human intuition with AI. *Nature, 600*(7887), 70–74.

DIN and DKE. (2020). Standardization roadmap artificial intelligence. Retrieved April 01, 2022, from https://www.din.de/resource/blob/772610/e96c34dd6b12900ea75b460538805349/normungsroadmap-en-data.pdf

Douglas, H. (2000). Inductive risk and values in science. *Philosophy of Science, 67*(4), 559–579.

Douglas, H. (2009). *Science, policy, and the value-free ideal*. University of Pittsburgh Press.

Durán, J. M., & Formanek, N. (2018). Grounds for trust: Essential epistemic opacity and computational reliabilism. *Minds and Machines, 28*(4), 645–666.

Elemento, O., Leslie, C., Lundin, J., & Tourassi, G. (2021). Artificial intelligence in cancer research, diagnosis and therapy. *Nature Reviews Cancer, 21*(12), 747–752.

European Commission, Directorate-General for Communications Networks, Content and Technology. (2019). Ethics guidelines for trustworthy AI. *Publications Office*. https://doi.org/10.2759/177365

Funk, C., Hefferon, M., Kennedy, B., & Johnson, C. (2019). *Trust and mistrust in Americans' views of scientific experts*. Pew Research Center.

Gamble, V. N. (1997). Under the shadow of Tuskegee: African Americans and health care. *American Journal of Public Health, 87*(11), 1773–1778.

Haakonsen, J. M. F., & Furnham, A. (2022). COVID-19 vaccination: Conspiracy theories, demography, ideology, and personality disorders. *Health Psychology*.

Hagendorff, T. (2022). A virtue-based framework to support putting AI ethics into practice. *Philosophy & Technology, 35*(3), 1–24.

Hobbs, R. (2021). Integrating ethically align design into agile and CRISP-DM. *SoutheastCon, 2021*, 1–8.

Holman, B., & Elliott, K. C. (2018). The promise and perils of industry funded science. *Philosophy Compass, 13*(11), e12544.

Hsu, A. L., Johnson, T., Phillips, L., & Nelson, T. B. (2022). Sources of vaccine hesitancy: pregnancy, infertility, minority concerns, and general skepticism. In *Open forum infectious diseases* (Vol. 9, No. 3, p. ofab433). Oxford University Press.

King, S. (1999). Vaccination policies: Individual rights v community health. *BMJ, 319*(7223), 1448–1449.

Mahese, E. (2021). Covid-19: WHO says rollout of AstraZeneca vaccine should continue, as Europe divides over safety. *BMJ 372*, n728. https://doi.org/10.1136/bmj.n728

Maxmen, A. (2019). Science under fire: Ebola researchers fight to test drugs and vaccines in a war zone. *Nature, 572*(7767), 16–17.

Momplaisir, F., Haynes, N., Nkwihoreze, H., Nelson, M., Werner, R. M., & Jemmott, J. (2021). Understanding drivers of coronavirus disease 2019 vaccine hesitancy among blacks. *Clinical Infectious Diseases, 73*(10), 1784–1789.

Nguyen, L. H., Joshi, A. D., Drew, D. A., Merino, J., Ma, W., Lo, C.-H., Kwon, S., et al. (2022). Self-reported COVID-19 vaccine hesitancy and uptake among participants from different racial and ethnic groups in the United States and United Kingdom. *Nature Communications, 13*(1), 1–9.

Privor-Dumm, L., & King, T. (2020). Community-based strategies to engage pastors can help address vaccine hesitancy and health disparities in black communities. *Journal of Health Communication, 25*(10), 827–830.

Resch, M., & Kaminski, A. (2019). The epistemic importance of technology in computer simulation and machine learning. *Minds and Machines, 29*(1), 9–17.

Rutjens, B. T., Sutton, R. M., & van der Lee, R. (2018). Not all skepticism is equal: Exploring the ideological antecedents of science acceptance and rejection. *Personality and Social Psychology Bulletin, 44*(3), 384–405.

Rutjens, B. T., Sengupta, N., van Der Lee, R., van Koningsbruggen, G. M., Martens, J. P., Rabelo, A., & Sutton, R. M. (2021). Science skepticism across 24 countries. *Social Psychological and Personality Science, 13*(1), 102–117.

Satariano, A. (2020). British grading debacle shows pitfalls of automating government. *The New York Times*, August 20.

Sikimić, V., Nikitović, T., Vasić, M., & Subotić, V. (2021). Do political attitudes matter for epistemic decisions of scientists? *Review of Philosophy and Psychology, 12*(4), 775–801.

Sikimić, V. (2022). How to improve research funding in academia? Lessons from the COVID-19 crisis. *Frontiers in Research Metrics and Analytics, 7*. https://doi.org/10.3389/frma.2022.777781

Sismondo, S. (2021). Epistemic corruption, the pharmaceutical industry, and the body of medical science. *Frontiers in Research Metrics and Analytics, 6*, 2. https://doi.org/10.3389/frma.2021.614013

Steinert, J. I., Sternberg, H., Prince, H., Fasolo, B., Galizzi, M. M., Büthe, T., & Veltri, G. A. (2022). COVID-19 vaccine hesitancy in eight European countries: Prevalence, determinants, and heterogeneity. *Science Advances, 8*(17), eabm9825.

Tso, R. V., & Cowling, B. J. (2020). Importance of face masks for COVID-19: A call for effective public education. *Clinical Infectious Diseases, 71*(16), 2195–2198.

Van Noorden, R. (2020). The ethical questions that haunt facial-recognition research. *Nature, 587*, 354–358.
Vučković, A., & Sikimić, V. (2022). How to fight linguistic injustice in science: Equity measures and mitigating agents. *Social Epistemology*, 1–17.
Wang, C. C., Prather, K. A., Sznitman, J., Jimenez, J. L., Lakdawala, S. S., Tufekci, Z., & Marr, L. C. (2021). Airborne transmission of respiratory viruses. *Science, 373*(6558), eabd9149.
Wonkam, A. (2021). Sequence three million genomes across Africa. *Nature, 590*, 209–211.

Science, Shame, and Trust: Against Shaming Policies

Sarah C. Malanowski, Nicholas R. Baima, and Ashley G. Kennedy

Abstract Scientific information plays an important role in shaping policies and recommendations for behaviors that are meant to improve the overall health and well-being of the public. However, a subset of the population does not trust information from scientific authorities, and even for those that do trust it, information alone is often not enough to motivate action. Feelings of shame can be motivational, and thus some recent public policies have attempted to leverage shame to motivate the public to act in accordance with science-based recommendations. We argue that because these shame policies are employed in non-communal contexts, they are both practically ineffective and morally problematic: shame is unlikely to be effective at motivating the public to behave in accordance with science-based policy, and shaming citizens is an unethical way to get them to comply. We argue that shame-based policies are likely to contribute to further distrust in scientific authority.

1 Introduction

In the context of a democratic society, scientific beliefs matter because they often influence the way that people act. If someone doesn't believe that, for example, a face mask is an effective tool in preventing the spread of infectious disease, they will likely balk at a rule that requires them to wear one. Because motivating public action requires that the public have certain beliefs, and because beliefs are notoriously resistant to change, some have recently suggested the application of epistemic paternalism, which involves interfering with the public's acquisition of information

S. C. Malanowski · N. R. Baima (✉) · A. G. Kennedy
Harriet L. Wilkes Honors College, Florida Atlantic University, 5353 Parkside Drive, Jupiter, FL 33458-2906, USA
e-mail: NBaima@fau.edu

S. C. Malanowski
e-mail: smalanowski@fau.edu

A. G. Kennedy
e-mail: kennedya@fau.edu

M. M. Resch et al. (eds.), *The Science and Art of Simulation*,
https://doi.org/10.1007/978-3-031-68058-8_10

for their own good. The idea is that, by withholding information or presenting it in a particular light, the public's biases can be avoided and potential misunderstandings circumvented before they have a chance to negatively affect the belief formation process. However, aside from being potentially morally problematic, it isn't clear that epistemic paternalism is a very effective strategy either. In our view, it is doubtful that information alone, even of the paternalistic sort, is enough to motivate people to act: people need to not only trust the information they are given, they must also accept the recommendation to act on this information. Thus, something above and beyond raw information is needed—something that will provide the motivation to act in accordance with policy recommendations. As we will see in what follows, certain emotions, such as shame, have been leveraged to play this motivating role in the public arena.

To further elaborate, shame has been utilized both implicitly and explicitly by communities, governments, and institutions as a way to alter the public's behavior. The idea is that by exposing an individual's or a group's undesirable behavior, their reputation is thereby damaged, and thus the individuals or groups in question are (presumably) motivated to change this behavior. However, while shame can be an effective motivator, we will argue that there are significant practical and moral costs that come with the use of shame as a means of motivating public behavior.

First, while shame may help to promote behaviors in some, it does not work when the individuals or groups being shamed do not trust or respect the ones who are doing the shaming. And often, the people who need the most convincing are those who are the least likely to have this level of trust and respect. Thus, the technique of shaming, in these instances, is likely only to create resentment for and backlash against the message being conveyed. In what follows, we will support this claim by drawing upon real-life examples of both effective and ineffective uses of shame in public policy. From these examples, we will develop two models of shaming via public policy, communal and non-communal, and argue that the latter cannot motivate behavioral change effectively. Because the shame-based policies put forth by Western liberal democracies are examples of non-communal shame, we argue that such policies will not have the desired practical effect.

Second, we will argue that the use of non-communal shame is morally problematic because it not only violates trust and respect but also involves an inaccurate appraisal of the person being shamed and disrupts democratic deliberation. As such, it violates some of the central aims of liberal democracy. In sum, we will argue that shame-based policies are not a practically or ethically sound way to motivate public behavior.

Before we begin, we'd like to emphasize that the paper's focus is not on the moral and political dimension of shaming in general, but rather on the use of shame in public policy by governmental agencies and local communities specifically. Accordingly, we will not discuss the issue of private citizens shaming other individuals, corporations, or governmental agencies.

2 Against Epistemic Paternalism

There are many reasons why scientific communication, and in particular, communication between scientific experts and the non-expert public, is difficult. A prominent example of this pervasive difficulty in communication involves the issue of anthropogenic climate change. Despite broad consensus in the climate-science community regarding such change, there is far from this level of consensus in the public realm. This example raises the more general question of why non-experts often do not trust experts when it comes to scientific matters. There are potentially many reasons for why this kind of mistrust occurs, such as a lack of scientific literacy and ideological bias (see Bardon, 2019, ch. 1). However, for many issues like climate change, deliberate misinformation campaigns are a substantial cause of the public's distrust of science. These campaigns are funded by industries that stand to suffer financial loss if the public were to understand the actual state of the science and change their behaviors as a result (Oreskes & Conway, 2010).

To combat mistrust and misinformation, some philosophers have suggested that it would be prudent to engage in some form of epistemic paternalism, that is, to filter the information the public receives in a way that promotes positive behavioral outcomes (Axtell & Bernal, 2020). The idea here is that if scientific information is presented in its original form, the public could misunderstand the data because they do not comprehend the scientific methods used in gathering it, leaving room for industry-endorsed skeptics to further encourage this misunderstanding. Hence, it is argued that, in non-ideal conditions like this, science communicators might need to disseminate information in a way that lacks transparency, openness, sincerity, and honesty in order to overcome epistemic obstacles and promote positive behaviors (John, 2018).

Nonetheless, there are at least three problems with this approach. First, many will find such an approach morally unsavory, arguing that it constitutes a form of deception (see Moore, 2018). Second, we are skeptical concerning the effectiveness of approaches to motivating public action that rely on a lack of openness and transparency. The United States' debacle surrounding the effectiveness of masks early on in the COVID-19 pandemic illustrates this concern: in order to prevent mask shortages for healthcare workers, Americans were told by public health officials that masks were not effective at preventing disease spread. When public mask use was later mandated, the juxtaposition to the earlier recommendations led to confusion, distrust, and backlash. Third, even on the assumption that this approach is morally acceptable and that the information would be well-received, it still might not provide the desired outcomes, as there is often a significant gap between information and motivation: as Bennett (2020) argues, people might accept the facts that justify a policy, and even in some sense judge that the policy is good, yet still lack the motivation to act according to the policy. Thus, in our view, a more preferred method of motivating public action should be one that is less morally problematic and more motivationally effective.

3 Shame and Policy

3.1 Shame Introduction

In the previous section, we saw that epistemic paternalism is morally problematic, and it isn't clear that it motivates action. A more ethically sound and effective solution to change public behavior may thus be to instead aim for a type of *emotional paternalism* in our science-based public policies and messages. Creating public policy that elicits moral emotions might avoid the ethical traps of epistemic paternalism and also more effectively motivate behavior. Emotions have a motivational force that pure information lacks—you can know about the statistics linking, say, cigarette smoking to lung cancer, but until you see the frightening images of cancerous lungs on the packet of cigarettes, you might not be motivated to quit.[1]

Thus, there are good prima facie reasons for believing that evoking the emotions of the public could effectively motivate them to comply with science-based policies. But which emotions should science communication attempt to elicit? Although there are several promising candidates, we will focus here on policies that attempt to elicit shame, as shaming tactics have become increasingly commonplace with the rise of social media (social media hashtags have become a common means of public shaming; see Ronson, 2015), and there have been several recent popular and philosophical defenses for the use of shame in enforcing policies related to public health and climate change (see Aaltola, 2021; Jacquet, 2015). Shame is also a particularly salient emotion to consider in the context of public policies and recommendations because it is often evoked by social norm violations, as will be discussed below.

Shame is an emotion that is felt when one feels exposed to public scrutiny and disapproval. There doesn't need to be anyone actually present to evoke it either—simply imagining how others view you can be enough. When we feel shame, we do not just feel bad about a particular behavior—shame makes us feel bad about *ourselves*.[2] We feel that we have been exposed to others as being deficient in some way. As psychologists Tangney et al. have put it: "In shame, an objectionable behavior is seen as reflecting, more generally, a defective, objectionable self…With this painful self-scrutiny comes a sense of shrinking or of 'being small' and feelings of worthlessness and powerlessness" (1996, 1257). Shame is considered to be one of the "moral emotions"—emotions that are tied to the "interests and welfare" of others (Haidt, 2003, 853). The moral emotions, in general, are thought to have played an important evolutionary role in promoting group living, as they help to motivate prosocial behavior (and provide disincentives for anti-social behavior) that is required for

[1] In moral philosophy, the idea that reason is motivationally inert without emotion is rooted in the philosophy of Hume and is developed by contemporary non-cognitivists (see Cohon, 2018). The influence that emotions have on reason is well-known by moral psychologists (see Haidt, 2001). With respect to this issue and smoking, see Amonini et al. (2015).

[2] The self versus behavior dichotomy is one of the main aspects that separates shame from guilt: when we feel guilty, we feel bad about something we did, while when we feel shame, we feel bad about ourselves (Tangney et al., 1996; Williams 1993).

people (or early hominids) to trust one another enough to live together and exchange resources. Shame, in particular, may have evolved because it can motivate an individual to abide by group norms: by exposing an individual's undesirable behavior, that individual's reputation is damaged, and thus individuals are motivated to abide by social norms in order to avoid such a punishment. When we feel shame, we are made aware that we are not seen in a positive way by others. We are thus motivated to either change our behaviors to be more acceptable to others or to hide the objectionable aspects of ourselves from others, so they are not out in the open.

A pertinent example of the motivational power of shaming can be seen in the recent trend of online shaming campaigns. Today, we frequently see examples of individuals (or groups of individuals) informally "calling-out" a particular transgressor online over behavior that is deemed unacceptable. This is typically done by individuals sharing the objectionable content with an accompanying expression of disapproval. Online shaming is also a way to signal to one's social media followers which norms the shamer adheres to. For example, recent online shaming about COVID-19 related behaviors (e.g. mask-wearing, social distancing) can be both an attempt to change the behavior of people who do not adhere to public health advice as well as a way of showing support for that public health advice. Shaming can thus serve both as a way of motivating others to adhere to norms and as a way of showing personal acceptance of those norms.

3.2 *Shame-Based Policy*

Given that shame can be a highly effective motivator, it is not surprising that authority figures sometimes use it to motivate the general public's behavior. Shame-based punishments, such as mandated brightly-colored license plates for drivers convicted of drunk driving violations, are currently used in many places for various crimes, as they are believed to function as deterrents to crime and as a way of conveying rules to the community. Shame is also sometimes used in order to motivate people to comply with science-based policies and recommendations, and it is easy to see why one might think this is an effective strategy: after all, such recommendations are an attempt to establish a norm, and shaming is a way to get people to comply with norms (cf. Harris & Darby, 2009).

Authority figures and agencies sometimes use direct shaming messages to motivate behavioral changes, but such outright shaming can be difficult to incorporate into policy. However, people do not need to be directly shamed in order to feel shame: simply drawing attention to the way others view an individual can be enough to arouse feelings of shame in that individual. This kind of indirect shame can thus work more subtly as a behavioral "nudge." Nudging involves structuring the environment in a way that predictably alters behavior without forbidding options—nudges simply make people more or less likely to choose a given option or behave a certain way (Thaler & Sunstein, 2008). An example of a nudge would be to put healthier food at eye level in a work cafeteria so that it is easier to see than the junk food options. In doing so, individuals are given a free choice as to whether they want to eat healthy foods, but the environment is structured in such a way as to influence their behavior. When someone is "nudged," the desired behavior is not achieved through rational persuasion or restriction; rather, it is achieved via changes to the "choice architecture" that might attach minor inconveniences or negative associations to a particular option.

One thing that can increase compliance with a recommendation is emphasizing that a person is not doing what the majority of people are doing—this functions as a nudge and is likely modulated by subtle feelings of shame. For example, some utility companies include a comparison of how a household's energy consumption compares with the average household in their neighborhood in order to reduce energy use and combat climate change. By calling attention to the discrepancy between one household's use compared to the norm, that household may feel ashamed about their overconsumption and thus take steps to reduce energy usage. Such a policy does not restrict people from using as much energy as they want, nor does it directly shame them for their usage—instead, it may nudge them into changing their behavior by attaching a minor negative association to overconsumption.

Antismoking campaigns and laws have also leveraged the motivational power of shame to nudge people into not smoking. Policies and laws that create designated "smoking zones" for smokers make smoking more physically inconvenient and thus can nudge individuals into smoking less. However, such policies may also have worked to nudge the public as a whole into smoking less by attaching shame to smoking: as Eyal (2014) has argued, such areas effectively banish smokers from the rest of the community, thus attaching a stigma to smoking and causing feelings of shame in those who are forced to physically distance themselves if they wish to smoke. Indeed, antismoking ads that call attention to the idea that smokers are viewed as "outsiders" work particularly well in motivating smokers to quit (Amonini et al., 2015). Such messages may work on viewers by evoking the shame smokers feel when isolated from others.

Direct shaming and the use of nudges that induce shame may thus be an effective way to get people to act in accordance with policies and science-based recommendations, as shame not only motivates, but motivates actions that adhere to norms—and norms are precisely what such policies and recommendations aim to establish. Shame-based tactics have an additional advantage in that they avoid many ethical issues raised by epistemic paternalism, as they do not involve withholding information from the public, but instead focus on drawing attention to how others view those who do not abide by the policy or recommendation. We'll now consider the ethical case for the use of shame in this way before arguing against it.

4 Shame and Normativity

As discussed above, shame is currently incorporated into some science-based public policy messaging, and there is some evidence that the use of shame in this way can be effective at motivating behavioral changes. In this section, we will discuss two arguments that can be used to support the claim that shame-inducing public policy messaging is an ethical way to motivate the public to act in accordance with science-based recommendations.

4.1 Shame and Positive Consequences

The most straightforward justification for the use of shame in public policy messaging is that it will promote positive consequences. In her defense of the use of shame to combat climate change, Aaltola (2021) argues that recent instances of "climate-shaming"—like those employed by Greta Thunberg in her address to the United Nations' 2019 Climate Action Summit—are not only a morally acceptable way to motivate individuals and corporations into changing their behaviors to be more climate-conscious, but also that we may in fact have a moral duty to shame *more* because shame is effective at motivating behavioral change. Aaltola points out that when we feel shame, we are confronted with the fact that others find our behavior objectionable. When we are "climate-change shamed," our self-centered ways of living in the world and disregarding the environment are placed under public scrutiny. Thus, shaming people for their harmful actions against the climate can motivate those people to critically examine their behaviors and the ways in which those behaviors affect others. When we hear, for example, Thunberg proclaim "How dare you!" (Thurnberg, 2019) about our dismissiveness of climate change, and when we see others on social media boycotting air travel because of the carbon footprint it leaves, Aaltola argues that we are led to ask important questions about how our behaviors reflect upon ourselves, and the shame we feel can redefine our attitude towards the world around us.

The use of shame in public policy messaging can thus be justified because it can bring about positive social change (Aaltola focuses on the specific case of climate change, but the argument can easily extend to other science-based practices that promote overall social good, like encouraging vaccines, lessening smoking and junk food consumption, and promoting good sanitation practices). The use of shame in such cases may also have the additional positive effect of promoting moral learning. Because shame is connected to social norm violations, when we feel shame, we are alerted to the fact that we are violating some communal value. Thus, feeling shame can teach us that our behaviors are not in accordance with the community's accepted values (Williams, 1993). The shamed agent is thus in a position where they can reflect upon their behavior and the reasons why society views it as a norm violation, and then internalize that norm, if, upon reflection, they deem it to be a good one.

Aaltola acknowledges that shame policies could potentially be counterproductive, since being shamed can cause people to become defensive and angry, and they could also cause psychological harm, as shame makes us feel bad about ourselves. However, she maintains that neither of these worries is sufficient to outweigh the positive consequences. The former worry can be overcome with moral maturity, as shame experiences can be viewed as opportunities for moral growth (Aaltola, 2021, 15; Velleman, 2001). The latter worry can be overcome with basic consequentialist reasoning: given the enormous costs of allowing climate change to continue uninhibited, we can justify the psychological harms of shame to motivate behavioral change (Aaltola, 2021, 17). Similarly for other science-based recommendations that have the potential to mitigate much suffering and loss of life, it can be argued that the benefits gained vastly outweigh any psychological harms caused by policies and messages that utilize shame to change the behavior of the public.

4.2 Shame and Fittingness

Shame can also be argued to be a morally acceptable means to motivate behavioral changes because it can help us understand that a particular situation is moral in nature in the first place. It is possible that situations exist in which we have a moral duty to do something, or there are moral violations that we should recognize, but the moral dimension of such situations, for whatever reason, fails to register with us. Climate change is a plausible candidate for such a situation (Gardiner, 2011), as is vaccination in the COVID-19 pandemic: failing to act in these situations contributes to the suffering of others, but there is widespread resistance to viewing these situations as moral in nature. Causing people to feel a moral emotion such as shame in response to these situations may thus help bridge the gap between the moral nature of the situation and the public's awareness of that moral nature, since moral emotions can play an important role in "alerting" us to the fact that there could be a moral dimension to the eliciting situation.

It is possible, then, that shame can be a *fitting* emotion to feel in response to one's behaviors. Feeling fitting emotions can demonstrate an appreciation of a normative fact, just as our aesthetic responses can demonstrate an appreciation of a fine piece of art (Srinivasan, 2018). Similarly, our experience of shame can be fitting when our behavior is a genuine moral violation, and the feeling of shame demonstrates an appreciation of the fact that our behavior was wrong. Thus, developing policies and messages that evoke shame may develop us epistemically by allowing us to properly understand the moral nature of the situation and the fact that certain behaviors are moral violations.

5 Against Shaming Policies

As noted in the introduction, the target of this paper is not shaming generally, but rather the implementation of shame in Western democratic scientific policy messaging. We can distinguish between two possible types of shaming via public policy, depending on the relationship between the policymakers/implementors and the people the policy is meant to regulate. The first type, communal shaming, occurs in a small group in which the people know each other, share values, and ideally, trust, and respect each other. In contrast, non-communal shame occurs in a large group in which the people don't know each other, don't necessarily share values, and intimate trust and respect are not possible or practical.[3] Because modern democratic societies involve large groups of people with a plurality of values and interests, it simply isn't possible for them to intimately connect with the people developing and implementing science-based policies; thus, the shame policies in such a society will mostly be non-communal. We argue that non-communal shaming faces two problems: (1) it is unlikely to be effective long-term, and (2) it is ethically problematic.

5.1 Practical Concerns

As discussed above, one of the major arguments in defense of shame-based policies is that they lead to significant positive consequences. However, we are skeptical that non-communal shame policies are effective long-term, and, despite the responses to the claim that shame leads to anger and defensiveness, we worry that the way shame is used in the cases at issue here—motivating people to behave in accordance with science-based policies in Western democracies—involves precisely the conditions under which we can expect shaming to be met with counterproductive backlash.

One reason to doubt the long-term effectiveness of shame-based policies comes from recent arguments that draw a distinction between the effectiveness of "pure nudges" versus "moral nudges." Pure nudges involve adjusting defaults or salience, whereas moral nudges leverage positive or negative emotions (e.g. fear, shame, pride) to encourage correct behavior (Carlsson et al., 2021). A pure nudge would be placing healthy food at eye level, while a moral nudge would be notifying individuals that they used more energy than their neighbors in similar-sized houses since this triggers a shame response. Speaking about moral nudges, economists Carlsson et al. write that moral nudges "trigger a conscious psychological response" that makes them "more prone to backlash… because the intended behavior is not in line with the preferences of the individual or because the individual objects to being nudged at all" (2021, 219). Thus, we can expect nudging people via shame to lack effectiveness, as the

[3] Community-Led Total Sanitation (CLTS) would be an example of communal shame; see (Sanitation Learning Hub, Retrieved 2022). However, because science-based policy in Western democracies is non-communal, our focus here is on non-communal shame.

individuals being nudged are likely to *notice* they are being nudged, and then attempt to counteract it.

If moral nudges, which are quite negligible and innocuous, can be counterproductive, it seems likely that more robust forms of shaming can backfire as well. We hypothesize that shame is likely to be more effective if a group member has internalized the group's set of values and respects and trusts the group. This is because the purpose of the shame isn't to develop a set of values from without; instead, the feeling of shame merely directs one's attention to how one failed to live up to one's own standards (which are shared amongst the group). In addition, if there is genuine respect and trust in the group, then there is less of a chance for the individual to feel stigmatized or ostracized, and thus they will be less likely to harbor resentment. Indeed, defenses of shame often argue that moral maturity, shared values, and trust are required (see Aaltola, 2021; Deonna et al., 2012, 239–243; Jacquet, 2015, ch. 6; Nussbaum, 2004, 212–213).

However, the problem with non-communal shame policies is that there can't be intimate trust and respect since the shame occurs in a large group of unknown people. Furthermore, in countries like the United States, there is often widespread disagreement over the very issue that shame policies would be adopted for, such as climate change and Covid-19, and these disagreements themselves have become tied to group identity. For example, the quality of being skeptical of what academic scientists advise is one among a cluster of qualities that helps define certain current subgroups in the United States, and so any attempt to motivate their behavior to be more in line with recommendations based on scientific data will be coming from outside of their ideological group. For these reasons, we doubt that shame policies will motivate long-term behavioral change; instead, we suspect that they will lead to resentment and counterproductive behavior.[4] Simply put, though one can leverage shame when there is respect, trust, and shared values, it isn't realistic—or perhaps even possible—for shame to promote respect, trust, and shared values when these qualities are not already present.

[4] Evidence suggests that trust in national public health authorities and scientists is an important factor in acceptance of COVID-19 vaccines (Lindholt et al., 2021), thus suggesting that trust must come before attempts to alter behavior will be effective. Although there are few studies directly linking shame policies with scientific distrust, we believe that related empirical evidence, as well as conceptual arguments, suggest the likelihood of such a link. Empirically, for example, patients who feel they are judged negatively by their healthcare providers about their weight are less likely to report trust in those healthcare providers (Gudzune et al., 2014). Conceptually, feeling shame causes individuals to want to hide, withdraw, and conceal their shameful behaviors from those that shame them—things one does when they do not trust others. Indeed, feelings of shame about COVID-19 infection correlate with lower intentions to comply with public health authorities' guidelines about distancing and reporting infection (Travaglino & Moon, 2021).

5.2 *Ethical Concerns*

Besides being skeptical about its long-term efficacy, we believe that non-communal shaming is ethically problematic for four reasons. First, non-communal shaming is likely to produce an inaccurate judgment of individuals—one that is overly negative. As noted earlier, the experience of shame involves feeling that some aspect of one's *self*, rather than just one's action, is unacceptable to others (this distinction is what separates shame from guilt). Thus, in non-communal shaming, the person's whole character, not just a single behavior, will be judged. Problematically, however, the person's entire character will not be known by those doing the shaming, since the shamers come from outside the person's community. Accordingly, it is unlikely that the person *should* feel shame over their whole character, so causing the individual to feel shame is an excessively negative way to motivate them to change their behavior—there is a mismatch between what is known about the individual and the "punishment" the individual is receiving (that is, shame about their self). In contrast, since individuals will know each other in communal shame, it is at least possible for the shame to match what is known about the person's character.

Second, since non-communal shame doesn't involve a close connection between individuals, there isn't anything to secure that the shame will involve the best interest of the person being shamed. In communal shame, the close connection between individuals makes it possible that the shame can reflect the best interest of the individuals: for example, when loving parents or a group of friends confront an individual about a problem, the parents or friends likely have the well-being of the individual in mind. However, since governmental agencies cannot know citizens on a personal level, it isn't clear that they can express genuine concern for the welfare of individual citizens while they shame them. More troubling, if the shame policy invites the public to shame an individual (by, say, exposing them for non-compliance), it is quite unlikely that the public will express sincere concern about the individual's welfare. Indeed, internet shaming largely functions like mob justice, and this is problematic because, as Nussbaum points out, "it invites the 'mob' to tyrannize over whoever they happen to like. Justice by the mob is not the impartial, deliberative, neutral justice that a liberal-democratic society typically prizes" (2004, 234). Though her point is about shaming punishments, it applies to shame policies more generally.

Third, shame policies risk corrupting democratic deliberation, which involves the public weighing of reasons. Democratic deliberation is thought to be essential to respecting citizens, protecting autonomy, and securing political legitimacy and justice (see Quong, 2022). In addition, some philosophers have argued that, besides having ethical value, public deliberation has epistemic value in both its procedure and truth conduciveness (see Estlund & Landemore, 2018). If citizens respond to shame, they are likely not responding to the policy's science-based reason; rather, their reaction probably reflects not wanting to be ostracized or made to feel bad. Not only does this mean that the deliberative process will not really involve the weighing of reasons, but it could also potentially create further distrust of scientific authority.

For instance, nudging "bypasses" reason by altering behavior in a way that is typically below the level of conscious awareness, using methods that are not themselves good reasons for changing behavior (e.g. the location of health food vs. junk food is not in itself a very good reason to choose one over the other) (see Schmidt & Engelen, 2020). The scientific community is granted the epistemic authority that it has in large part because its methods for acquiring and analyzing evidence deliberately attempt to adhere to epistemic values. Thus, circumventing rational deliberative processes by arousing shame to motivate public behavior could cause the public to distrust the scientific community. Why should the public trust that the methodology science uses to get to their conclusions adheres to epistemic values if the methodology they use to get the public to accept their conclusions does not? Thus, although shame policies are not as clear violations of these epistemic values as the forms of epistemic paternalism discussed in Sect. 2, they are still in tension with these values. Motivating citizens to comply with a policy by inducing a negative emotion neither encourages understanding of the science-based reasons behind the policy nor encourages scientific understanding generally. Indeed, it will likely encourage distrust of the scientific community.

Fourth, there is something troubling about how shame policies will likely be applied. Shame policies can operate as an alternative to legal restrictions: where a law might be seen as overly imposing or where its effective implementation is unlikely given the complexity of the issue, moral emotions, such as shame, are leveraged to motivate compliance. For example, rather than making vaccines a legal requirement in the United States, the government could implement policies and messages that shame individuals who are unvaccinated. But in this case, the government would be sending a perplexing message by shaming an individual for an action they have no intention of making illegal and doing so for reasons the individual doesn't themselves accept (see Nussbaum, 2004, 246).

Communal shame can avoid some of these problems because the connection between the individuals in the community can ensure that the policy's reasons are communicated clearly, and the shaming is implemented with compassion. That said, communal shaming poses its own dangers. Losing face in one's inner community carries significant weight; thus, communal shaming done poorly could be seriously problematic. For this reason, alternatives that utilize positive emotions would likely be preferable.[5]

[5] For instance, it has been found that dignity and pride are a strong source of motivation in some CLTS programs; see Venkataramanan et al. (2018). Another example, could be littering programs that appeal to pride; such as, (Keep America Beautiful, Retrieved 2022) or (Don't Mess with Texas, Retrieved 2022).

6 Conclusion

Scientific beliefs are important because they shape society's policies and behavior. However, scientific communication faces various hurdles, ranging from simple ignorance to ideological bias. Some have defended epistemically paternalistic policies as a way to overcome these impediments. However, epistemic paternalism faces problems of its own: questions about its efficacy and moral status loom large. Shame policies have an advantage in that they are emotionally laden, making them more likely to be motivational, and they also avoid some of the moral missteps of epistemic paternalism, as they do not involve directly withholding information. Thus, it is thought that shame policies might offer an alternative to epistemic paternalism when more forthright forms of scientific communication fail. However, in this paper, we have distinguished between communal and non-communal shame and argued that this distinction matters when it comes to the acceptability of implementing shame-based policies. Communal shame occurs in groups where the members know each other and share values and interests, thus making genuine respect and trust possible, whereas non-communal shame occurs in groups that lack those features. As we have argued in this paper, the non-communal context in which shamed-based policies are being proposed—Western liberal democracies—makes it such that these policies are practically and ethically objectionable: when trust and respect are not present, shame policies are prone to backfiring and creating further distrust in scientific authorities. In sum, we should avoid shame policies.

References

Aaltola, E. (2021). Defensive over climate change? Climate shame as a method of moral cultivation. *Journal of Agricultural and Environmental Ethics, 34*, 1–23.

Amonini, C., Pettigrew, S., & Clayforth, C. (2015). The potential of shame as a message appeal in antismoking television advertisements. *Tobacco Control: An International Journal, 24*, 436–441.

Axtell, G., & Bernal, A. (Eds.). (2020). *Epistemic paternalism: Conceptions, justifications and implications*. Rowman & Littlefield Publishers.

Bardon, A. (2019). *The truth about denial: Bias and self-deception in science, politics, and religion*. Oxford University Press.

Bennett, M. (2020). Should I do as I'm told? Trust, experts, and COVID-19. *Kennedy Institute of Ethics Journal, 30*, 243–264.

Carlsson, F., Gravert, C., Johansson-Stenman, O., & Kurz, V. (2021). The use of green nudges as an environmental policy instrument. *Review of Environmental Economics and Policy, 15*, 216–237.

Cohon, R. (2018). Hume's moral philosophy. In E. N. Zalta (Ed.), *The Stanford encyclopedia of philosophy*. Retrieved December 07, 2022, from https://plato.stanford.edu/archives/fall2018/entries/hume-moral/

Deonna, J. A., Rodogno, R., & Teroni, F. (2012). *In defense of shame: The faces of an emotion*. Oxford University Press.

Don't Mess with Texas. Retrieved February 15, 2022, from https://www.dontmesswithtexas.org/

Estlund, D., & Landemore, H. (2018). The epistemic value of democratic deliberation. In A. Bächtiger, J. Dryzek, J. Mansbridge, & M. Warren (Eds.), *The Oxford handbook of deliberative democracy* (pp. 113–131). Oxford University Press.

Eyal, N. (2014). Nudging by shaming, shaming by nudging. *International Journal of Health Policy and Management, 3*, 53–56.
Gardiner, S. (2011). *A perfect moral storm: The ethical tragedy of climate change*. Oxford University Press.
Gudzune, K., Bennett, W., Cooper, L., & Bleich, S. (2014). Patients who feel judged about their weight have lower trust in their primary care providers. *Patient Education and Counseling, 97*, 128–131.
Haidt, J. (2001). The emotional dog and its rational tail: A social intuitionist approach to moral judgment. *Psychological Review, 108*, 814–834. https://doi.org/10.1037/0033-295X.108.4.814
Haidt, J. (2003). The moral emotions. In R. J. Davidson, K. R. Scherer, & H. H. Goldsmith (Eds.), *Handbook of affective sciences* (pp. 852–870). Oxford University Press.
Harris, C., & Darby, R. (2009). Shame in physician-patient interactions: Patient perspectives. *Basic and Applied Social Psychology, 31*, 325–334.
Jacquet, J. (2015). *Is shame necessary? New uses for an old tool*. Pantheon Books.
John, S. (2018). Epistemic trust and the ethics of science communication: Against transparency, openness, sincerity and honesty. *Social Epistemology, 32*, 75–87.
Keep America Beautiful. Retrieved February 15, 2022, from https://kab.org/
Lindholt, M., Jørgensen, F., Bor, A., & Petersen, M. (2021). Public acceptance of COVID-19 vaccines: Cross-national evidence on levels and individual-level predictors using observational data. *British Medical Journal Open, 11*, e048172. https://doi.org/10.1136/bmjopen-2020-048172
Moore, A. (2018). Transparency and the dynamics of trust and distrust. *Social Epistemology Review and Reply Collective, 7*, 26–32.
Nussbaum, M. (2004). *Hiding from humanity: Disgust, shame, and the law*. Princeton University Press.
Oreskes, N., & Conway, E. M. (2010). *Merchants of doubt: How a handful of scientists obscured the truth on issues from tobacco smoke to global warming*. Bloomsbury Press.
Quong, J. (2022). Public reason. In E. N. Zalta (Ed.), *The Stanford encyclopedia of philosophy*. Retrieved September 5, 2024, from https://plato.stanford.edu/entries/public-reason/
Ronson, J. (2015). *So you've been publicly shamed*. Riverhead Books.
Sanitation Learning Hub. Retrieved February 15, 2022, from https://sanitationlearninghub.org/
Schmidt, A., & Engelen, B. (2020). The ethics of nudging: An overview. *Philosophy Compass, 15*, e12658. https://doi.org/10.1111/phc3.12658
Srinivasan, A. (2018). The aptness of anger. *Journal of Political Philosophy, 26*, 123–144.
Tangney, J. P., Miller, R. S., Flicker, L., & Barlow, D. H. (1996). Are shame, guilt, and embarrassment distinct emotions? *Journal of Personality and Social Psychology, 70*, 1256–1269.
Thaler, R., & Sunstein, C. 2008. *Nudge: Improving decisions about health, wealth, and happiness*. Yale University Press.
Thurnberg, G. (2019). United Nations Action Summit September 23, 2019. Retrieved February 15, 2022, from https://www.usatoday.com/story/news/2019/09/23/greta-thunberg-tells-un-summit-youth-not-forgive-climate-inaction/2421335001/
Travaglino, G., & Moon, C. (2021). Compliance and self-reporting during the COVID-19 pandemic: A cross-cultural study of trust and self-conscious emotions in the United States, Italy, and South Korea. *Frontiers in Psychology, 12*, 565845. https://doi.org/10.3389/fpsyg.2021.565845
Velleman, J. D. (2001). The genesis of shame. *Philosophy & Public Affairs, 30*, 27–52.
Venkataramanan, V., Crocker, J., & Karon, A., & Bartram, J. (2018). Community-led total sanitation: A mixed-methods systematic review of evidence and its quality. *Environmental Health Perspectives, 126*(2) CID: 026001. https://doi.org/10.1289/EHP1965
Williams, B. (1993). *Shame and necessity*. University of California Press.

Decision Making, Values and (Dis)Trust in Science: Two Cases from Public Health

Elena Popa

Abstract This paper examines trust in science in relation to public health and decisions affecting different groups. I employ philosophical work on trust and distrust, particularly how the perpetuation of injustices leads to warranted distrust, and literature on science and values to argue that trust in science can increase if justice and equity are taken into account in the decision-making process. As illustration, I discuss the case of lockdowns during the COVID-19 pandemic, and protecting maternal and fetal health from metylmercury poisoning by removing fish from the diet. Both examples involve measures that take a disproportionate toll on the most vulnerable: the economically less-off and historically discriminated groups. More just, and thus trust conducive, decisions would consider the provision of additional support for those affected, alongside incorporating broader concerns about welfare, or environmental safety.

1 Introduction

As trust in science is receiving increasing philosophical attention, it is important to investigate the drivers of trust or distrust. Problems regarding skepticism about science or vaccine hesitancy are often framed solely as questions of educating the public. On such views, failure to comply with recommendations by scientists is due to the public lacking relevant knowledge. However, solutions along the lines of supplying more information overlook instances where distrust is justified. Goldenberg's (2021) work on vaccine hesitancy points out that medical professionals' dismissal of queries about vaccine safety by patients, as well as information campaigns framed as a clash between the knowledgeable experts versus the ignorant public breed justified distrust. This paper will look at a related possibility, which has been subject to less scrutiny so far, namely whether certain practices within science itself may also lead to warranted distrust. If scientists do not take into account the

E. Popa (✉)
Interdisciplinary Centre for Ethics, Jagiellonian University, Kraków, Poland
e-mail: elena.popa@uj.edu.pl

M. M. Resch et al. (eds.), *The Science and Art of Simulation*,
https://doi.org/10.1007/978-3-031-68058-8_11

situations or interests of specific segments of the public when making policy relevant decisions, this may lead to justified distrust among the respective groups. If this is right, then restoring trust in science would require a change in the practices of scientific decision making, particularly seeking policies that are fair for everyone involved. In what follows, I will explore this in relation to public health decisions.

First, I rely on work in science and values, also drawing on an expansion of the value ladenness thesis in relation to public health (Sect. 1). Secondly, I look at philosophical approaches to trust and distrust, singling out justice and equity as values central to trust. Thirdly, I argue that building public trust is essential for the success of future public health interventions. The theoretical part of the paper concludes that for successful public health interventions, research and policy recommendations should be guided by justice and equity (Sects. 2 and 3). I then illustrate the proposed framework with two case studies: the first looks at public health measures during the COVID-19 pandemic and the neglect of vulnerable groups (Sect. 4), and the second looks at effects of maternal and child nutrition policies on Native American women whose diet is affected by pollution (Sect. 5). One common thread here is that public health decisions have disproportionately affected the most vulnerable, subsequently leading to warranted distrust. To counter such effects, public health decisions should be guided by justice and equity.

2 Values in Science and the Public Health Context

A discussion on how specific values can shape decisions made by scientists requires a broader overview of value influences in science. Two key questions in the literature are whether values play a role in decisions by scientists, and, on a normative level, whether they *should* play any role. Before exploring the relevant answers, two clarifications are needed. The first is the distinction between cognitive values, connected to strictly epistemic aspects, such as accuracy or coherence, and contextual values, which can be economic, ethical, social, political, or aesthetic. Contextual values (henceforth, values) are at the focus of the controversies in the philosophy of science, because accepting that non-epistemic factors influence scientific decisions can undermine the credibility of science. Secondly, when discussing value influences on decisions by scientists, the focus is on decisions whether to accept or reject a theory or a hypothesis, or whether to consider a type of evidence viable or not, which raises deep questions about scientific objectivity.

A review of the science and values literature singles out two main stances (Reiss & Sprenger, 2020):

- *The value-free ideal*—gathering evidence and accepting theories or hypotheses should be independent from values; in relation to the questions above, this stance supports the normative claim that science ought to be value-free.
- *The value ladenness thesis*—value influences are inevitable, even in processes central to science, such as theory acceptance or evidence assessment; in relation

to the questions above, this thesis makes the descriptive claim that it is impossible for science to be value-free.

In the following, I take the value ladenness thesis to be true, as defended in the literature (e.g., Douglas, 2009; Elliott, 2017; Longino, 1996). Particularly relevant here is Longino's (2002) argument, building upon the feminist critique of the concept of objectivity inherent to the value-free ideal. Starting from examples such as gender bias in scientific research, Longino highlights that since scientific knowledge is social, objectivity should include the possibility of transformative criticism, enabling approaches more sensitive to gender, or other past patterns of discrimination (2002: ch. 6). Objectivity understood in this sense helps highlight that science should not reflect only perspectives from the more well-off, but be open to criticism and accept a plurality of views. Since Longino's view includes contextual values, such as a commitment to equality of intellectual authority, alongside shared cognitive values, this defense of the value ladenness thesis brings together both types of values under a view of scientific knowledge as a social product.

Still, my argument should also be of interest to defenders of the value-free ideal. Evidence-based policy rests on scientific contributions when designing policies. Even if scientists are only concerned about epistemic aspects, what they take to be legitimate knowledge, or what they omit, can have ethical consequences. For example, choosing to study a certain illness only on men, then presenting the findings and patterns as valid for everyone could lead to overlooking manifestations of the illness in women. Here, even if the scientists do not take a stance in the conflict between supporting gender equality versus the interests of men of high socio-economic status, they should at least inform policy makers about the limitations of their research. As will be shown below, public health policies can have negative effects if the knowledge they rely upon excludes the situations of discriminated groups.

I will now extend the discussion of values to an approach looking beyond past or current practices, also including future developments within science. The prospective approach by Russo (2021) holds that concepts, values, and norms rarely work in isolation. Under this view, opting for a certain concept or method can influence what kinds of values will be promoted. In what follows, I rely on a question brought forward by Russo to analyze the cases: '*If* we conceptualize X such-and-such, *then* what actions should follow? One can replace X by their favourite concept: health, evidence, adverse drug reaction, causality, etc.' (2021: 6).

To sum up, if decisions by scientists are unavoidably value laden, and furthermore, if the choice of methods and concepts leads to certain values being promoted, then a question arises regarding which values should influence decisions in public health. In the next section, I will make the case for values that foster trust. By contrast, focus on other values, particularly those that tend to exacerbate injustice and inequalities, can decrease public trust. An example is prioritizing the economic interests of pharmaceutical companies against the interests of the public, as shown by the scarcity of COVID-19 vaccines in low-income countries (Hassan et al., 2021). Henceforth, I look at philosophical analyses of trust and distrust and their connection to the values of justice and equity. The gist of the argument is that if research and policy making in public health are to foster trust, then they should employ approaches in line with justice and equity.

3 Trust and Distrust in Science

Philosophical analyses have singled out trust as an attitude, and trustworthiness as a property. One point of agreement across different accounts is that trust is more than mere reliance (Goldberg 2020; Hawley, 2014; McLeod 2021). What the additional component of trust is beyond reliance has been subject to controversy. Without seeking an exhaustive overview, relevant attempts have analyzed trust as follows:

- Taking a risk, or placing oneself in a position of vulnerability from the part of the trustor (Baier, 1986; Becker, 1996; Dasgupta 2000; Luhmann, 2000). Trust as reducing the complexity of social interactions, and enabling more advanced forms of human cooperation (Luhmann, 1977/2018) also fits here, as not checking on others involves taking risks.
- Trustworthiness is viewed as competence plus the right kind of motivation for the trustee. An example is the 'encapsulated interests' view, where maintaining the trusting relationship with the trustor is in the trustee's self-interest (Hardin, 2002) or goodwill (Jones, 1999). These are also known as motivational accounts.
- The trustor taking the participant stance, i.e., treating the trustee as a person responsible for one's actions (Holton, 1994). These are also known as non-motivational accounts.

While these views are often in competition, with various objections raised, I will not pursue these debates. Rather, I will investigate how these features may apply to public health. My interest lies in the connection between trust and likelihood of complying with public health recommendations. Successful public health interventions require compliance from a large part of the population, while insufficient compliance can undermine effectiveness. Immunity levels and vaccine uptake are an example: under a high rate of vaccination a disease may no longer be a threat, but if vaccination rates fall below a certain threshold new outbreaks emerge. How does compliance come together with reliance and/or trust? Regarding reliance, I use Holton's view, focusing on relying on someone to do something from a motivation that stems from them (1994: 65–66). One may not comply to guidelines from institutions or professionals if their motivation does not align with one's interests and expectations. An extreme example are people whose access to medical services is completely cut off. Why would it make sense for one to bear the costs of complying with various measures if one cannot access the medical system? This can also connect to the risk component above. Similarly, if one cannot draw benefits, why expose oneself to risks associated with public health measures? Less extreme examples can include precarious workers with no health coverage and/or paid sick leave, or people from regions with underfunded medical systems, unable to offer assistance to everyone. This situation can be explained through Hansson's (2017) framework for risk analysis including risk takers, those who benefit, and the decision makers. Marginalized groups who receive little benefits from the healthcare system would be the risk takers, while other groups would be those benefiting, and/or making decisions. Benefits here do not only include health as a common good, but also economic goods: while precarious workers may lose their income or other sources of livelihood as a result of specific public health measures, other members of the society can continue to work.

The other components of trust highlight values: goodwill and responsibility involve commitment to a set of values common to the trustor and trustee. A question here is whether goodwill and responsibility can be present in a context of deep structural injustices and inequity, e.g., in societies leaving large groups unable to access healthcare services or basic welfare. While the accounts above do not discuss justice explicitly, it is implied by both motivational and non-motivational approaches. Discussions of distrust can help highlight the importance of justice, which I will now review.

As philosophical work on distrust is still emerging, different views have been proposed agreeing on several features (D'Cruz 2020; Hawley, 2014; McLeod, 2020):

- Distrust and trust are exclusive, but not exhaustive.
- Distrust is not merely non-reliance, though it is a kind of non-reliance.
- Distrust has a normative dimension; knowing that someone distrusts you is perceived as bad.

As with trust, the disagreement concerns what it is that characterizes distrust in addition to non-reliance. Of particular interest here are instances 'when distrust is warranted by people who experience oppression' (McLeod, 2020). One such account is Krishnamurthy (2015), discussing distrust as a means of resisting injustice. Krishnamurthy focuses on the civil rights movement by African-Americans, and the warranted distrust towards white moderates with regard to taking action against racial injustice. Distrust is defined as 'a confident belief that other individuals, groups, or institutions will not act as justice requires' (2015: 397).

Discussing distrust in relation to public health raises the issue whether groups that have been neglected by the public health authorities, or have previously born most of the risks while reaping little benefits are justified in distrusting the said authorities. On a first glance, the issue is non-reliance, but the problem goes beyond that. Given a history of oppression and neglect of certain groups, simply supplying reliable access to medical services is not going to address distrust. In line with Krishnamurthy's approach, trust would also require addressing the existing injustices.

Work on epistemic injustice in science is consistent with the points above. Grasswick has pointed out the links between social conditions, particularly injustice, and how this can permeate scientific practice: 'because scientific knowledge production is tightly intertwined with social needs and goals (…), social injustices can push science in certain directions such that it creates new forms of understanding that can then serve as sources of further injustices' (2017: 315). Particularly relevant is the concept of epistemic trust injustice, which points to harms caused by justified distrust in science among oppressed groups (2017: 319). This helps explain how the unjust social conditions that subject certain groups to oppression lead to further injustice, alienating the said groups from scientific knowledge and expertise. Grasswick focuses on race and gender, but the categories can be expanded. This is also connected to broader work on justice and epistemology by Scheman, taking warranted distrust into account: 'the credibility of science suffers, and, importantly, ought to suffer (…) when its claims to trustworthiness are grounded in the workings of institutions that are demonstrably unjust—even when those injustices cannot be shown to be responsible for particular lapses in evidence gathering or reasoning' (Scheman, 2001: 36). In response to epistemic trust injustice, Grasswick calls for more accountability of scientific institutions, particularly in relation to those who have good reasons for distrust. As Grasswick does not expand on the sense of accountability, my proposal can contribute to this line of argument. A revision of epistemic practices so that marginalized groups are not left out when designing studies or making policy recommendations is a good start for science to become more inclusive and to reduce the perpetuation of injustices. It would also provide alternatives to current deficient practices, to be discussed in the case studies.

Before moving on, a further question is worth considering: how can scientists incorporate concerns about justice, or combat injustice in their research? Accepting the value ladenness thesis discussed out above brings forth the possibility of choosing between values. On the view defended here, when such choices arise, justice and equity should be prioritized over values that would deepen inequalities. This can be spelled out through Hansson's (2017) framework for risk analysis, namely making sure that there are no groups that only take risks, with other groups reaping the benefits, or making decisions. Hansson criticizes utilitarianism in particular for allowing certain groups to be the least well-off as long as the overall welfare is maximized (2017: 1826). While agreeing with Hansson's point that deontology can help, I believe that public health especially provides a good reason to rethink the utilitarian approach. As low trust brings about low compliance with public health requirements, the overall welfare is likely to decrease in the long run, particularly for long lasting public health threats, such as pandemics. As such, public health approaches that promote trust can also be defended from a consequentialist viewpoint, and if scientists are not convinced about the intrinsic value of trust, there is an instrumental case to be made as well. An example of how equity has been explicitly incorporated in research, is the aim of producing a 'vaccine for the world' against COVID-19 set by Oxford University and AstraZeneca. This has taken into account that storage and transportation facilities are not the same across the world, and that less demanding conditions make the vaccine accessible to low-income countries. The case studies will provide a more in-depth perspective on this.

4 Lockdowns During COVID-19

Despite having been described as a great equalizer, at the moment of writing this paper the COVID-19 pandemic is linked to the exacerbation of inequalities (e.g. Cash-Gibson et al., 2021; Chen & Wang, 2021). While some of these effects are due to various pre-existing socio-economic conditions, or the pandemic itself, my interest lies in how specific public health measures in response to COVID-19 fare from the point of view of justice and equity. I will focus on lockdowns as means of preventing COVID-19 infections.

One difficulty with analyzing lockdowns is the variation between rules enforced in different parts of the world. A lockdown may involve limitations of movement, curfews, a ban on gatherings, closure of services deemed non-essential etc. Assuming some of these features are in place, and lead to decreased social contact and economic activity, the question is whether lockdowns affect everyone equally. Relevant contrasts include:

- Ability to work remotely versus the need to be physically present at work.
- Formal work and ability to access unemployment benefits versus informal work without benefits.
- Living conditions with sufficient space, sanitation etc. versus overcrowded conditions, without access to sanitation.

These contrasts have been present throughout the pandemic, with the sharpest instances noticeable between affluent people in Global North countries able to work remotely or access welfare, and the least well-off people in the Global South depending on informal work for their livelihood. I stress that this is not meant to overlook contrasts within these contexts too, as will be explained shortly.

Philosophical work on the COVID-19 pandemic has criticized the assumption that similar measures are appropriate across the world, particularly lockdowns, because it neglects local conditions, and can cause more harm than good (Broadbent & Smart, 2020; Broadbent & Streicher, 2022). Particularly relevant is the egalitarian argument against lockdowns (Broadbent et al., 2020a, 2020b): through their economic effects, lockdowns take the heaviest toll on the poor, or those in already precarious situations, who are more likely to lose their livelihoods, suffer from food shortages, and have their access to other medical or welfare services cut off. These have been discussed in the context of South Africa (Broadbent et al., 2020a, 2020b; Smart et al., 2021). Further examples include Kenya (Onditi et al., 2021), or countries in South Asia, with focus on informal workers (Naz et al., 2021).

Empirical work has shed light on the situation of marginalized groups, and for illustration I use a study on challenges faced by inhabitants of slums in Dhaka, Bangladesh during the lockdown (Zaman et al., 2021). Interviews with informal workers living in slums show an exacerbation of pre-existing inequalities, particularity the division between those living in the city and in the slums, as an answer from a tea seller shows: 'When a city dweller sees me around them, they tell us to go away and stand far away. It seems we are not people' (Zaman et al., 2021:

30). Interestingly, when asked about the major problems they face, residents of two different slums have reported different problems, which can be explained in terms of their work conditions (2021: 30–31). In the Sadek Kahn slum, where most residents are informal workers, access to fuel and unemployment were the most frequent problems reported. COVID-19 was also on the list, but when asked to clarify, the respondents did not mention infection, but the effects of the policies, such as loss of income or limited mobility. By contrast, residents of the Gabtali slum, who work for the city corporation, mainly as cleaners, and have kept their jobs throughout, listed problems they have faced previously: lack of sanitation and lack of access to schools. COVID-19 does not appear in the list, and when asked about it, residents did not perceive it to be a problem. The seemingly paradoxical finding here is that although cleaners are directly exposed to waste that may cause infection, they did not identify the pandemic as a problem, possibly because they have not lost their income as a result of it, unlike the informal workers.

One point to stress is that many of the problems pointed out above about lockdown effects are not confined to Global South context. Examples of vulnerable groups in Global North countries include migrants, who may also live in overcrowded conditions, have scarce sources of income, rely on family as the main source of social support, and are more exposed to COVID-19 infection (e.g., Finell et al., 2021; Ojwang, 2020 for Finland). Being subject to historical discrimination, they continue to be alienated when failing to comply to lockdown rules.

Seeing how blanket lockdowns take a higher toll on those already worse off, and deepen existing inequalities and injustices, the question is how to address comparable public health problems in the future. I rely on the prospective approach discussed above and the question '*If* we conceptualize X such-and-such, *then* what actions should follow?' (Russo 2021). In the current case study, health was understood as the absence of COVID-19 infection. As limiting social contact is an effective way of preventing new infections, lockdowns appear to be good measures. However, if we conceptualize health as absence of COVID-19 infection plus absence of other negative health effects associated with unemployment or lack of food, a different picture emerges. On this view, either lockdowns would be implemented differently, according to the local context, or would be supplemented with support for the less well-off, particularly informal workers.

The problems above have been stressed in philosophical discussions of the pandemic: the need for a broader range of evidence, particularly from the social sciences (Caniglia et al., 2021; Lohse & Bschir, 2020; Lohse & Canali, 2021), weighing other effects of lockdowns, particularly psychological ones (Popa, 2021), and the critique of epistemic ignorance regarding the global poor (Timmermann, 2020). From a political perspective, the role of welfare as a way of resolving the tension between saving lives and livelihoods has been emphasized (Lecamwasam, 2021).

In sum, the way lockdowns were implemented was likely to lead to distrust in science and the public health authorities, particularly among marginalized groups. Facing recommendations impossible to follow, the said groups have also been subject to blame in addition to other difficulties. A more just approach would take their situations and interests into consideration. This is both an epistemic point, regarding what perspectives to consider, and an ethical one concerning the health impact of economic and social issues.

5 Maternal and Child Health and Diet in Indigenous Populations

Work on environmental epigenetics has shown that fetal development can be affected by factors such as pollution. Public health responses to these findings have consisted in lifestyle and nutrition advice for women who are pregnant or of reproductive age (Kenney & Müller, 2017). I focus on one instance here: the connection between nutrition, metylmercury poisoning, and neurodevelopmental damage in children. Metylmercury is a toxic organometallic compound that emerges after mercury is released in aquatic systems and interacts with bacteria, making its way into the food chain. Exposure to even small doses of metylmercury during pregnancy can negatively affect the child's later neural development (Mansfield, 2012). Guidelines by the US Environmental Protection Agency and the Food and Drug Administration have advised pregnant women to limit their consumption of fish.

As earlier, one may ask whether this public health advice is neutral. As stressed by Mansfield, 'owing to racial disparities in fish consumption, not only do the advisories have greater impact on women of color, but they change the problem from contamination itself to the abnormal diets of these women' (2012: 352). The first problem is that for certain groups, such as Native-American women, fish is part of the traditional diet, and changing that would amount to significant effort and additional expenses. Secondly, framing this as a question of diet choice places the responsibility on pregnant women, overlooking the role of the pollutants that have enabled metylmercury to enter the food chain (Lock & Palsson, 2016). Analyzed through the perspective above, such measures are far from just: they shift responsibility onto those affected by pollution, and away from the polluters. They are not equitable either: it is more difficult for women of color to comply with these measures. While Mansfield's discussion has focused on race, similar issues have been pointed out in relation to socio-economic status: poorer women find it more difficult to comply with recommendations regarding diet, being subject to blame (Singh, 2012).

Employing the prospective approach shows that if fetal health is conceptualized as the individual responsibility of the mother, then interventions targeting the mother's diet are appropriate. If, however, fetal health is conceptualized as the joint responsibility between the mother and institutions ensuring environmental safety and access to food, then the need to reduce pollution in the long term, and providing the mother with alternative food sources would also be highlighted. The latter conceptualization would be more just and equitable, as it stresses the need for additional support for communities that are heavily affected by contamination. Looking at trust now, while in the short term women may struggle to meet the recommendations, they are left on their own, particularly if sources of food keep dwindling, and would be justified in distrusting future public health mandates.

Both examples show how public health interventions can leave vulnerable people behind, neglecting specific challenges they are facing. In both cases there appears to have been an argument from urgency: these measures can help avert some of the immediate damage. Still, the effectiveness of such measures in the future is brought into question by their impact on trust in science. These deficiencies can also be singled out from the perspective of past scholarship on ethics and public health. For instance, Kass has noted that ‘it is hard to find a more powerful predictor of health than class and it is thus an appropriate, if not obligatory, function of public health to reduce poverty, substandard housing conditions, and threats to a meaningful education—if for no other reason than to reduce the incidence of disease’ (2001: 1781). Measures such as the ones discussed above are likely to have the opposite effects, increasing the gap between the the well-off and the poor. More broadly, water pollution and the contamination of food sources is linked to further health problems, which individual responsibility alone is unlikely to address. This perspective can change by focusing on problems faced by historically discriminated groups.

6 Conclusions

This paper has investigated trust and values in relation to public health decisions. Using the value ladenness thesis and philosophical investigations of trust, I have argued that trust in science can be fostered by choosing approaches in line with justice and equity. Scientific approaches and policies that support an uneven distribution of burdens, taking a higher toll on the most vulnerable, can lead to warranted distrust, especially where a history of injustice is present. Countering that requires more than focus on scientific communication and education, namely changing the decision making process in public health, emphasizing the situation of the most vulnerable. This, in turn, calls for research methods and conceptualizations that are sensitive to structural injustices, drawing from a plurality of perspectives. The case studies on lockdowns during COVID-19 and maternal and child health have shown that current public health decisions can breed distrust by neglecting the interests of disadvantaged groups. The framework proposed here shows how to move towards better decisions, to increase trust in science. I have taken a philosophical perspective, with the hope that the normative conclusions can inform policy. Nevertheless, the methods and sources of evidence that helped highlight the problems draw on different research areas such as sociology, anthropology, and feminist scholarship. Thus, a further point is that addressing these problems also requires a more interdisciplinary engagement with public health.

Acknowledgements This research is part of the project No. 2021/43/P/HS1/02997 co-funded by the National Science Centre and the European Union Framework Programme for Research and Innovation Horizon 2020 under the Marie Skłodowska-Curie grant agreement no. 945339. I am grateful to Maria Temmes for discussions on the second case study and the organizers and audience of the Trust in Science conference at the University of Stuttgart, particularly Dunja Šešelja and Andreas Kaminski, for helpful questions and suggestions.

References

Baier, A. (1986). Trust and antitrust. *Ethics, 96*(2), 231–260.

Becker, L. C. (1996). Trust as noncognitive security about motives. *Ethics, 107*(1), 43–61.

Broadbent, A., & Smart, B. T. (2020). Why a one-size-fits-all approach to COVID-19 could have lethal consequences. *The Ethics of Pandemics, 78.*

Broadbent, A., & Streicher, P. (2022). Can you lock down in a slum? And who would benefit if you tried? Difficult questions about epidemiology's commitment to global health inequalities during Covid-19. *Global Epidemiology*, 100074.

Broadbent, A., Combrink, H., & Smart, B. (2020a). COVID-19 in South Africa. *Global Epidemiology, 2*, 100034.

Broadbent, A., Walker, D., Chalkidou, K., Sullivan, R., & Glassman, A. (2020b). Lockdown is not egalitarian: The costs fall on the global poor. *The Lancet, 396*(10243), 21–22.

Caniglia, G., Jaeger, C., Schernhammer, E., Steiner, G., Russo, F., Renn, J., ... Laubichler, M. D. (2021). COVID-19 heralds a new epistemology of science for the public good. *History and Philosophy of the Life Sciences, 43*(2), 1–6.

Cash-Gibson, L., Pericàs, J. M., Martinez-Herrera, E., & Benach, J. (2021). Health inequalities in the time of COVID-19: The globally reinforcing need to strengthen health inequalities research capacities. *International Journal of Health Services, 51*(3), 300–304.

Chen, D. T. H., & Wang, Y. J. (2021). Inequality-related health and social factors and their impact on well-being during the COVID-19 pandemic: Findings from a national survey in the UK. *International Journal of Environmental Research and Public Health, 18*(3), 1014.

Dasgupta, P. (2000). Trust as a commodity. *Trust: Making and Breaking Cooperative Relations, 4*, 49–72.

D'Cruz, J. (2020). Trust and distrust. In *The Routledge handbook of trust and philosophy* (pp. 41–51). Routledge.

Douglas, H. (2009). *Science, policy, and the value-free ideal.* University of Pittsburgh Press.

Elliott, K. C. (2017). *A tapestry of values: An introduction to values in science.* Oxford University Press.

Finell, E., Tiilikainen, M., Jasinskaja-Lahti, I., Hasan, N., & Muthana, F. (2021). Lived experience related to the COVID-19 pandemic among Arabic-, Russian-and Somali-speaking migrants in Finland. *International Journal of Environmental Research and Public Health, 18*(5), 2601.

Goldberg, S. C. (2020). Trust and reliance 1. In *The Routledge handbook of trust and philosophy* (pp. 97–108). Routledge.

Goldenberg, M. J. (2021). *Vaccine hesitancy: Public trust, expertise, and the war on science.* University of Pittsburgh Press.

Grasswick, H. (2017). Epistemic injustice in science. In *The Routledge handbook of epistemic injustice* (pp. 313–323). Routledge.

Hansson, S. O. (2017). Ethical risk analysis. In *The ethics of technology: Methods and approaches* (pp. 157–172).

Hardin, R. (2002). *Trust and trustworthiness*. Russell Sage Foundation.

Hassan, F., Yamey, G., & Abbasi, K. (2021). Profiteering from vaccine inequity: A crime against humanity? *BMJ, 374.*

Hawley, K. (2014). Trust, distrust and commitment. *Noûs, 48*(1), 1–20.

Holton, R. (1994). Deciding to trust, coming to believe. *Australasian Journal of Philosophy, 72*(1), 63–76.

Jones, K. (1999). Second-hand moral knowledge. *The Journal of Philosophy, 96*(2), 55–78.
Kass, N. E. (2001). An ethics framework for public health. *American Journal of Public Health, 91*(11), 1776–1782.
Kenney, M., & Müller, R. (2017). Of rats and women: Narratives of motherhood in environmental epigenetics. *BioSocieties, 12*(1), 23–46.
Krishnamurthy, M. (2015). (White) Tyranny and the democratic value of distrust. *The Monist, 98*(4), 391–406.
Lecamwasam, N. O. (2021). Lives or livelihoods? The erosion of welfare in Sri Lanka's COVID-19 response. *Is the Cure Worse Than The Disease?*, 87. https://www.cpalanka.org/wp-content/uploads/2021/08/08-Chapter-4.pdf
Lock, M. M., & Palsson, G. (2016). *Can science resolve the nature/nurture debate?* John Wiley & Sons.
Lohse, S., & Canali, S. (2021). Follow* the* science? On the marginal role of the social sciences in the COVID-19 pandemic. *European Journal for Philosophy of Science, 11*(4), 1–28.
Lohse, S., & Bschir, K. (2020). The COVID-19 pandemic: A case for epistemic pluralism in public health policy. *History and Philosophy of the Life Sciences, 42*(4), 1–5.
Longino, H. E. (1996). Cognitive and non-cognitive values in science: Rethinking the dichotomy. In *Feminism, science, and the philosophy of science* (pp. 39–58). Springer, Dordrecht.
Longino, H. E. (2002). *The fate of knowledge*. Princeton University Press.
Luhmann, N. (2000). Familiarity, confidence, trust: Problems and alternatives. *Trust: Making and breaking cooperative relations, 6*(1), 94–107.
Luhmann, N. (2018). *Trust and power*. John Wiley & Sons.
Mansfield, B. (2012). Race and the new epigenetic biopolitics of environmental health. *BioSocieties, 7*(4), 352–372.
McLeod, C. (2021). Trust. In E. N. Zalta (Ed.), *The Stanford encyclopedia of philosophy*. https://plato.stanford.edu/archives/fall2021/entries/trust/
Naz, F., Ahmad, M., & Umair, A. (2021). COVID-19 and inequalities: The need for inclusive policy response. *History and Philosophy of the Life Sciences, 43*(3), 86.
Ojwang, F. (2020). Deconstructing socially constructed subtle prejudices during the first wave of Covid-19 pandemic among immigrant population in Finland. *Sociology, 10*(6), 267–279.
Onditi, F., Nyadera, I. N., Obimbo, M. M., & Muchina, S. K. (2021). How urban 'informality' can inform response to COVID-19: A research agenda for the future. *History and Philosophy of the Life Sciences, 43*(1), 1–5.
Popa, E. (2021). Loneliness and negative effects on mental health as trade-offs of the policy response to COVID-19. *History and Philosophy of the Life Sciences, 43*(1), 1–5.
Reiss, J., & Sprenger, J. (2020). Scientific objectivity. In E. N. Zalta (Ed.), *The Stanford encyclopedia of philosophy*. https://plato.stanford.edu/archives/win2020/entries/scientific-objectivity
Russo, F. (2021). Value-promoting concepts in the health sciences and public health. http://philsci-archive.pitt.edu/19287/
Scheman, N. (2001). Epistemology resuscitated: Objectivity as trustworthiness. In N. Tuana & S. Morgen (Eds.). (2001). *Engendering rationalities* (pp. 23–52). Suny Press.
Singh, I. (2012). Human development, nature and nurture: Working beyond the divide. *BioSocieties, 7*(3), 308–321.
Smart, B., Combrink, H., Broadbent, A., & Streicher, P. (2021). *Direct and indirect health effects of lockdown in South Africa*. Center for Global Development.
Timmermann, C. (2020). Epistemic ignorance, poverty and the COVID-19 pandemic. *Asian Bioethics Review, 12*(4), 519–527.
Zaman, S., Hossain, F., Ahmed, S., & Matin, I. (2021). Slums During COVID-19: Exploring the Unlocked Paradoxes, *BIGD Working Papers*. https://bigd.bracu.ac.bd/wp-content/uploads/2021/10/BIGD-Working-Paper_Slums-During-COVID-19.pdf

Trust and Funding Science by Lottery

Jamie Shaw

Commentators have recently proposed novel modifications of traditional peer review as a method of evaluating grant proposals. One that has gained increased attention is introducing *lotteries*, or elements of random chance, into funding allocation methods. The primary arguments for the use of lotteries concerns its purported cost efficiency, openness to innovative research, and ability to minimize the marginalization of particular groups through traditional peer review. Considerations of *trust* are nearly entirely absent from the discussion, although they are extremely important when considering practical implementations. The purpose of this paper is to outline a series of plausible hypotheses about how lotteries may impact the trust of various publics and scientists that could be subjected to tests. This provides the beginnings of a research program for analyzing the relationship between lotteries in science funding policy and trust.

1 Preamble

Funding institutions have an interesting place in the fiduciary networks of scientists, science, and the public. On the one hand, skepticism of funding sources is routinely used to undermine trust in science funded from that source. However, the reliability of the funding source rarely safeguards the trustworthiness of the science it funds (Sprecker, 2002). Skepticism of funding sources also functions differently in the private and public sphere (Bubela et al., 2009). Private sources of funding, with some notable exceptions, are almost automatically targets of skepticism. They are mistrusted from the start and must *earn* their trust. This is largely due to a history of vested private interests distorting scientific research for reasons that go against

J. Shaw (✉)
Institut für Philosophie, Leibniz Universität Hannover, Hannover, Germany
e-mail: jshaw222@uwo.ca

M. M. Resch et al. (eds.), *The Science and Art of Simulation*,
https://doi.org/10.1007/978-3-031-68058-8_12

any viable conception of the public good, as have been revealed in numerous meta-analyses on sponsorship bias (see e.g., Fabbri et al., 2018). Whether it be pharmaceutical companies, whose funded research reaches suspiciously convenient results (Sismondo, 2008), the Pioneer Fund's support of race and intelligence research while having connections to white supremacist movements (Tucker, 2002), tobacco companies research on the health effects of smoking (Bero et al., 1994), or conservative think tanks research on climate change (Oreskes & Conway, 2011)—there is a seemingly endless array of publicly prominent instances in which private entities have betrayed public trust for financial or political gains.

Public patrons also have their own fiduciary upkeep that is necessary for public support of science funding. Recently, Matthew Motta has highlighted how public American science funding bodies must stimulate interest in the sciences to continue to support government spending (Motta, 2019).[1] Throughout history, public funding bodies have undergone many challenges to their legitimacy, thus forcing them to establish themselves as institutions with the public good at heart (Muñoz et al., 2012).[2] While some of these events will be canvassed in this paper, my primary concern is the impacts that proposed *changes* of funding allocation mechanisms might have on trusting relationships between scientists, funding bodies, and various publics. Science funding policy is currently undergoing several reforms. Some agencies, such as the International Development Research Council, incorporate stakeholders into decision-making procedures, while others (e.g., the Center for AIDS Prevention Studies, the National Institute of Mental Health) introduced pre-screening feedback systems. Moreover, numerous funding allocation methods that augment or circumvent traditional peer review such as peer-to-peer networks, crowdsourcing, or a universal basic income for scientists are being practiced and discussed. One proposed change is to introduce *lotteries* into funding allocation methods, and little has been written about how they may impact trust. For example, the *Powering Discovery* document, recommending alterations to Canada's science funding policy, suggests experimenting with lotteries and does not mention how it may impact public trust at all (Tilghman et al., 2021). This omission is extremely unfortunate since lotteries are likely to introduce new fiduciary relationships and disrupt those that have been established with traditional peer review. Detailed investigations on how lotteries might impact trust is desperately needed.

Unfortunately, there is little theoretical or empirical research on trust and lotteries in science funding policy. This is to be expected, as the initiative is fairly new. Because of this, my ambition in this paper is not to argue that we should or should not trust lotteries. Rather, I will enumerate and motivate some hypotheses about how trust *might* be impacted by the practice of lotteries that can be refined and tested in future research.

---

[1] Although I cannot help but express the worry that, in the final paragraph, Motta suggests that this could be done through early education which sounds close to indoctrination.

[2] This has also been the case for private funding bodies. This functions slightly differently for philanthropic bodies such as the Rockefeller foundation than private companies like the British Petroleum company funding research on the ecological impacts of oil spills (see O'Connor, 2007).

2 Background on Funding Science by Lottery

Traditionally, funding for scientific research relies on peer reviewers. While peer review has been "troubled from the start" (Csiszar, 2016), there has been a renewed and reinvigorated skepticism that has been building since the 1970s (Mitroff & Chubin, 1979). The most common worries are that it is unduly expensive, discriminates against marginalized groups, and disincentivizes innovative research (Fang & Casadeval, 2016). This motivates novel funding allocation methods that reduce the use of peer review. One allocation method that has gained significant attention is *funding science by lottery.*

Lotteries introduce elements of random chance into funding decisions. While there are several ways that lotteries have been practiced, every serious theoretical or practical proposal integrates lotteries with traditional peer review within a more limited scope. The Volkswagen Foundation, for example, uses peer review to winnow a pool of applicants by accepting and rejecting the top- and bottom-ranked proposals respectively and using a lottery for the remaining grants (Roumbanis, 2019). Thus, funding science by lottery does not depart from conventional methods as radically as it may seem. Still, this new method for allocating funds raises new problems and prospects that must be thought through carefully.

One issue that has received little attention is how the implementation of lotteries can transform fiduciary relationships. Three, in particular, seem especially important:

(1) Scientist's trust of funding bodies
(2) The public's trust of funding bodies.
(3) Trust in the scientific research itself.

While lotteries have been implemented in Canada, Switzerland, England, Germany, Austria, and New Zealand, little data exists on how these policy changes have impacted trust or how increased and more publicly noticeable uses of lotteries may transform trusting relationships many have with funding agencies. This may be a mere oversight, or that there is not much data to be had—these policy changes are fairly new (the first modern use of lotteries in funding policy is from 2013) and have received modest attention in popular media. This omission is extremely noteworthy, as trust plays an increasingly central role in science and its place in society (de Melo-Martín & Intemann, 2018; Fernandez-Pinto, 2020; Goldenberg, 2021). If we are to move forward with lotteries, trust needs to be considered more explicitly.

Trust functions differently in (1), (2), and (3) and, for that matter, within each category. Scientist's trust in funding bodies can impact the production of scientific knowledge in several ways. First, scientists may not *request* funds from bodies whose methods they don't trust. While this is possible, it is unlikely to pose a problem for lotteries in particular. Preliminary polls suggest that lotteries are popular amongst scientists (Liu et al., 2020), and applications for lottery-based calls have not decreased in number. Moreover, since external sources of funding are a necessity of academic life, there are few other options scientists have. However, in some areas of research, several funding sources are open for scientists to support their research.

If scientists do not trust lotteries, *ceteris paribus*, they will seek funding elsewhere (see Collins, 2000). In some cases, this is a mundane point. However, if increasing numbers of scientists sought funds from *private* sources or governmental agencies with (primarily) non-scientific agendas (e.g., military agencies), then this may have a more drastic impact on scientific research (see Elder 2012; Laudel, 2006; Solovey, 2013). Therefore, while the possibility of researchers switching funding sources seems low, we must keep an eye on it—especially with 'mission-oriented' research which is closer to applications and more likely to be funded by exogenous sources (Mazzucatto, 2018).

As for (2), it is worth repeating that there is no such thing as 'the public'—there are many publics—and science-society relations function differently depending on the science and its social milieu. Public trust can be essential for several reasons. First, without trust, uptake of science, at policy levels or in citizen decision making, is less likely. Second, loss of public trust can also lead to a lack of citizen willingness to support science. Third, a lack of trust of scientists can increase social polarization.

3 Quick Preliminaries on Trust

The philosophical literature on trust has focused almost entirely on questions about when trust is warranted or conceptualizing the idea of 'trust.' Of course, funding policy involves normative considerations. However, the approach taken here does not presume *whether* trust is warranted. Rather, the question is purely descriptive/explanatory—about whether lotteries might undermine or promote trust—with normative questions in mind. Callous trust or foolish distrust remain issues to be resolved regardless of their epistemic warrant (see Govier, 1992). As Nick Fury of the Avengers once put it, "we take the world as it is, not as we want it to be."

Second, I take it—perhaps controversially—that trust, at least in this case, reduces to other virtues a funding allocation method may have. Say, for example, I am warranted in trusting funding bodies if they reliably identify the best scientific research proposals (by whatever definition of 'best' I hold). Or imagine I trust a funding agency because I perceive the director to be honorable and competent. Then insofar as the allocation method is reliable or the director is honorable and competent, then it is to be trusted. I don't see much that can be added to understanding the proper functioning of funding bodies by evaluating policy in this manner. This way of investigating trust is just a roundabout and confused way of talking about the virtues funding agencies should possess. However, if we attempt to understand how trust will be impacted by implementing lotteries, then several normative consequences arise. Whether we like it or not, trusting dispositions, whatever they may be, matter for what policies we can responsibly enact.

Before hypothesizing, it is worth detailing what is meant by trust. Trust, in this paper, requires the *acceptance of the results of a process regardless of what the particular results may be.* To trust a funding method is not tantamount to trusting the science it funds—this must be established through other means. Rather, trust requires accepting the results of the funding procedures as fair or justified, whatever they may be. Analogously, people trust democratic procedures not because the elected officials are inherently trustworthy but because the procedures themselves are fair or reliable. We can accept that Justin Trudeau won a fair election while not accepting him as the best candidate for Prime Minister of Canada. Trust, to be worthy of the title, must be relatively *robust*. By this, I mean that someone's trust does not waver too much in the face of exogenous changes. For example, before the 2020 election, the vast majority of Americans claimed to trust the electoral system—that is, they thought their vote and everyone else's votes would be counted and reported fairly.[3] However, by a few well-placed and easily refutable words, Donald Trump was able to manipulate the fiduciary relationships many Americans had towards their electoral system. On my view, we can say that those who 'changed their minds' *never* trusted the electoral system—or, at least, they never trusted it very much. Their trust could be undermined even though the electoral system did not undergo any significant changes. Of course, robustness comes in degrees, and I have no intention on fixing the level of robustness necessary for trust here.

Some degree of robustness is necessary for trust to be tractable enough to be studied. There are numerous changes constantly surrounding funding bodies, and many of them have the potential to impact trust. If trust depends on the quality of one's morning coffee, whether the Moon is in the seventh house, or who won the Stanley Cup most recently, then it would fluctuate too much to be predictable. For now, we will presume that trust in funding is not so volatile as to undergo major changes as a result of minor changes in funding structures or independent events. This somewhat reasonable hope is necessary unless we want to merely wait and see what impacts lotteries have on trust.

4 Hypotheses on Trust and Lotteries

As mentioned, we know very little about lotteries in science funding policy and trust. My purpose here, therefore, is not to aggrandize what little we know during what is likely a transitional phase in funding policy. Such an assessment will likely become out-of-date by the time it hits the presses. Rather, my goal is to get the ball rolling on a research program that investigates the ways in which lotteries may impact trust. This will be accomplished by formulating *hypotheses* about how trust might be impacted by the implementation of lotteries. I will motivate why these hypotheses might turn out to be roughly true but will not pretend that they are entirely unambiguous. Hopefully, as we take more of a detailed empirical look at trust and lotteries, these hypotheses will become more precise, and we can figure out whether these hunches are correct.

[3] Although distrust in the American electoral system had been steadily increasing since the early 2000s (Norris et al., 2014).

To put all the cards on the table, here are four hypotheses about how trust will be impacted, positively or negatively, by funding science by lottery:

Hypothesis #1: Lotteries will decrease incentives to hype research, which will impact longitudinal trust in funding agencies.
Hypothesis #2: Research funded by lottery will be more vulnerable to doubt.
Hypothesis #3: Accusations of systemic bias will be blunted.
Hypothesis #4: Lotteries will decrease public trust in science as a whole insofar as scientific research projects emerge independently from the social sphere.

Let's go over some background on each hypothesis in turn.

Hypothesis #1: *Lotteries will decrease incentives to hype research, which will impact longitudinal trust in funding agencies.*

Many scientists hype their research to make it seem like it will be more impactful than can be reasonably expected. While scientists often hype their research to other scientists—this is much more pronounced when the promises are remedies to social ills. When the Human Brain Project (HBP) accrued over 1.2 billion Euros from the European Research Council (ERC) and more than double that from a consortium of private investors, it was the promise of discovering neuromarkers for Alzheimer's that caught the eye of the public and was responsible for its price tag. While the HBP led to some interesting theoretical discoveries and the development of novel technologies, it's failure to attain this goal is what have led many to label this project as a failure (see Mullin, 2021; Shaw, 2018). While I am not aware of studies on this particular project, it wouldn't be hard to imagine that public trust in the ERC (and particular scientists) was unaffected by this highly public disappointment.[4]

Hyping rarely involves straightforward lies (Intemann, 2022). This is because hype usually concerns *future* performance, and overly optimistic guesses are not the same thing as committing fraud. Moreover, the well-known fallibility of anticipating future success allows for plenty of wiggle-room to maneuver out of direct accusations of misleading portrayals of research potential. In other words, we "soon reach a point where the probable results of research, as well as the possible practical consequences of a variety of possible outcomes, become so uncertain and so diverse that the only honest thing we can do is to fall back on a generalized faith in the average beneficence of new knowledge" (Brooks, 1978, 181). Moreover, even if a research project does not provide its promised results on time, it is almost always possible to dispel frustrations by pushing the deadline back a little bit. This makes it difficult to test hype claims or hold hypers accountable. Scientists also may believe their own hype, and this is to be expected when scientists are proposing new research that they are excited about. The energy that goes into grant writing will likely, psychologically speaking, blur the line between hype and realistic assessment. So even if hype is not deliberate,

[4] It might be possible that trust wasn't damaged at the public at large as evidenced by the little press coverage of the HPB after its initial controversy. Most people may have simply forgotten about the project. However, trust amongst those who had vested interests in the HBP obtaining its goals (e.g., families of those who suffer from Alzheimer's) may be more impacted.

and the grantees have the most honest intentions, hype may still be a by-product of the grant writing process. Moreover, hyping is often perceived as necessary for obtaining funding by convincing peer reviewers, and sometimes the public, that the hype is real (Caulfield, 2018).

As hyper-competition increases, scientists are under increasing pressure to make their grant application stand-out. This can be difficult to accomplish by wading through technical waters, so highlighting eye-catching consequences of research are incentivized. This can be more pronounced by popular news media, whose profit motives move them towards sensationalizing research projects. As Harvey Brooks astutely observes,

> The incentive to oversell basic research on grounds of its utility becomes greater as the competition for research funding becomes keener. At the same time, the excesses of a few entrepreneurs may ultimately discredit the whole exercise of forecasting the consequences of research. The problem is exacerbated by the attitude of the news media which tend to overemphasize highly speculative or uncertain applications of basic research (Brooks, 1978, 181).

This point has become more pronounced in the twenty-first century. Despite the surprising accuracy of popular science journalism when disclosing ongoing science to the public, separating hype from realistic excitement or idealism is an arduous and, perhaps, impossible task. As a result, the phenomenon of hype seems intrinsic to science in an age of hyper-competition (Caulfield, 2004). However, to complicate the issue, sometimes hype plays an important function in scientific progress. The blockbuster *Jurassic Park*, for example, hyped theories that birds evolved from dinosaurs and that dinosaurs were warm-blooded, communal in nature, used forms of sophisticated communication, and nurtured their young (Kirby, 2003). These theories turned out to be true and were marginalized before the release of *Jurassic Park*. Hype aided and was arguably essential for transforming a niche research program into an accepted body of knowledge in paleontology. More bombastically put, propaganda can be helpful to allow room for a new research program that might otherwise be discredited prematurely (Feyerabend, 1975/1993, 114).

Hyping can have important consequences for trust. When scientists hype their research, publics whose taxes fund this science become frustrated with unrealized hype, it is reasonable to expect publics to become disillusioned not only with the scientists themselves but the funding body that constantly disappoints its patrons. This will likely only be the case with science that is more visible to publics capable of influencing funding policy. While most public funding bodies advertise research they have supported, little of this will catch the public eye. Research like space travel, which captures public attention more than research on the mechanisms by which miRNAs regulate their targets, is likely to be under increased scrutiny. Since journalists rarely disclose funding sources in popularizations of science (Cook et al., 2007), it is reasonable to expect this effect to be localizable to more public-facing scientific projects. Moreover, for the effect to be notable it will likely require repeated failures over a stretch of time unless a particular failure is especially catastrophic.

There are good reasons to think that lotteries will decrease the need to hype research, although they will not eliminate it entirely. A primary purpose of hype in grant proposals is to convince reviewers. With less reviewers to convince, it seems reasonable to expect that hype would decrease. Of course, the grant proposal will be read by very few and the public display of the research project is a separate entity altogether. Since hype may still be necessary to promote one's work—whether academic or not, commercial or not, private or public—it is plausible that the incentive to hype in grant applications will be minimal at more important scales.

Hypothesis #2: *Research funded by lottery will be more vulnerable to doubt.*

Regardless of the faults peer review may have, it is still largely perceived as a gatekeeper or guarantor of rigor (Wing & Chi, 2011). In hotly politicized debates surrounding climate change, nuclear power, drug efficacy, or vaccination, reporters, journalists, public figures, and public-facing scientists routinely use peer review as a shorthand for establishing the validity of a claim or rejecting a claim as 'pseudo-science.' For example, in response to Tom Coburn's (R-Oklahoma) report *The National Science Foundation: Under the Microscope*, Patrick Clemins (then-director of the R&D Budget and Policy Program at the American Association for the Advancement of Science) attempted to reassure the public by referencing the NSF's use of peer review: "Just as we rely on legislators to make legislative decisions, we should rely on scientific experts to make the scientific decisions on where the next big innovation might occur" (quoted in Boyle, 2011). NSF spokesperson Maria Zacharias similarly singled out peer review to assure that the NSF was not funding frivolous research when pressed at a Congressional budget meaning (Pappas, 2010). While peer review is more frequently cited in its capacity to ensure the legitimacy of *publications*, it is likely that research funded by lotteries will be considered with added skepticism. That is, it could be perceived as research that could not pass traditional peer review. As Philipps (2022) reports, scientists were only in favor of lotteries once it was clear that traditional peer review still played an essential role in its functioning. Not all scientists and certainly many members of the public will pay attention to the details and this reassurance could be muted in broader discourses.

To be clear, the hypothesis is not that research funded by lotteries will be dismissed or treated with suspicion from the get-go. As mentioned, funding sources are usually brought to the fore *after* a controversy has arisen. Some lotteries do not request researchers to mention their funding source so this could be easily hidden. However, a wide variety of sources can foster doubt on a research project—from legitimate scientific criticisms, to jealousy, or political controversy. Research funded by lotteries could be *especially vulnerable* to doubt should it arise, which can have drastic implications for its uptake.

Since lotteries are relatively recent, it probably isn't too late to engage in a public relations campaign to relabel 'funding by lottery' as a version of peer review—which it is—which may elicit less blowback (however this relabeling might have negative consequences as well, see below). However, it isn't too hard for any journalist, scientist, or political figure who is peeved with a particular research project or funding body to exploit the rhetoric of lotteries to reinstate the aforementioned connotations. It's because of this that I doubt that public justifications of lotteries will stymie the trust lost from truncating peer review. Regaining the trust that comes with peer review may take sustained, decade long marketing (or propaganda?) campaigns for lotteries.

Hypothesis #3: *Accusations of systemic bias will be blunted.*

Accusations of systemic bias amongst scientific communities or sub-communities are extremely difficult to address. Worries that scientific communities do not take viable criticisms or interests into account come from across the political spectrum. Many conservatives see academic scientists as mere pawns in the progressive bastion of universities (Mede & Schäfer, 2020). Many progressives see some scientific groups as covertly protecting power interests (see Reiter, 2021). These concerns often stem from the worry that dissenters, usually who accept a dissenting opinion that is in-line with the accusers, are not likely to thrive in a field that is hell-bent on rejecting their approach for unfair reasons. If a community is perceived as containing systemic bias, it will likely be perceived as less trustworthy.

Of course, funding policy only represents one part of the organization of scientific communities (Pierson & Millum, 2017). Even well-funded science, if it were systemically discriminated against, would struggle to survive. It would be difficult, though not impossible, to get presented at conferences, published, and cited on a consistent basis. Moreover, scientists almost always require institutional credentials to apply for funds and, therefore, would need to be trained and accredited which would leave open some charges of systemic biases. If systemic bias truly existed, changes in funding policy on their own are unlikely to remove it. However, it seems plausible to think that lotteries will *blunt* accusations of systemic bias in one direct and one indirect way. The direct way is: Lotteries are will be perceived as fairer since they allow everyone an equal chance to obtain funding (see Guildenhuys, 2020). While this isn't strictly true, since all lotteries still use peer review to eliminate *some* grants from consideration, the marketing of 'funding by lottery' could bypass these details. In other words, this may give pause to accusations of systemic biases making them less likely to permeate discourse and become widely accepted.

There is another, more longitudinal and indirect manner in which lotteries may blunt accusations of systemic bias. One of the primary promises of lotteries is that they increase the prospects of receiving funds for marginal groups (Philipps, 2020, 2022). As a result, we should expect lotteries to play a role in gradually changing the composition of scientific communities. As communities become more diverse, it is more difficult to accuse them of containing systemic prejudices. As has been argued extensively elsewhere, people tend to trust scientists when they believe that their interests are aligned. This can be, and often is, correlated with perceptions of social identities and a sense of group solidarity (Goldenberg, 2021). The more diverse communities become, the more likely some publics will trust the science since it is more likely to be produced by *people like them.*

Finally, there is a wildcard consideration about what future studies on bias and lotteries will reveal and, given these results, how these findings might seep into public consciousness. The current thinking is that lotteries decrease the power biases have over access to funds,[5] but whether this is actually true is unknown and how publics may come to think of bias and lotteries in light of currently unconducted studies is fairly difficult to predict.

Hypothesis #4: *Lotteries will decrease public trust in science as a whole insofar as scientific research projects emerge independently from the social sphere.*

In the 1970s, American Senator William Proxmire announced his 'Golden Fleece' awards which were granted to publicly funded scientific research that was perceived as frivolous expenditures of taxpayer money (see Solovey, 2020, Chap. 5). This motivated the Bauman amendment which would allow Congressional vetoes over NSF budget proposals. While the amendment never passed the Senate, it remains an instructive historical episode about how the autonomy of science can be challenged. If scientists frequently engage in research that the public sees no value in, then the promise of taxpayer dollars stands on wobbly legs. While the decrease of hype will likely partially nullify this worry, publicly funded science is constantly required to justify its necessity (see Gauchat, 2012).

As previously mentioned, lotteries are partially motivated by their propensity to support innovative research. This comes alongside of more *freedom* for scientists to pursue what fits their fancy, which will allow for a wider range of research projects. Even if social impact metrics or citizens were incorporated into lotteries, which has yet to be done, there is less control to direct the flow of funds to desired projects. This would increase the chance that projects perceived to be frivolous (or actively harmful) will be funded, thus leaving science funding as a whole less secure. If history were to repeat itself (and it did in 2011, see above), the freedom scientists enjoy and the *increased* freedom they could possess due to lotteries might make it more difficult to assure the public that their research is worth the investment. There is, of course, a chance that things could go in the opposite direction. Perhaps more freedom for scientists will incline them to pursue research that is more personal to them and, perhaps by virtue of how they are situated, is more personal to members of the broader public. This might increase sympathies with the choice of research priorities. My hunch, though, is that the odds of this are low and insofar as it occurs, will likely be outweighed by the projects scientists pursue out of curiosity that are stimulated by their professional training. Regardless, this may be tempered by recent calls to make public engagement a condition of receiving research (e.g., Salmon & Priestley, 2015), which could be included in lottery calls.

5 Concluding Remarks

As is frequently the case, empirical work on the relationships between trusting bodies and funding institutions that utilize lotteries will refine the hypotheses offered here. Until the relevant empirical research is conducted, these hypotheses are mostly based

[5] This has been resisted by Lee et al. (2020).

on educated (hopefully) intuitions. However, I hope to have offered the start of a research program whose realization is essential for a broader assessment of the viability of funding science by lottery. We know very little about this topic as of now, so much more research needs to be done.

Acknowledgements I greatly appreciate the constructive feedback from Michael Herrmann, Manuela Fernandez-Pinto, Carlos Santana, and Mark Solovey.

References

Bero, L. A., Glantz, S. A., & Rennie, D. (1994). Publication bias and public health policy on environmental tobacco smoke. *JAMA, 272*(2), 133–136.

Boyle, A. (2011). Funny science sparks serious spat. *Cosmic Log*. https://web.archive.org/web/20110530002942/http://cosmiclog.msnbc.msn.com/_news/2011/05/26/6724606-funny-science-sparks-serious-spat.

Bubela, T., Nisbet, M. C., Borchelt, R., Brunger, F., Critchley, C., Einsiedel, E., Geller, G., Gupta, A., Hampel, J., Hyde-Lay, R., & Jandciu, E. (2009). Science communication reconsidered. *Nature Biotechnology, 27*(6), 514–518.

Brooks, H. (1978). The problem of research priorities. *Daedalus, 170*(2), 171–190.

Caulfield, T. (2004). Biotechnology and the popular press: Hype and the selling of science. *Trends in Biotechnology, 22*(7), 337–339.

Caulfield, T. (2018). Spinning the genome: Why science hype matters. *Perspectives in Biology and Medicine, 61*(4), 560–571.

Cook, D., Boyd, E., Grossmann, C., & Bero, L. (2007). Reporting science and conflicts of interest in the lay press. *PLoS ONE, 2*(12), e1266.

Collins, J. (2000). Preemptive prevention. *The Journal of Philosophy, 97*(4), 223–234.

Csiszar, A. (2016). Peer review: Troubled from the start. *Nature, 532*(7599), 306–308.

de Melo-Martín, I., & Intemann, K. (2018). *The fight against doubt: How to bridge the gap between scientists and the public.* Oxford University Press.

Edler, J. (2012). Toward variable funding for international science. *Science, 338*(6105), 331–332.

Fabbri, A., Lai, A., Grundy, Q., & Bero, L. (2018). The influence of industry sponsorship on the research agenda: A scoping review. *American Journal of Public Health, 108*(11), e9–e16.

Fang, F., & Casadevall, A. (2016). Research funding: The case for a modified lottery. *mBio, 7*(2), 1–7.

Feyerabend, P. (1975/1993). *Against method* (3rd ed.). Verso Books.

Gauchat, G. (2012). Politicization of science in the public sphere: A study of public trust in the United States, 1974 to 2010. *American Sociological Review, 77*(2), 167–187.

Goldenberg, M. (2021). *Vaccine hesitancy: Public trust, expertise, and the war on science.* University of Pittsburgh Press.

Govier, T. (1992). Distrust as a practical problem. *Journal of Social Philosophy, 23*(1), 52–63.

Gildenhuys, P. (2020). Lotteries make science Fairer. *Journal of Responsible Innovation, 7*(2), S30–S43.

Intemann, K. (2022). Understanding the problem of "Hype": Exaggeration, values, and trust in science. *Canadian Journal of Philosophy, 52*(3), 279–294.

Kirby, D. (2003). Science consultants, fictional films, and scientific practice. *Social Studies of Science, 33*(2), 231–268.

Laudel, G. (2006). The art of getting funded: How scientists adapt to their funding conditions. *Science and Public Policy, 33*(7), 489–504.

Lee, C., Grant, S., & Erosheva, E. (2020). Alternative grant models might perpetuate black-white funding gaps. *The Lancet, 396*(10256), 955–956.

Liu, M., Choy, V., Clarke, P., Barnett, A., Blakely, T., & Pomeroy, L. (2020). The acceptability of using a lottery to allocate research funding: A survey of applicants. *Research Integrity and Peer Review, 5*(1), 1–7.

Mazzucato, M. (2018). *Mission-oriented research & innovation in the European Union*. European Commission.

Mede, N., & Schäfer, M. (2020). Science-related populism: Conceptualizing populist demands toward science. *Public Understanding of Science, 29*(5), 473–491.

Mitroff, I., & Chubin, D. (1979). Peer review at the NSF: A dialectical policy analysis. *Social Studies of Science, 9*(2), 199–232.

Motta, M. (2019). Explaining science funding attitudes in the United States: The case for science interest. *Public Understanding of Science, 28*(2), 161–176.

Mullin, E. (2021). How big science failed to unlock the mysteries of the human brain. *MIT Technology Review*. https://www.technologyreview.com/2021/08/25/1032133/big-science-human-brain-failure/

Muñoz, A., Moreno, C., & Luján, J. (2012). Who is willing to pay for science? On the relationship between public perception of science and the attitude to public funding of science. *Public Understanding of Science, 21*(2), 242–253.

Norris, P., Frank, R., & Coma, F. (2014). Measuring electoral integrity around the world: A new dataset. *PS: Political Science & Politics, 47*(4), 789–798.

O'Connor, A. (2007). *Social science for what?: Philanthropy and the social question in a world turned rightside up*. Russell Sage Foundation.

Oreskes, N., & Conway, E. M. (2011). Merchants of doubt: How a handful of scientists obscured the truth on issues from tobacco smoke to global warming. Bloomsbury Publishing USA.

Pappas, S. (2010). Scientists: Call for citizen review of funding is misleading. *Science on MSNBC*. https://web.archive.org/web/20110601043811/http://www.msnbc.msn.com/id/40574700/ns/technology_and_science-science/t/scientists-call-citizen-review-funding-misleading/

Philipps, A. (2020). Science rules! A qualitative study of scientists' approaches to grant lottery. *Research Evaluation, 30*(1), 102–111.

Philipps, A. (2022). Research funding randomly allocated? A survey of scientists' views on peer review and lottery. *Science and Public Policy, 49*(3), 365–377.

Pierson, L., & Millum, J. (2017). The limits of research institutions in setting research priorities. *Journal of Medical Ethics, 43*(12), 810–811.

Pinto, M. F. (2020). Commercial interests and the erosion of trust in science. *Philosophy of Science, 87*(5), 1003–1013.

Reiter, B. (2021). *Decolonizing the social sciences and humanities: An anti-elitism manifesto*. Routledge Press.

Roumbanis, L. (2019). Peer review or lottery? A critical analysis of two different forms of decision-making mechanisms for allocation of research grants. *Science, Technology, & Human Values, 44*(6), 994–1019.

Salmon, R., & Priestley, R. (2015). A future for public engagement with science in New Zealand. *Journal of the Royal Society of New Zealand, 45*(2), 101–107.

Shaw, J. (2018). Why the realism debate matters for science policy: The case of the human brain project. *Spontaneous Generations: A Journal for the History and Philosophy of Science, 9*(1), 82–98.

Sismondo, S. (2008). Pharmaceutical company funding and its consequences: A qualitative systematic review. *Contemporary Clinical Trials, 29*(2), 109–113.

Solovey, M. (2013). *Shaky foundations: The politics-patronage-social science nexus in cold war America*. Rutgers University Press.

Solovey, M. (2020). *Social sciences for what? Battles over public funding for the "Other Sciences" at the National Science Foundation*. MIT Press.

Sprecker, K. (2002). How involvement, citation style, and funding source affect the credibility of university scientists. *Science Communication, 24*(1), 72–97.

Tilghman, S., Bailey, J., & Bourguinon, J.-P. (2021). *Powering Discovery*. https://cca-reports.ca/reports/international-practices-for-funding-natural-science-and-engineering-research/

Tucker, W. (2002). *The funding of scientific racism: Wickliffe draper and the pioneer fund*. University of Illinois Press.

Wing, J. M., & Chi, E. H. (2011). Reviewing peer review. *Communications of the ACM*, *54*(7), 10–11.

Sociological, Communicative and Media Aspects of Trust in Science

Establishing Trust in Algorithmic Results: Ground Truth Simulations and the First Empirical Images of a Black Hole

Paula Muhr

Abstract When the first empirical images of a black hole's shadow were released in April 2019, they transformed this defining black hole feature from a theoretical into an explorable physical entity. But although derived from empirical measurements, the production of these images relied on the deployment of the algorithmic pipelines designed specifically for this purpose to enable the selection of optimal imaging parameters. How could the researchers involved trust their imaging pipelines to deliver faithful reconstructions of unknown images from the noisy, fragmentary measurements? This article analyses the media-specific operations through which the researchers generated sufficient evidence of the trustworthiness of their algorithmic outputs by constructing and deploying a specifically tailored ground truth dataset that comprised simulated model images and simulated measurement data. The article also argues that the epistemic trust established through such operational procedures is relational and contingent on the adequacy of the media-specific operations that generated it.

1 Introduction

Black holes are massive cosmic objects of enormous density whose gravitational pull is so strong that everything that crosses their boundary, called the event horizon, disappears inside them (EHTC, 2019a, 1). Yet some of the largest specimens—termed supermassive black holes—are extremely luminous objects due to their accretion discs. Such discs are created by the matter that, under the black hole's gravitational pull, is forced to orbit around the event horizon close to the speed of light, slowly spiralling inward. In the process, the swirling matter heats up, emitting radiation.

P. Muhr (✉)
Institute of Philosophy, History of Literature, Science, and Technology, Technische Universität Berlin, Hardenbergstr. 16-18 10623, Berlin, Germany
e-mail: paula@paulamuhr.de

Institute for Implementation Science in Health Care, Faculty of Medicine, University of Zurich, Zurich, Switzerland

M. M. Resch et al. (eds.), *The Science and Art of Simulation*,
https://doi.org/10.1007/978-3-031-68058-8_13

According to the theory of general relativity, when viewed by a distant observer, in the centre of the bright accretion ring, there is a dark circular "shadow" cast by the event horizon. Due to gravitational lensing, the shadow's radius is several times larger than the event horizon's (Falcke, 2017, 1–4).

Initially predicted by general relativity, the existence of black holes has been supported by growing indirect empirical evidence since the 1970s (Goddi et al., 2019, 25–26). But until recently, there was no empirical evidence that event horizons exist. This changed in April 2019, when the Event Horizon Telescope (EHT) Collaboration, an international project with more than 200 scientists, released the first empirical images of the supermassive rotating black hole M87*, which is some 55 million light years away from Earth (EHTC, 2019a, 2019b, 2019c, 2019d, 2019e, 2019f) (Fig. 1). In astrophysics, all previous visualisations of black holes' accretion disks were based on calculations or computer-generated simulations of theoretical models, including recent 3D general relativistic magnetohydrodynamic (GRMHD) simulations derived from coupling general relativity with hydrodynamic models of ideal accretion flows (EHTC, 2019e, 15). In contrast, the four empirical EHT images released in April 2019 were reconstructed from petabytes of radio signal data acquired on four nights in April 2017 (EHTC, 2019d).

All four EHT images show a blurred ring of shiny matter with a dark circular patch in the centre, which was interpreted as a black hole shadow. These images thus provide the first "strong [visual] evidence for the existence of an event horizon" (Goddi et al.,

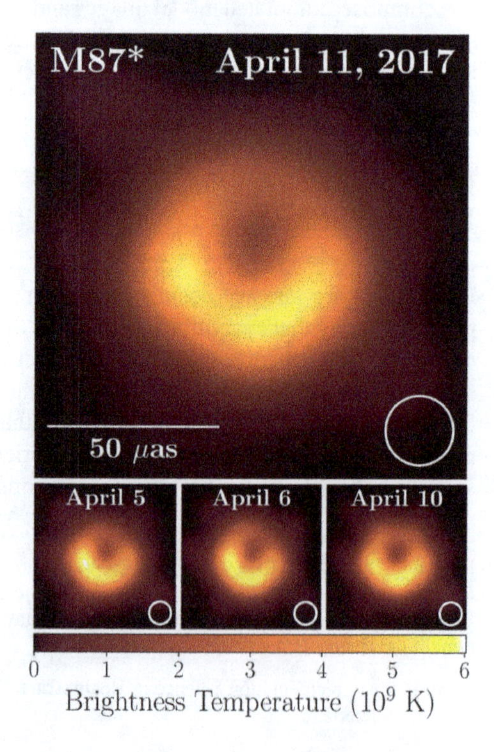

Fig. 1 EHT images of M87* reconstructed from measurement data collected on four nights in April 2017 (From EHTC, 2019a)

2019, 34), effectively "transforming this elusive boundary from a mathematical to a physical entity" (EHTC, 2019a, 9). Moreover, these images were created to fulfil further epistemic functions, such as estimating M87*'s physical properties (e.g., mass and spin direction) (EHTC, 2019e, 2019f) and, more generally, for exploring "gravity at its most extreme limit" (EHTC, 2019a, 1).

Elsewhere, I have argued that to fulfil these epistemic functions, the EHT images had to be reconstructed from the non-visual measurements as "empirically authentic," meaning that they must have a demonstrably tenable referential relation to the black hole they visualise (Muhr, 2024). Deploying Bruno Latour's (1999) concept of the *chain of transformations* and Ludwig Jäger's (2006) concept of *discursive evidence*, I have claimed that this referential relation was established through a multibranched network of mutually coordinated operations that gradually translated the noisy non-visual measurements into empirically authentic images of M87* (Muhr, 2023, 2024). The main stages of this process included: (1) acquiring radio signals with an array of telescopes (EHTC, 2019b); (2) processing noisy data through correlation and calibration (EHTC, 2019c); (3) assessing the adequacy of different statistical image-reconstruction models through "blind imaging" by human experts (EHTC, 2019d, 9); (4) reconstructing the images from the processed data using scripted algorithmic pipelines developed specifically for the EHT project (EHTC, 2019d, 9–18); and (5) validating the imaging results (EHTC, 2019d, 18–29). Whereas elsewhere (Muhr, 2023, 2024), I analysed the mutual coordination of these different stages, here I focus on the pipeline-based stage of image reconstruction, exploring how the EHT team grounded their epistemic trust (Rolin, 2020) in the results of their novel algorithmic pipelines.

In the philosophy of science, there is an intense discussion on the epistemic status of outputs of computational modelling in scientific settings and how scientists can trust such outputs. Much of this discussion has focused on identifying general principles that justify scientists' beliefs in algorithmic outputs as valid sources of knowledge (Baumberger et al., 2017; Duede, 2022; Durán & Formanek, 2018; Hubig & Kaminski, 2017; Symons & Alvarado, 2016, 2019). Due to this focus, little attention has been paid to step-by-step procedures through which a context-specific epistemic justification is established in a concrete scientific setting, as I analyse here. Following Rolin (2020, 354–55), who defines trust within science as epistemic when scientists' beliefs in research results are grounded on adequate evidence,[1] I will examine how the EHT team generated evidence to justify their belief in the trustworthiness of their algorithms' performances. Trustworthiness is understood here as the algorithms' capability under given conditions to faithfully reconstruct unknown images of M87*

[1] According to Rolin (2020, 355), epistemic trust is a specific form of reliance that "arises only in a relation of epistemic dependence." Fittingly, the EHT team epistemically depended on the algorithms to reconstruct empirically authentic images from their data (EHTC, 2019d, 9). But whereas Rolin discusses epistemic trust we place in individual scientists, scientific communities, and institutions, I extend it to algorithms and their results. While many scholars claim that relations of trust apply only to human beings, my approach is aligned with Starke et al. (2021), who persuasively argue that we can also trust non-human agents such as algorithms.

from sparse and noisy non-visual data without introducing epistemically deleterious errors.

I will approach the above question from a media-theoretical perspective by deploying the German media theorist Ludwig Jäger's concept of *transcription*. Jäger (2010, 51) defines transcription as the fundamental operational logic of all sign systems through which the production of meaning—e.g., trustworthiness—is dynamically organised. He argues that when a sign's meaning is problematic for some reason—e.g., due to its novelty—its new meaning cannot be constructed by referring to a trans-semiotic world. Instead, the problematic sign's new meaning must be constituted processually through intramedial and intermedial transcriptive operations that produce semantic effects by forging targeted references among different signs (Jäger, 2006). Intramedial operations interconnect signs within a single medium (e.g., comparing images to images), whereas intermedial operations set up references across different media (e.g., relating images to numerical data) (Jäger, 2010, 49–50). Drawing on Jäger, I will show that the EHT team established their algorithmic results as trustworthy by constructing a complex network of visual, mathematical, and discursive references between their measurement data and a simulated dataset they developed for this purpose. I will thus claim that instead of being an intrinsic property of the algorithmic outputs, trustworthiness, in this case, is a product of context-dependent transcriptive processing. In short, this article does not examine if the EHT team were justified in trusting the results of their imaging algorithms but how their justification was processually enacted through media-specific operations.

2 Ground Truths and Epistemic Challenges

With the increasing use of automated computer algorithms for knowledge production, much has been written lately about the algorithms' potentially problematic epistemic opacity (Duede, 2022; Ferreira, 2021; Humphreys, 2009; Ziewitz, 2016). Since the computational operations through which such algorithms map the input onto the output data are essentially inscrutable to humans, a scientist cannot determine how an algorithm solved a particular task and if the underlying operations were adequate. To circumvent the problem of opacity, scientists quantitatively assess how well an algorithm performs the task it was designed for by examining the accuracy of its outputs (Jaton, 2021).

Several recent studies have foregrounded the central role of what is called *ground truths* in delimiting and assessing the performance of automated algorithms (Henriksen & Bechmann, 2020; Jaton, 2017, 2021). In the context of computer algorithms, ground truth denotes an external referential dataset that, by relating input data to pre-defined output targets, establishes a "yardstick" with which a new algorithm's performance is both shaped—by setting its parameters—and evaluated (Jaton, 2021, 4). As shown by Jaton (2017), what constitutes an adequate ground truth in a particular context depends on the task a new algorithm was designed for but also, as we will see shortly, on the characteristics of the intended input data. For this reason, although often derived from real-world data (Henriksen & Bechmann, 2020),[2] adequate ground truth datasets must be specifically constructed for a new algorithm, the tasks at hand, and the particular data type. Thus, before we can analyse the media-specific operations through which the EHT team designed and used their ground truth dataset in order to, as I will claim, establish the trustworthiness of their algorithmic results, we must examine the challenges their image reconstruction faced.

The first major challenge lay in the EHT measurements' intrinsic characteristics. To generate the data, the team deployed very long baseline interferometry (VLBI). This method entails a GPS-based synchronisation of multiple widely spaced telescopes that simultaneously record radio signals in the same frequency band emitted by the astronomical object of interest (EHTC, 2019b). An interferometric dataset is obtained through pairwise superposition of the signals recorded by individual telescopes and consists of spatial frequencies or Fourier components of the target object's brightness distribution in the sky (EHTC, 2019c). In VLBI, the interferometric data are called visibilities. In standard VLBI, which has been used in astronomy since the 1960s, a spatially resolved image is algorithmically reconstructed from such visibilities through well-tested methods (EHTC, 2019d, 4). However, the EHT team used a new telescope configuration tailored to observing M87* and recorded signals at a lower wavelength than in standard VLBI. Consequently, their data had novel characteristics and could not be unproblematically translated into images through the standard VLBI algorithm called CLEAN. Not only were the EHT data much noisier than standard VLBI data (EHTC, 2019c, 1–2), but because very few telescopes were suitable for the measurement, the EHT data were extremely sparse. This meant that the inverse problem of reconstructing an image from this data was ill-posed and required "information, assumptions, or constraints beyond the interferometer measurements" (EHTC, 2019d, 4).

To minimise the reliance on potentially biasing expert judgments in resolving these issues,[3] the EHT team built three imaging pipelines: DIFMAP, SMILI, and eht-imaging (EHTC, 2019d, 4–5). One pipeline used the classical CLEAN algorithm adapted to deal with the characteristics of the EHT data. The other two used the new image reconstruction algorithms (Akiyama et al., 2017; Chael et al., 2018). These were developed specifically for the EHT by deploying a statistical modelling approach less commonly used in VLBI (EHTC, 2019d, 4). Whereas CLEAN is based on inverse modelling, the other two algorithms take a forward-modelling approach called the regularised maximum likelihood method (EHTC, 2019d, 4–5). Each pipeline was designed to perform multiple iterative cycles of image generation followed by self-calibration, during which the trial images were fitted to the data, before computing the final image reconstruction (EHTC, 2019d, 12–14). Whereas

---

[2] Initially, the term ground truth was introduced in the 1960s in remote sensing to designate "the information obtained by direct measurement at ground level," which serves as a benchmark for verifying remotely obtained data through comparison (Woodhouse, 2021).

[3] On the vital role of human judgment in the preceding phase of 'blind imaging,' which I will not discuss here, see Muhr (2024).

each pipeline had "some choices that are fixed," many of the statistical models' parameters could take a range of values, thus influencing the pipeline's performance (EHTC, 2019d, 11). Hence, another major challenge was that, with numerous parameter values, the pipelines were expected to deliver untrustworthy results, i.e., "produce poor [image] reconstructions" (EHTC, 2019d, 11).

It should be noted that even before acquiring the EHT data, the team organised a series of "imaging challenges" to assess how the algorithms they were developing performed on test data derived from heterogeneous ground truth images ranging from GRMHD simulations to astronomical and real-world pictures (Bouman, 2017, 133–184). These challenges allowed the researchers to "better understand each of the algorithms' strengths and weaknesses, and even develop stronger methods" (Bouman, 2017, 133), including the decision to script the pipelines. But the challenges did not allow them to determine the parameter values that could be generalised to the EHT data. To identify the parameters with which their three algorithmic pipelines could be trusted to reconstruct physically credible and empirically authentic images of M87* from the EHT measurements—and to justify these choices—the team needed an adequately tailored ground truth dataset. But since "there are no previous VLBI images of any source at 1.3 mm wavelength, and there are no comparable black hole images on event-horizon scales at any wavelength" (EHTC, 2019d, 9), the researchers lacked real-world data they could use as their ground truth. To solve this problem, they reverted to using *simulated* data.

3 Constructing the Ground Truth Dataset

Before creating their ground truth dataset, the EHT team first had to define what it meant for their algorithmic pipelines to perform sufficiently well to be trustworthy and how to measure the performance quality. In their definition, the pipelines performed sufficiently well if they produced quantifiably "faithful image reconstructions" of M87* from the sparse and noisy interferometric data acquired on four nights in April 2017 (EHTC, 2019d, 10). To be regarded as faithful, an algorithmically reconstructed image had to satisfy two criteria. First, the image had to be consistent with the data from which it was computed (EHTC, 2019d, 15). However, since the EHT measurements were extremely sparse, each algorithm had to statistically infer the missing data from the existing noisy data to reconstruct a complete image. Depending on the choice of its parameters, while 'filling in' the missing data, a pipeline could introduce spurious artefacts, compromising the reconstructed image's fidelity. Thus, the second criterion of fidelity was that the uncertainty introduced by the pipeline's free parameter choices should not exceed the intrinsic measurement uncertainty of the EHT data (EHTC, 2019d, 21–22). Whereas the first criterion could be assessed by comparing the reconstructed images to the EHT data, testing the second criterion required a specifically tailored ground truth dataset. Drawing on Jäger (2010), in what follows, I will show how the construction and subsequent deployment of the ground truth dataset, which, in turn, established the trustworthiness of the algorithmic results, hinged on targeted transcriptive operations.

To enable the researchers to assess their second criterion of fidelity, the ground truth dataset had to contain two major components. The first component consisted of the simulated interferometric ground truth data that could be fed to the pipelines during parameter surveys to obtain various image reconstructions with different combinations of free parameters. The second component consisted of the known simulated target images—i.e., ground truth images—to which the algorithmic image reconstructions could be compared to quantify the accuracy of each reconstructed image and thus determine the best performing parameter combinations. The researchers first focused on creating the simulated target images and then derived the simulated visibilities from these images. Understanding the individual steps in this process is crucial for our discussion.

To begin with, the team chose four simple geometric models as their ground truth images. All four models had a compact structure and lacked fine-scale features, yet their global morphologies differed. These were: (1) a ring of 44 μas diameter; (2) a crescent of the same diameter with enhanced brightness in the south; (3) a full disk of 70 μas diameter; and (4) two asymmetrically positioned circular components (EHTC, 2019d, 10) (Fig. 2). Importantly, the researchers selected these geometric models because the amplitudes of their Fourier components—the spatial frequencies into which each model could be mathematically decomposed via a Fourier transform—were similar to the amplitudes of the actually observed M87* data (Fig. 3). The salient similarity between the models' and the actual data's Fourier components included: "(1) a large decrease in flux density on baselines between 0 and 1 Gλ, indicating extended structure, (2) visibility nulls at ~3.4 and ~8.3 Gλ, and (3) a high secondary peak between the nulls at ~6 Gλ" (EHTC, 2019d, 10). Put simply, the four model images were designed to replicate the salient characteristics of the EHT data closely enough that each of these images could be considered a potentially plausible reconstruction of the actual sparse data. Hence, the ground truth images were created to adequately simulate the unknown image of the black hole that the researchers aimed to reconstruct from their data.

Two aspects are thereby of particular interest. First, it transpires from the above that establishing the adequacy of the four geometric models as ground truth images relied on a transcriptive operation (Jäger, 2010). By setting up a mathematically defined referential relation of salient similarity between the geometric models and the non-visual EHT data, the researchers justified their choice of these geometric models as their ground truth images. In other words, through such intermedial transcription, the choice of particular ground truth images was based not on their visual but on their underlying non-visual components.

Second, it was epistemically significant that the team decided to use not one but four different geometric models, none of which, despite their salient similarity, "reproduces all features seen in the M87 data" (EHTC, 2019d, 10). In doing so, the team defined a range of possible outcome targets for their algorithmic pipelines instead of imposing a single pre-defined imaging solution. Their aim thereby was to avoid biasing the subsequent image reconstruction through prior expectations. Thus,

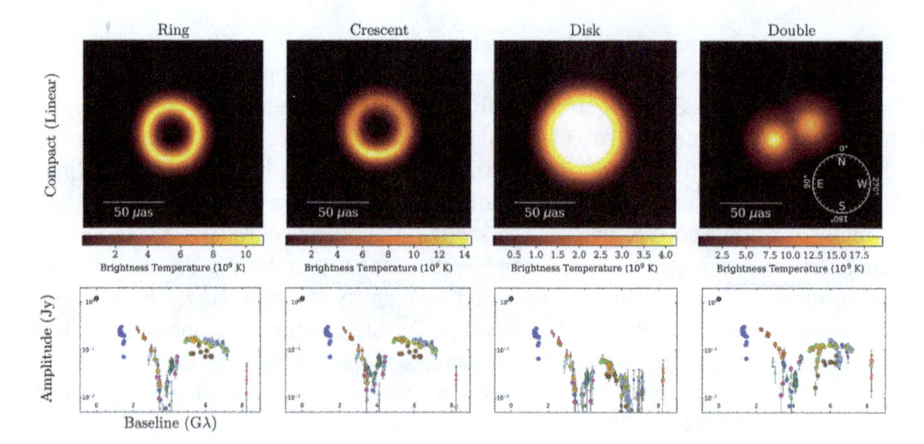

Fig. 2 Four simple geometric models used as ground truth images (*upper row*) and interferometric data generated from these models to simulate the EHT data from April 11, 2017 (*lower row*) (From EHTC, 2019d)

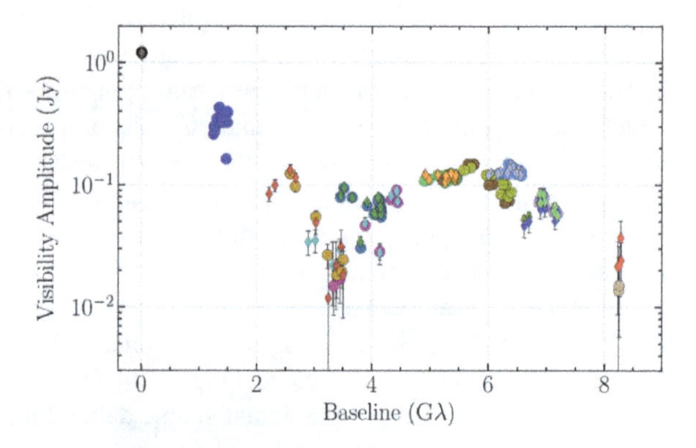

Fig. 3 Interferometric data of M87* for April 11, 2017 after amplitude and network calibration (From EHTC, 2019d)

two of the four geometric models showed a bright shape with a circular shadow in its centre, which according to general relativity, the reconstructed EHT images should look like (EHTC, 2019d, 1). The other two geometric shapes that did not fit the theoretical predictions were meant to safeguard the researchers from unduly constraining the possible outcomes of the algorithmic image reconstruction through their expectations. Also in this case, the adequacy of the team's choices was justified through transcriptive operations. These entailed setting up the intramedial relations of visual differences among the individual geometric models and the intermedial relations of selective dissimilarity between the geometric models and the M87* data. Just as importantly, the additional intermedial relations of similarity and difference were transcriptively established between the ground truth images and theoretical predictions about black holes.

Having constructed the ground truth images and justified their adequacy through the transcriptive operations, the team used these images to generate the simulated ground truth interferometric data. First, they deployed one of their three algorithmic pipelines to compute the Fourier transform of each model image (EHTC, 2019d, 38). The resulting simulated visibilities were generated with the same sparsity of coverage and sensitivity as the actual M87* data. However, the thus obtained simulated data were uncorrupted, whereas the M87* data were riddled with noise. Previously, the researchers had analysed, quantified, and mathematically modelled the different types of noise in their M87* data, which stemmed from various sources of measurement errors (EHTC, 2019d, 7–8). Apart from the well-understood thermal noise, other more insidious sources of systematic uncertainty included station-based errors from each of the seven telescopes with which the M87* data were collected. The station-based errors had a time-varying component that fluctuated with changing atmospheric conditions both during each measurement and across measurements taken on different days.

Next, the researchers drew on their analysis of noise patterns in the EHT data to computationally add corresponding noise patterns to the simulated interferometric data. The statistical modelling and inclusion of the time-varying noise components were crucial since the aim of the simulation was to create "realistic" ground truth visibilities that closely reproduced the salient features of the M87* data, including their errors (EHTC, 2019d, 7). To account for the changing atmospheric conditions across the measurement days, the team added a different pattern of simulated noise derived from the actual data to the ground truth visibilities for each measurement day. Thus, to construct 'realistic' data simulations, it was necessary to establish the complex statistically defined transcriptive relations of similarity between the simulated and actual data.

Finally, the EHT team performed another transcriptive operation to verify that the mathematically modelled noise in the simulated data was not just similar to the noise in the actual data but also that the modelled noise was physically plausible. For this purpose, they fed one of their model images into another algorithmic pipeline to generate simulated interferometric data. This pipeline used "a more physical approach" by adding noise based on actually "measured weather parameters and antenna pointing offsets" (EHTC, 2019d, 11) instead of statistical analysis of the M87* data. The comparison of the simulated data generated from the same model image by the two different pipelines showed that both datasets were broadly consistent (EHTC, 2019d, 40). Owing to this intramedial transcription, the researchers could argue that their initially simulated interferometric data were 'realistic' in a double sense—not only did their statistical characteristics match the M87* data, but they were also physically plausible simulations of the actual measurement.

In sum, I have shown that to synthesise the ground truth images and to compute the adequately simulated interferometric data from these images, the EHT team had to forge a network of intramedial and intermedial transcriptive references across different images, actual and synthetic non-visual datasets, statistical models, noise patterns, and theoretical predictions. Besides the direct source-output relation between the simulated model images and the simulated interferometric data, the researchers also established the relations of salient similarity between the simulated and measured data. In the latter case, this included an elaborate mathematical and physically plausible replication of the original data's sparsity and specific time-varying noise patterns. It is through these meticulously organised transcriptive operations that the four model images and the corresponding simulated interferometric data acquired the epistemic status of the adequate ground truth in our case study.

4 Deploying the Ground Truth Dataset

Once adequately constructed, the ground truth dataset could then be deployed to assess how well and under which conditions the novel image reconstruction pipelines performed their task, thus delivering the epistemic evidence for the trustworthiness of the algorithmic outputs. This process entailed two stages. First, we will examine how the team relied on the ground truth to determine the combination of parameters that, for each pipeline, resulted in quantifiably faithful image reconstructions from the M87* data. Second, we will discuss how the researchers subsequently used their ground truth to quantify the uncertainties of select visual features in the final image reconstructions. As I will show, both stages entailed a range of transcriptive operations.

To begin with, it was during the image reconstruction that the ground truth dataset had a crucial epistemic function since the EHT team used it as a benchmark for performing parameter selection and thus shaping their pipelines' performances. With it, they explored and quantitatively assessed how the performance quality of their pipelines varied across different choices of free parameters. For each pipeline, up to tens of thousands of parameter combinations were tested separately on the simulated and the actual interferometric data (EHTC, 2019d, 17).[4] These parameter surveys resulted in image reconstructions whose different fidelity levels were quantified through multiple comparisons that established a complex web of transcriptive relations.

Applying their first criterion of image reconstruction fidelity, the team transcriptively related the reconstructed images to the EHT data. Using the chi-squared statistical test to quantitatively assess the fit between the reconstructed images and the EHT data, they identified the parameter combinations for each pipeline that produced images consistent with the data (EHTC, 2019d, 15). Next, they narrowed down these parameter combinations by deploying the second criterion of image reconstruction fidelity. To do so, the team tested with which of the surviving parameter combinations the algorithms adequately statistically inferred the missing data from the sparse, noisy data by comparing the images reconstructed from the simulated interferometric data to the corresponding ground truth images (EHTC, 2019d, 15–17). To establish the transcriptive relation of similarity between these two sets of images, the researchers

[4] As the EHT emphasised, "the ensemble of results from these parameter surveys do not correspond to a posterior distribution of reconstructed images. Our surveys are coarse-grained and do not completely explore the choices in the imaging process" (EHTC, 2019d, 10).

computed the normalised cross-correlation between the voxels of each reconstructed image and the corresponding ground truth images.

Based on this statistical measure, the different parameter combinations were ranked according to the quality of their performance. Next, a quantitative cut-off value was defined below which the image reconstructions were no longer acceptable (EHTC, 2019d, 17). All parameter combinations—for each algorithm separately—that survived this threshold were designated as the "Top Set" parameters. These resulted in "acceptable" image reconstructions as they fulfilled both fidelity criteria (EHTC, 2019d, 14). First, when applied to the EHT data, the Top Set parameters produced images of M87* that were consistent with the data. Second, when applied to the simulated data, they produced images that were sufficiently similar to the corresponding ground truth images.

However, the Top Set contained between 30 and 1572 parameter combinations per pipeline (EHTC, 2019d, 15). Moreover, the quality of each pipeline's performance with a particular parameter combination varied across the simulated interferometric data from the different ground truth images. Through an additional transcriptive comparison, the team then identified a single parameter combination from each pipeline's Top Set with which the particular pipeline uniformly produced the most faithful reconstructions across the simulated interferometric data from all four ground truth images. Termed "fiducial [i.e., trustworthy] parameters," this single parameter combination for each pipeline delivered the best average performance on all simulated data (EHTC, 2019d, 17). And because they performed uniformly and measurably well on all ground truth interferometric data—which, as shown earlier, were constructed to simulate the EHT data's salient features—the fiducial parameters were also deemed to deliver faithful image reconstructions from the actual EHT data.

To further reinforce the fiducial parameters' trustworthiness, the researchers performed two more transcriptions. Using synthetic data generated from a snapshot of a GRMHD model of an ideal black hole, they showed that the fiducial parameters delivered faithful image reconstructions also from data not used for determining these parameters (EHTC, 2019d, 16). Moreover, by evaluating the parameter combination for each pipeline that performed best on three of the four ground truth datasets, the researchers used it to reconstruct images from the fourth dataset that was withheld from this evaluation. For all three pipelines, the resulting reconstructions were faithful to the respective ground truth images and did not resemble "any of the other images in the training set" (EHTC, 2019d, 17). Importantly, the latter transcription demonstrated that the ground truth images were not biasing the algorithms' performances.

In sum, based on the evidence generated by transcriptively deploying the ground truth dataset, the EHT team could claim that, when used with the corresponding fiducial parameters, their novel algorithmic pipelines could be trusted to produce faithful image reconstructions from their measurement data. I have shown that this evidence was gradually constructed by systematically forging a network of multidirectional, dynamic transcriptive references across different sets of images and data. First, the statistically defined relation of consistency was established between the algorithmic image reconstructions and the EHT data. This was followed by quantitatively establishing the intramedial relations of sufficient visual similarity between the reconstructed images and the corresponding ground truth images. The final step entailed creating the comparative relations across the different segments of the ground truth dataset to identify the fiducial parameters. All these complex and, in part, mutually intertwined media-specific operations and the resulting transcriptive references were necessary to determine and empirically demonstrate with which parameter combinations the pipelines could trustworthily perform the task for which they were designed.

After determining the respective fiducial parameters, the EHT team used these to compute the "fiducial [i.e., trustworthy] images" of M87* (EHTC, 2019d, 18). They did so separately for each pipeline and each measurement day, thus obtaining 12 fiducial images (Fig. 4). Despite some variations in peak intensities and other visual details, all 12 images had a "prominent, asymmetric ring feature, approximately 40 μas in diameter with enhanced brightness toward the south" and a dark circular shadow in the centre (EHTC, 2019d, 18–19). Finally, to generate a single conservative image for each measurement day, the researchers averaged the three fiducial images obtained for that day by the different pipelines after blurring them to a joint resolution (see Fig. 1).

Having computed the pipeline-averaged images, the researchers then returned to their ground truth dataset. This time their aim was to assess the levels of uncertainty

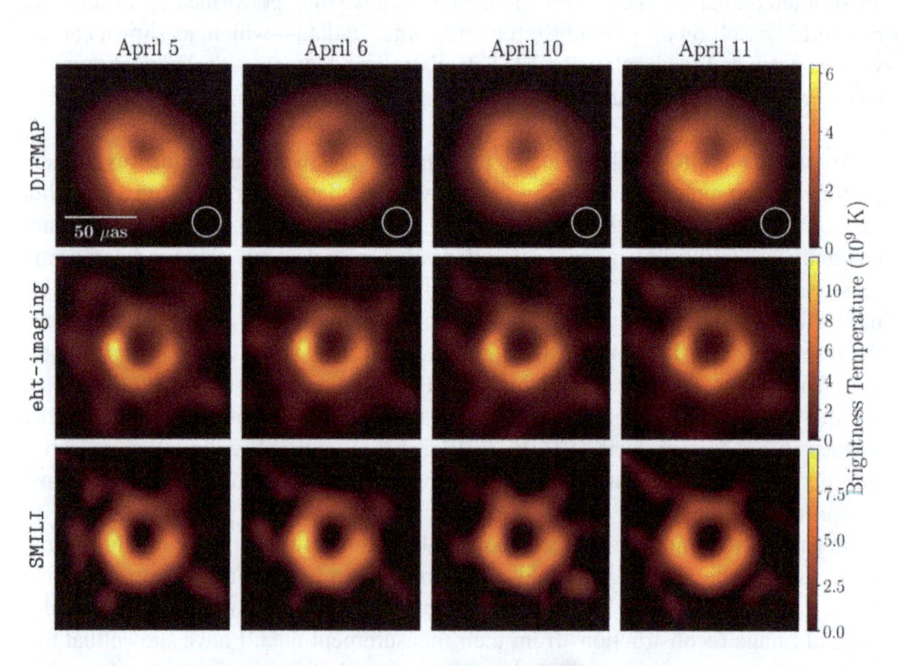

Fig. 4 Fiducial (i.e., trustworthy) images of M87* separately reconstructed by three independent algorithmic pipelines for four measurement days (From EHTC, 2019d)

with which different visual features of interest in their fiducial images were reconstructed. To this end, they first identified the five salient visual features of the asymmetric ring in the reconstructed images. These included the ring's diameter, width, orientation angle, azimuthal brightness asymmetry, and fractional central brightness (EHTC, 2019d, 26). After defining the mathematical procedures for measuring the uncertainty of these features, the team applied them to a subset of the ground truth dataset from the crescent model image.

Through these procedures, the researchers forged multiple targeted transcriptive relations between the simulated interferometric data, different algorithmic image reconstructions derived from these data, and the crescent ground truth image. First, they quantified and compared how each of the five salient ring features varied between the images reconstructed from the simulated data with the fiducial parameters and the ground truth image (EHTC, 2019d, 27–28). Then they calculated how much the measurement of each salient feature varied across the images reconstructed with different Top Set parameters. These two transcriptive operations allowed the team to disentangle "two distinct sources of error: the intrinsic measurement uncertainty on each quantity from a single image, and the uncertainty in the quantity from varying imaging parameters across the Top Set" (EHTC, 2019d, 27). The researchers thus established that for all the imaging pipelines, the ring diameter was the visual structure in their fiducial images that was reconstructed with the lowest statistical uncertainty since the variation in the choice of parameters across the Top Set did not significantly affect its value. However, they also determined that the reconstructions of the other four salient ring features were less certain since their error budgets were "more evenly divided between intrinsic uncertainty in a single image and the scatter across the Top Set" (EHTC, 2019d, 28).

Hence, although the simulated ground truth dataset enabled the researchers to demonstrate through transcriptive procedures that their novel algorithmic pipelines produced trustworthy fiducial images of M87* on the whole (i.e., the general ring morphology), not all visual details of the resulting ring structure were reconstructed with equal trustworthiness. We have also seen that the subsequent transcriptive deployment of the simulated ground truth allowed the team to quantify the varying levels of uncertainty with which the fiducial images' different visual structures were reconstructed. They thus identified which ring features were relationally 'more trustworthy' and could be justifiably used to estimate M87*'s physical properties. In effect, it can be argued that not just the overall trustworthiness of the fiducial and pipeline-averaged images but also the varying levels of trustworthiness of these images' different visual features were direct semantic effects of complex, context-dependent transcriptive procedures.

5 Conclusion

This article has shown that one of the crucial aspects of producing the first empirical images of a black hole consisted in establishing epistemic trust in the performance of the new algorithmic pipelines, without whose mediation the translation of the measurements into empirically authentic images would not have been possible. In this situation of epistemic dependence, the EHT team had to produce sufficient evidence that their epistemically opaque algorithmic pipelines could compute unknown images from the sparse and noisy data without introducing spurious reconstruction artefacts that would have jeopardised the images' ability to be used for generating new scientific insights into black holes. I have shown that to produce such evidence, the researchers relied on a ground truth dataset they constructed from simulated model images and simulated interferometric data. This specifically tailored simulated dataset played a crucial epistemic function as it provided an external referential framework with which the EHT team could assess and constrain the performance of their algorithmic pipelines through scripted parameter selection.

I have also demonstrated that the construction and deployment of this ground truth dataset hinged on the targeted establishment of a complex network of transcriptive references (Jäger, 2010) of salient similarity, selective dissimilarity, and image-data consistency. Often multidirectional and mutually interwoven, such references were systematically forged between the actual and simulated interferometric data, between image reconstructions and pre-defined model images, across non-visual data and different images, and among empirical measurements, mathematical models, and theoretical predictions. Jointly, these transcriptive operations enabled the EHT team to produce evidence that justified the trustworthiness of the algorithmic results obtained with the fiducial parameters and assess the different levels of statistical uncertainty of the various visual details in the final reconstructed images. However, instead of being normative, the epistemic trust grounded in the evidence generated through such transcriptive operations is context-specific and relational. This trust is contingent on the quality and stability of the transcriptive relations that justified it and is limited to the algorithms' performances with the fiducial parameters on the particular M87* data to which the simulated ground truth dataset and the respective transcriptive operations were tailored. Therefore, for each new dataset, a different image reconstruction task, or a new algorithmic pipeline, the context-specific transcriptive processing and the epistemic trust that emerges as its semantic effect must be enacted anew.

References

Akiyama, K., Kuramochi, K., Ikeda, S., Fish, V. L., Tazaki, F., Honma, M., Doeleman, S. S., et al. (2017). Imaging the Schwarzschild-radius-scale structure of M87 with the Event Horizon Telescope using sparse modeling. *The Astrophysical Journal, 838*, 1. https://doi.org/10.3847/1538-4357/aa6305

Baumberger, C., Knutti, R., & Hirsch Hadorn, G. (2017). Building confidence in climate model projections: An analysis of inferences from fit. *WIREs Climate Change, 8*, e454. https://doi.org/10.1002/wcc.454

Bouman, K. L. (2017). *Extreme imaging via physical model inversion: Seeing around corners and imaging black holes*. PhD Diss., Massachusetts Institute of Technology.

Chael, A. A., Johnson, M. D., Bouman, K. L., Blackburn, L. L., Akiyama, K., & Narayan, R. (2018). Interferometric imaging directly with closure phases and closure amplitudes. *The Astrophysical Journal, 838*, 23. https://doi.org/10.3847/1538-4357/aab6a8

Duede, E. (2022). Instruments, agents, and artificial intelligence: Novel epistemic categories of reliability. *Synthese, 200*, 491. https://doi.org/10.1007/s11229-022-03975-6

Durán, J. M., & Formanek, N. (2018). Grounds for trust: Essential epistemic opacity and computational reliabilism. *Minds and Machines, 28*, 645–666. https://doi.org/10.1007/s11023-018-9481-6

Event Horizon Telescope Collaboration (EHTC). (2019a). First M87 Event Horizon Telescope results. I. The shadow of the supermassive black hole. *The Astrophysical Journal Letters, 875*, L1: 1–17. https://doi.org/10.3847/2041-8213/ab0ec7

Event Horizon Telescope Collaboration (EHTC). (2019b). First M87 Event Horizon Telescope results. II. Array and instrumentation. *The Astrophysical Journal Letters, 875*, L2: 1–28. https://doi.org/10.3847/2041-8213/ab0c96

Event Horizon Telescope Collaboration (EHTC). (2019c). First M87 Event Horizon Telescope results. III. Data processing and calibration, *The Astrophysical Journal Letters, 875*, L3: 1–32. https://doi.org/10.3847/2041-8213/ab0c57

Event Horizon Telescope Collaboration (EHTC). (2019d). First M87 Event Horizon Telescope results. IV. Imaging the central supermassive black hole. *The Astrophysical Journal Letters, 875*, L4: 1–52. https://doi.org/10.3847/2041-8213/ab0e85

Event Horizon Telescope Collaboration (EHTC). (2019e). First M87 Event Horizon Telescope Results. V. Physical origin of the asymmetric ring. *The Astrophysical Journal Letters, 875*, L5: 1–31. https://doi.org/10.3847/2041-8213/ab0f43

Event Horizon Telescope Collaboration (EHTC). (2019f). First M87 Event Horizon Telescope Results. VI. The shadow and mass of the central black hole. *The Astrophysical Journal Letters, 875*, L6: 1–44. https://doi.org/10.3847/2041-8213/ab1141.

Falcke, H. (2017). Imaging black holes: Past, present and future. *Journal of Physics: Conference Series, 942*. https://doi.org/10.1088/1742-6596/942/1/012001

Ferreira, M. (2021). Inscrutable processes: Algorithms, agency, and divisions of deliberative labour. *Journal of Applied Philosophy, 38*, 646–661. https://doi.org/10.1111/japp.12496

Goddi, C., Crew, G., Impellizzeri, V., Martí-Vidal, I., Matthews, L. D., Messias, H., Rottmann, H., et al. (2019). First M87 Event Horizon Telescope results and the role of ALMA. *The Messenger, 177*, 25–35. https://doi.org/10.18727/0722-6691/5150

Henriksen, A., & Bechmann, A. (2020). Building truths in AI: Making predictive algorithms doable in healthcare. *Information, Communication & Society, 23*, 802–816. https://doi.org/10.1080/1369118x.2020.1751866

Hubig, C., & Kaminski, A. (2017). Outlines of a pragmatic theory of truth and error in computer simulation. In M. Resch, A. Kaminski, & P. Gehring (Eds.), *The science and art of simulation I* (pp. 121–136). Springer.

Humphreys, P. (2009). The philosophical novelty of computer simulation methods. *Synthese, 169*, 615–626. https://doi.org/10.1007/s11229-008-9435-2

Jäger, L. (2006). Schauplätze der Evidenz. Evidenzverfahren und kulturelle Semantik. Eine Skizze. In M. Cuntz Barbara Nitsche, I. Otto, & M. Spaniol (Eds.), *Die Listen der Evidenz* (pp. 37–52). DuMont.

Jäger, L. (2010). Transcriptivity matters: On the logic of intra- and intermedial references in aesthetic discourse. In L. Jäger, E. Linz, & I. Schneider (Eds.), B. Pichon & D. Rudnytsky (Trans.), *Media, culture, and mediality: New Insights into the current state of research* (pp. 49–76). Transcript.

Jaton, F. (2021). Assessing biases, relaxing moralism: On ground-truthing practices in machine learning design and application. *Big Data & Society, 8*, 1–15. https://doi.org/10.1177/20539517211013569

Jaton, F. (2017). We get the algorithms of our ground truths: Designing referential databases in digital image processing. *Social Studies of Science, 47*, 811–840. https://doi.org/10.1177/0306312717730428

Latour, B. (1999). *Pandora's Hope: Essays on the reality of science studies*. Harvard University Press.

Muhr, P. (2023). The "cartographic impulse" and its epistemic gains in the process of iteratively mapping M87's black hole. *Media+Environment, 5*. https://doi.org/10.1525/001c.88163

Muhr, P. (2024). "What We Thought Was Unseeable": Die mediale Konstruktion der ersten authentischen empirischen Bilder eines Schwarzen Lochs. In A. Bahr & G. Fröhlich (Eds.), *"Ain't Nothing Like the Real Thing?": Formen und Funktionen medialer Artefakt-Authentifizierung* (pp. 19–49). Transcript.

Rolin, K. (2020). Trust in science. In J. Simon (Ed.), *The Routledge handbook of trust and philosophy* (pp. 354–366). Routledge.

Starke, G., van der Brule, R., Elger, B. S., & Haselager, P. (2021). Intentional machines: A defence of trust in medical artificial intelligence. *Bioethics, 36*, 154–161. https://doi.org/10.1111/bioe.12891

Symons, J., & Alvarado, R. (2016). Can we trust Big Data? Applying philosophy of science to software. *Big Data & Society, 3*. https://doi.org/10.1177/2053951716664747

Symons, J., & Alvarado, R. (2019). Epistemic entitlements and the practice of computer simulation. *Minds & Machines, 29*, 37–60. https://doi.org/10.1007/s11023-018-9487-0

Woodhouse, I. H. (2021). On "ground" truth and why we should abandon the term. *Journal of Applied Remote Sensing, 15*. https://doi.org/10.1117/1.jrs.15.041501

Ziewitz, M. (2016). Governing algorithms: Myth, mess, and methods. *Science, Technology & Human Values, 41*, 3–16. https://doi.org/10.1177/0162243915608948

Trust, Primary Source Knowledge, and Science Communication in the Internet Era: The Case of Mainstream Climate Blogging

George Zoukas

Abstract Much of the scholarly discourse on the relation between science and society concerns the potential of fostering public trust in science, especially regarding socially salient fields of science. In this article, to provide insight into the association between trust and science in the Internet era, I focus on the communication of climate science and climate change through the blogging platform. Following an interview-based qualitative case study approach, and using the concept of "primary source knowledge", initiated in the sociology of scientific knowledge by Harry Collins and Robert Evans in 2007, I examine the character of mainstream scientist-produced climate blogs and the knowledge communicated through them. I suggest that "mainstream climate blogging" is an authoritative and trustworthy niche of science communication, an example of direct communication between mainstream scientists and an interested and dedicated audience willing to obtain a deeper understanding of science.

1 Introduction

Much of the recent scholarly discussion about the relationship between science and society is aimed at understanding the public's attitudes towards scientists and the institution of science as a whole, especially in view of the past two years' Covid-19 pandemic outbreak (e.g. Agley, 2020; Kreps & Kriner, 2020). Factors relating to trust in governments, communities, and scientists seem to play an important role in how individuals react to recommendations about public health (Agley, 2020). Meanwhile, communicating science "carefully" is generally considered an essential way to maintain and raise public support for scientifically grounded policies (Kreps & Kriner, 2020). However, distrust of scientists with respect to public health is not only evident but also analogous to public responses to issues that arise within other

G. Zoukas (✉)
Department of History and Philosophy of Science, National and Kapodistrian University of Athens, Athens, Greece
e-mail: gzoukas@phs.uoa.gr

M. M. Resch et al. (eds.), *The Science and Art of Simulation*,
https://doi.org/10.1007/978-3-031-68058-8_14

science-related fields, such as issues concerning the environment and, most notably, climate change (Hamilton & Safford, 2021; Sarathchandra & Haltinner, 2020).

Trust *within* science generally means that trust is vital to a variety of activities and interactions that are, in one way or another, integral to scientific research, from peer reviewing, data sharing, replicating research results, and recruiting research participants to facilitating communications between scientists and universities, funding agencies, scientific journals, and other relevant organizations and institutions (Resnik, 2011). Public trust *in* science typically refers to the trust that the public, or society, puts in scientific research (Resnik, 2011); or alternatively, the extent to which one, despite their limited understanding of science, is confident that science and scientists offer reliable evidence and knowledge (Wintterlin et al., 2022).

Researching, and conceptually analyzing, public trust in science is a challenging endeavour for many reasons, including not only the impossibility of describing "the public" as a homogeneous entity but also the various ways the meaning of scientific expertise is experienced, perceived, and constructed within different social settings (Wynne, 1995). Notwithstanding, enhancing the communication between expert scientists and the public and engaging the public with science and scientific research have long since been regarded as efficient methods of restoring public trust in science (Haerlin & Parr, 1999; Wynne, 2006).

According to Oreskes and Conway (2010), a large body of sociological, anthropological, and communication studies has been concerned with how direct communication between scientists and the public, as well as between scientists and the press or policy-makers, could be improved. Using the example of climate change, the same authors have suggested that expanding science outreach could be a positive step towards rebutting "organized, sophisticated and persistent" efforts to undermine science (Oreskes & Conway, 2010: 687). Certainly, attempts to subvert science have been expressed increasingly through different types of media, both traditional and online (Cook, 2022; Treen et al., 2020). This is so especially with respect to issues relating to climate change, which, similarly to those about public health, stem cells, or nuclear energy, to name but a few, have a "socio-scientific" character, as they permeate considerably people's everyday lives (Hendriks et al., 2016). Even though a complete public understanding of science might not be feasible, a better understanding of some of the scientific principles that underlie such issues could be (Hendriks et al., 2016).

Nowadays, the potential of immediately accessing different types of online sources to be informed about climate change, as well as other science-related areas, has increased; yet so does the need to evaluate such sources and ultimately trust science and scientists (Hendriks et al., 2016). Social media platforms are probably some of the most obvious, and indeed popular, examples of online communication. Based on recent studies (Huber et al., 2019; Xiao et al., 2021), the relationship between social media use and trust in science and scientists can be considered two-sided. For instance, while Huber et al. (2019) suggest that social media can play a rather positive role in cultivating public trust in science, Xiao et al. (2021) argue that some of the dangers associated with social media use should not be overlooked, particularly the high possibility of misinforming and disinforming[1] the public.

In this article, to provide insight into the interrelation between science communication and public trust in science in the Internet era, I have drawn upon the Sociology of Scientific Knowledge (SSK), specifically, the concept of "primary source knowledge" (Collins, 2014, 2018; Collins & Evans, 2007; Collins et al., 2017). By looking at the example of blogs, one of the oldest and most well-received social media platforms, and concentrating on what I describe as "mainstream climate blogging", I explore the nature of the knowledge communicated through a selected group of scientist-produced blogs about climate science and climate change. My analysis focuses on the experiences and perceptions of both the scientist-bloggers and the non-expert readers, putting a special emphasis on how the readers feel about the communicated knowledge. Before I illustrate the methodology I followed and my case study, I present an overview of the climate change issue and the notion of communicating science online, specifically through the blogging platform.

2 The Climate Change Issue

An inherently complicated and multifaceted natural phenomenon, climate includes the interrelationship and interaction among the five main components that constitute the climate system, that is, the atmosphere, the ocean, the cryosphere, the land surface, and the biosphere, as well as various climatic and, often extreme, weather events, such as droughts, floods, hurricanes, and tornadoes (Goosse, 2015; Schmittner, 2021). A sub-discipline of the environmental sciences, climate science, or climatology, concentrates on the study of climate through the combination of different fields of research, from atmospheric chemistry and physics to paleoclimatology, ecology, and computer science, to mention but only a few, and involves the study of a changing climate (Farmer, 2015).

According to many relevant science textbooks (e.g. Eggleton, 2013; Houghton, 2004), human-induced increases in greenhouse gasses, especially carbon dioxide (CO_2), have been occurring since the eighteenth century's Industrial Revolution, resulting in global warming, a rapid growth in the Earth's average surface temperature. Global warming causes in turn climate change, which is briefly described as variations in atmospheric and oceanic behaviour (Eggleton, 2013). Given that such variations happen at unprecedented rates, they create environmental damages with adverse health, economic, and social impacts (Houston & Capalbo, 2021).

Since the development of the notion of "dangerous climate change" within the context of the United Nations Framework Convention on Climate Change

[1] The terms "misinformation" and "disinformation" differ as to why false information about a subject is provided. While misinformation is information presented initially as truthful, turning out later on to be incorrect though, disinformation typically means the deliberately and propagandistically disseminated false information (Lewandowsky et al., 2013).

(UNFCCC) in 1992, it has become increasingly evident that the climate change issue reaches beyond the natural sciences (Lorenzoni et al., 2005; Oppenheimer, 2005). As such, different studies have focused on its political and economic implications (e.g. Giddens, 2009; MacKenzie, 2007), as well as the ethical and social ones (e.g. Gardiner et al., 2010; Jasanoff, 2011; Nagel et al., 2008).

Despite the overwhelming consensus among scientists on the anthropogenic nature of global warming and climate change (e.g. Cook et al., 2013; Oreskes, 2004), the issue remains contentious on a social, political, and economic level (Eggleton, 2013; Hulme, 2009). What is typically described as "the global warming controversy" or "the climate debate" generally refers to the debate between mainstream scientists, as well as whoever concurs with their evaluations, and those who dispute, doubt, or reject the mainstream scientific assessments[2] (Van Rensburg & Head, 2017), typically undermining the role of human activities in impacting the climate.

Having implications for the natural and the social world, the climate change issue has stimulated widespread interest and concern within society (Kvaløy et al., 2012; Mostafa, 2016), with longstanding discussions and debates taking place in both official and unofficial contexts. The role that the Internet and its technologies play in offering online spaces for people to interact and communicate, through social networking, blogging, and similar platforms, is notable. As discussed below, blogs appear to be one of the most remarkable examples of communicating science online.

3 Science Communication and the Blogging Platform

The term "science communication" is commonly used to describe the communication between scientists and a broader audience, which can take place with or without mediators (Bultitude, 2011). Science communication is a considerably variegated endeavour, revealing and reinforcing different cultural orientations (Medin & Bang, 2014). As such, it can take different forms, from science journalism and a variety of live events like public lectures and science festivals to various interactions happening online (Bultitude, 2011).

Not long after the emergence and expansion of the World Wide Web, its capacity to foster independent and direct online communication models between scientists and a broader audience, transcending the typically journalist-mediated mass communication patterns, has drawn much scholarly attention (e.g. Peters et al., 2014; Weigold, 2001). Arguably, the new possibilities of the Internet and thc Web were first materialized, at least on a large scale, through the blogging platform, the initiation of which in 1997 is often associated with the beginning of the so-called "Social Web" period (Golbeck, 2013).

[2] The terms "climate skeptics", "climate contrarians", and "climate deniers" are often used to refer to those who belong to this latter category.

Defined as web pages which can be modified easily, posted in reverse temporal order, commented on by readers, and which may combine both written and non-written material, such as videos, images, and links to other blogs or websites (McGeehin Heilferty, 2012), blogs (or weblogs) have been used widely during the past two decades or so. Accordingly, their dynamics have been examined within various examples, including different cases of science communication.

The so-called "science blogs" seem to be a distinguishing case of online science communication, with the term "science blogging" generally referring to "writing, sharing, and discussing scientific subject matter online" (Mehlenbacher, 2019: 108). Depending on both the science blogger's identity (scientist, science writer, science enthusiast, science journalist, and so on) and the intended audience, science blogging may relate to activities and practices of communication among scientists or public engagement with science (Mehlenbacher, 2019). Those can vary, for example, from the review and critique of recent scientific studies, the analysis of science-related news, and the explanation of current fields of research interest to the documentation of scientists' experiences and the provision of venues for student writing (Shanahan, 2011).

A complete survey of studies on science blogging which fall within the social science area is beyond the scope of this article (for an extensive review, see Zoukas, 2019). However, it should be pointed out that the analysis of different cases of blogging as a form of online climate communication, in particular, that is, the communication of scientific information and knowledge relating specifically to climate science and climate change (see e.g. Somerville & Hassol, 2011), appears to be considerably limited. Some notable examples include studies focusing on the structure of the so-called "climate blogosphere" and the communication patterns constructed in it (e.g. Elgesem et al., 2015; Sharman, 2014); the character and meaning of the textual content communicated through blog posts (e.g.Fløttum et al., 2014; Poberezhskaya, 2017); or the role of scientist-bloggers as communicators of climate science (e.g. Thorsen, 2013).

Nonetheless, what seems to be missing from the relevant literature is a more "symmetrical", so to speak, examination of the blogging platform and the science communication models unfolding through it; in other words, a study that would take into account the experiences, perspectives, and behaviours of *both* the bloggers (and indeed the scientist-bloggers) and the readers. Such an approach is essential when examining the interplay between science communication and trust in the era of the Internet, exactly because it could reveal how trust in science can be constructed between scientists and a broader audience. Accordingly, this article constitutes an attempt to contribute some relevant knowledge by looking at mainstream climate blogs, a subcategory of science blogs, which are written and administered by mainstream scientists and address subjects relating specifically to climate science and climate change.

This article relies primarily on data collected through qualitative interviews with the scientist-bloggers and the blog readers (emphasizing, in this case, the reader's perceptions), constituting the first, to my knowledge, study of the specific type of climate blogs based on interviews with both groups of users (see also Zoukas, 2019).

4 Methodology

The research presented here relies upon the interpretation of data originally collected between 2015 and 2016 (see Zoukas, 2019). Additional data, which have been analyzed within the context of a recently finished two-year study on the history of the appropriation of the Internet by environmental scientists and engineers (February 2020–January 2022), have been utilized too.

The section providing background information about the history and "prehistory" of climate blogging derives mainly from the latter study, which is partially based on the examination of an assortment of written documents, in particular, articles published in scientific journals and magazines, professional magazines, as well as opinion pieces and news articles, collected between April and June 2021. More precisely, that section relies on forty-five articles published in scientific journals and magazines[3] between 1986 and 2001, which I have coded and analyzed thematically (Ayres, 2008; Guest et al., 2012). The rationale behind this choice was to understand how scientists felt about Usenet, an online platform which, as I explain later in this article, could be seen as the blogs' predecessor, at least in terms of the online climate- and generally science-related discussions.

The section about the character of the blogs and the knowledge communicated through them is based on a qualitative case study approach, which involves semi-structured in-depth interviews with thirty-eight users of twelve mainstream climate blogs written in English. I identified and selected the blogs through Google, BlogSeacrhEngine.org,[4] and Blogspot Blog Search,[5] between March 2015 and September 2015, after entering the term "climate blog" in the search bars (see also Sharman, 2014). I collected the blogs manually from the first ten result pages of each of the aforementioned search engines and subsequently, when applicable and using the snowball technique, from the so-called "blogroll", the list of blogs recommended by the bloggers. Particularly, to identify the blogs' relevance to climate science and climate change, I read their title, their "about" section, which generally includes illuminating information about the blogs' main subject and character, and individual posts. At that stage, I was able to collect a total of one hundred and three blogs, as well as blog-like services, relating to climate.

My main aim was to select blogs about climate administered independently by mainstream scientists. Additional criteria of selection were that such blogs should be written in English and distinguished clearly as blogs, having the main features of blogs mentioned earlier. As such, they also had to allow comments, specifically given

---

[3] Including *Annals of Mathematics*, *BioScience*, *Bulletin of the Ecological Society of America*, *Change*, *Child Development*, *Econometrica*, *Educational Technology*, *Environmental Health Perspectives*, *Environmental Science and Technology*, *Journal of Symbolic Logic*, *Mathematics Magazine*, *Nature*, *Publications of the Astronomical Society of the Pacific*, *Review of Agricultural Economics*, *Science*, *Science Teacher*, *Scientific American*, and *Transactions of the American Mathematical Society*.

[4] http://www.blogsearchengine.org/, accessed on 14/01/2023.

[5] https://www.searchblogspot.com/, accessed on 14/01/2023.

that I was also interested in the interactions taking place in the comment sections. Finally, they had to be active (see also Zoukas, 2019). Thirty seven of the initially selected blogs fulfilled those criteria. I am therefore confident that the twelve blogs used for this case study represent a good sample of the specific type of climate blogs.

In addition to the twelve mainstream scientist-bloggers, I interviewed twenty-six readers. I approached the scientist-bloggers by email, while the readers were "enrolled" via five of the selected blogs, the bloggers of which posted on my behalf an invitation where I described the subject and overall aim of my study (see also Baumer et al., 2011). Each of the readers had to read at least one of the blogs of this case study, while each blog was known to no less than one of the readers.

The interviews with both subgroups were conducted between April 2015 and June 2016, thirty-six of them orally (online, by phone, or in person), while two readers were interviewed through written questions sent by email. The oral interviews were, with the interviewees' permission, digitally recorded and transcribed in full, while all thirty-eight interviews have been coded and analyzed thematically.[6] To better understand the nature of the selected blogs, thematic coding and analysis has been similarly applied to more than three hundred and twenty blog posts.[7]

Let it be stressed that, even though I am aware of the longstanding, crucial, and controversial issue of climate change, in this research, no opinion regarding the interviewees' views and claims about the subject per se is expressed on my behalf. Rather, my aim has been to understand and reflect on why and how people with certain beliefs, educational backgrounds, and levels of expertise appropriate the blogging platform on the basis of their common interests, motivations, and goals; how they engage with the communicated knowledge; and how they feel about it. To that end, my interview questions focused on the experiences, perceptions, and incentives of the blog users.

5 The Case Study

At the time of the interviews, almost all scientist-bloggers were active scientists (two of them recently retired). Ten of them were involved professionally in climate science, one in physics, another one in chemistry. They were motivated to blog about climate as they felt that scientifically-grounded angles were noticeably absent from the public climate discussions and disputes. Like they explained, their main aims were to address climate mis/disinformation, especially considering the contrarian views increasingly expressed publicly, as well as to fill what they described as information or knowledge gaps in the public discussions and debates about climate science and climate change. They intended to do so not simply by adding some extra information,

[6] In this article, random codes, based on the AB, BC, CD, etc. sequence, are assigned to all interview quotes.

[7] The coding and analysis of data has been assisted by the MAXQDA qualitative data analysis software.

but rather contributing their science-based perspectives and knowledge. That kind of knowledge is mirrored on the character of the analyzed blogs. That is, while the scientist-bloggers, blogging independently, could of course approach any topic as they wish, they principally emphasize the scientific aspects of the climate change issue. In this sense, science-focused articles, especially those distinguished by their technical complexity, constitute the core material of the mainstream climate blogs, a material the quality of which has, by and large, attracted the interest of the readers.

Virtually all the readers that I interviewed have been academically educated at a postgraduate level and/or an undergraduate one. Most of them have gained degrees in the natural or applied sciences, some in the humanities or social sciences, even though there are cases where there is no relation between the readers' actual occupations and their educational backgrounds. Approximately one fourth of them have been academically educated in some of the earth sciences' sub-disciplines, such as climate science/climatology, geophysics, geology, and ecology. Even so, the whole group of the twenty-six readers is rather heterogeneous regarding their specific backgrounds and occupations. What is important to note is that twenty-three of them are aligned with the mainstream scientific assessment on global warming and climate change, while only three are skeptical about the issue. This is not surprising, given that they were enlisted through blogs administered and written by mainstream scientists, which, most probably, attract a like-minded audience. The reasons why most of the readers are interested in the particular type of climate blogs could be outlined as their wish to obtain a deeper understanding of the science behind a stimulating but concerning and contentious subject so that they can better communicate and argue about it. Before I present how the readers feel about the knowledge communicated through the mainstream climate blogs, a brief background note on the history, and indeed the "pre-history", of climate blogging is in order.

5.1 Historical Note

Despite being a niche of science communication, in the sense that it is undertaken by a small number of scientists and it appeals to a relatively narrow audience, mainstream climate blogging is a rather longstanding example of online climate communication. The first climate blogs appeared in 2004 (Zoukas, 2019), while online climate-related conversations can be traced back to Usenet, an online discussion system developed in the late 1970s.

Based on a plethora of topic-focused groups, the newsgroups, which were arranged thematically under different hierarchies (Fisher, 2003), Usenet seems to have been a particularly helpful tool for the communication of science, specifically through newsgroups under the sci.hierarchy. Aside from being a way for scientists to communicate scientific knowledge, information, ideas, or even research results among them (e.g. Abate, 1995; Bulman-Fleming et al., 1992), several sci.newsgroups appealed to a lay audience too, enabling them, for example, to interact with "top experts", who could offer both technical and lay explanations of different scientific topics, from physics and chemistry to astronomy and relativity (Kaiser, 1999: 139).

One of the most longstanding newsgroups of the sci.hierarchy was sci.environment, already active in the 1980s, discussing topics about the environment, sustainability, ecology, and associated areas (D' Souza & Walton III, 1997; Makulowich, 1993). Sci.environment deserves a special mention in this article, not only because it was one of the very first cases of climate communication online, as both scientists and a general, yet interested and informed, audience engaged in the newsgroup, but also because it can be regarded as the "prehistory" of climate blogging.

Partly due to Usenet's principally unmoderated environment, and also as a result of the platform's increasing openness to an ever-expanding public since the mid 1990s, Usenet began to deteriorate, rendering by the late 1990s/early 2000s most of its newsgroups unviable (Donath, 2014). To maintain an informative and constructive conversational context, and safeguard the quality of the communicated knowledge, some of the sci.environment's core participants, most of them scientists, moved their climate-related discussions to the moderated blogging platform.

Usenet and blogs might represent two quite diverse cases of online communication, as the former had a rather communal character, whereas the latter are essentially personal websites, administered typically by one person, or a small group of people. Notwithstanding, the importance of the scientists' transition to blogs consists in that it suggests that there has been a historical trend towards the utilization of online platforms by scientists, indicative of their wish to communicate their knowledge directly to a wider audience. Primary source knowledge, originally developed within the SSK field, is a good conceptual tool to analyze, and comprehend better, the knowledge that constitutes the nucleus of the material communicated through the mainstream climate blogs of this case study.

5.2 *Mainstream Climate Blogging and "Primary Source Knowledge"*

The notion of primary source knowledge (PSK) was initiated within the SSK field by Collins and Evans (2007), as part of their analysis and systematization of different types of science-related expertise and their explanation of what it means to be an expert when it comes to science. A straightforward and illustrative definition suggests that PSK can be obtained by "struggling" through professional journals or very technical material found on the Internet,[8] and that such a process could render the layperson able to "get the drift" of, for example, a physics paper, without necessarily understanding the equations (Collins, 2014: 67). According to a more recent definition, PSK can be gained by non-experts and citizens by "assiduously reading the primary scientific journal literature without being embedded in the cultural life of

[8] Described elsewhere as "primary or quasi-primary literature" (Collins & Evans, 2007: 22).

the corresponding technical specialty" (Collins et al., 2017: 1105). Three constitutive elements of PSK deserve special attention here, especially as they denote the relevance of that type of knowledge to this case study.

First, PSK is largely defined based on how it is experienced and perceived by individuals who engage with it, taking into account their educational backgrounds and what could be described as their intellectual capacity. As seen above, terms such as the "layperson", the "non-expert", or the "citizen" have been used as part of PSK's description. It is important nonetheless to consider that, to obtain that knowledge, someone needs to be fluent in the language, educated at a university level, and able to understand libraries as well as other sources (Collins, 2018). This means that PSK could be attainable by a certain category of citizens, non-experts, laypersons, and so on. Second, the audience to which PSK appeals is specified not only by the above characteristics but also by the same audience's common interests and concerns. Motivations to engage with PSK could vary from particular health needs and local conditions to imperative political agendas, while it is not rare that citizens, led by some combination of that sort of interests and incentives, become involved in "scientific circles", exposing themselves to "a deeper understanding" of the matter at hand (Collins & Evans, 2007: 36). Third, being a kind of knowledge typically derived from scientists and characterized by technical intricacy, PSK requires perseverance on behalf of those interested in acquiring it (Collins, 2014, 2018; Collins et al., 2017).

The analogy between PSK and the knowledge communicated through the mainstream climate blogs is considerable. The blogs' knowledge comes from scientists, is often distinguished by some sort of scientific complexity, and seems to appeal mainly to a certain group of readers, whose educational backgrounds, interests, concerns, and motivations are similar to those mentioned above. Even more so, that analogy has been demonstrated in different cases in my study, particularly when the non-expert readers explained that, by reading technical blog posts and persevering in comprehending them, they can gain a better understanding of some of the science underlying the issue of global warming and climate change, an issue they find both interesting and concerning.

One of the several indicative examples I examined is that of a person with a master's degree in information technology (IT), who said:

> Anything that's got an equation in it, a lot of people will just turn off straight away. And sometimes there has to be an equation, because, you know, in the hard sciences and the natural sciences, mathematics is a prerequisite for understanding what's going on (AB, oral interview, February 11, 2016).

A similar sentiment was expressed by another IT professional, with higher education in engineering, who explained that, when discussing, for instance, the lapse rate, he needs to understand "what the lapse rate is", often by studying some additional material (BC, oral interview, February 15, 2016). That kind of dedication was likewise emphasized by one more reader, who had a science degree and also worked in IT, when explicating:

> As I have a very particular set of blogs that I read, the majority of them have content which sometimes is quite challenging, and requires a degree of determination, and research, and all

> that sort of stuff, and a bit of practice sometimes [...]. Do I understand it? Sometimes; and sometimes not; which would then require reading or doing some more research or whatever (CD, oral interview, February 15, 2016).

Indeed, being more esoteric than the knowledge gained, for example, through popular science books, science magazines, or broadsheets (Collins & Evans, 2007), PSK can render people capable of having a better understanding of science. However, determination and technical comprehension are not enough to assess the *credibility* of that knowledge, especially when it comes to areas that are deeply contested (Collins & Evans, 2007; Collins et al., 2017). Like Collins and Evans (2007: 22) simply put it, to understand a scientific dispute, "one has first to know what to read and what not to read", which requires a great amount of experience and training. This is so because PSK could, for example, come from a published scientific paper not taken seriously within the corresponding community of working scientists, because it might have been written by an unreliable scientist or a scientist with no good research reputation within mainstream science (Collins et al., 2017). Perceiving the meaning of a published scientific paper or, for that matter, the meaning of a communication piece containing PSK, requires an understanding of how such a material is located within "the social milieu of the relevant technical domain" (Collins, 2018: 72). That understanding can be obtained through immersion in, and interaction with, the relevant community of expert scientists, a kind of social contact that would allow the non-expert to comprehend what is the current scientific consensus on the relevance and value of that paper/knowledge (Collins & Evans, 2007; Collins et al., 2017). Accordingly, although the Internet is considered "a powerful resource" for PSK by Collins and Evans (2007: 22), gaining knowledge from the Internet is unreliable "without further back-up" (Collins, 2018: 72), that is, if no additional assistance is provided through some sort of interaction with scientists.

The argument I put forward in this paper is that such interaction can be achievable through the mainstream climate blogs and the way their communicative environment has been shaped by their users. A few more empirical examples demonstrate this point.

One of them refers to the case of a photographer, educated at a university level in mathematics, who explained that he does not consider necessary for someone, including himself, "to follow the specifics of the equation" in order to comprehend the central concepts and arguments in the discussions unfolding through the blogs' comment sections. After clearly differentiating his knowledge from the "really in-depth knowledge" possessed by a field expert, the same interviewee touched upon the notion of trust and its relation to the climate blogs as follows:

> I think there's also a question of building trust. So, when you read certain writers, you develop a sense of whether they're trustworthy or not, both by what they say and what other people say about them (DF, oral interview, February 28, 2016).

For sure, obtaining PSK through the mainstream climate blogs could suggest that its quality and trustworthiness are, at least to some extent, taken at face value among the majority of the interviewed readers, particularly those aligned with mainstream science. It is true that some of them implied, directly or indirectly, that, being produced by scientists with well established reputations in mainstream climate science, they consider the blogs they read to be authoritative and credible. However, like the above passage connotes as well, trust is also something that can be developed by engaging with the blogs and the scientist-bloggers.

It is very essential for this article to stress that what virtually all the readers further explained is that one of the things they value the most is that, through the blogs, information and knowledge can become subject to discussions, critiques, and evaluations, typically by other readers who comment, many of whom are often working scientists themselves. For example, a university professor of sociology said that the bloggers are "very responsive to sincere feedback", often providing additional explanations themselves, whenever they are asked, or directing readers to other relevant sources of information, adding afterwards:

> And what you may also find is that readers chime in [the comments] with additional information. Sometimes they have technical knowledge that the blog owner does not, and that can be very helpful (FG, oral interview, February 12, 2016).

The technical knowledge the commenters often demonstrate was also acknowledged by another reader, a postdoctoral researcher with a PhD in ecology, when explained that he would realize the scientific education of many of the commenters, who were frequently willing to help others "understand the nuances of what's being discussed" (GH, oral interview, July 1, 2015). As a journalist with an undergraduate degree in chemistry explained, such interactions are especially characteristic of what he described as "boutique" or "niche" blogs, that is, blogs with a "dedicated readership", where not only real expert scientists but also "informed amateurs" offer explanations, clarifications, corrections, and so on (HI, oral interview, March 21, 2016). To use the words of a retired journalist with a degree in English literature, it is that kind of "virtual proximity, if not [...] actual direct communication, with people who really know what they're talking about" (IJ, oral interview, October 6, 2015), which is afforded by the blogs and appreciated by the readers.

As the readers are able to improve, through their interactions with the blogs' knowledge and environment described in this paper, not only their technical understanding but also their comprehension of, like some of them put it, "the politics" of climate change or "the discussions about the discussion", what is ultimately enhanced, I suggest, is their trust to the communicated science. To be clear, I do not make any strong claim here that such interactions are the same to those Collins and Evans (2007) consider necessary for citizens to be able to acculturate to the relevant community of expert scientists, that is, continuous, almost everyday, interactions among fully-fledged members of such a community. But what I do argue is that such interactions constitute an important step towards a better understanding of science and its underlying issues, which is why the character of PSK should be reconsidered as more trustworthy when it becomes part of the mainstream climate blogs' communicative environment.

6 Conclusion

Generally conceptualized within sociological research as a property of "interpersonal relations" (Barbalet, 2019), the notion of "trust" has taken different forms during its history, including, for example, trust based on beliefs (*cognitive trust*); "judgments, decisions, intentions and resolutions" (*conative trust*); or emotional states (*affective* trust) (Simpson, 2012: 564). Such forms may frequently overlap when it comes to trust in institutions, particularly those with a distinctive social character, like the institution of science. It could, for instance, be argued, even from our own experiences, that our trust, or distrust, in scientists working within fields intersecting, so to speak, with our everyday lives might result from the interplay among our viewpoints, motives, and emotions. This is why both the nature and the role of society's trust in science needs to be clarified and, subsequently, facilitated (Whyte & Crease, 2010).

As discussed earlier, science communication practices, especially methods supporting direct communication between the public and scientists, can be useful for restoring and/or enhancing public trust in science. However, science communication is also characterized by longstanding challenges, which pertain mainly to issues relating to information quality, credibility, reliability, and the role of experts (Bucchi, 2019). Such issues merit even greater consideration when we examine the communication of science within Internet-based environments, which are very often characterized by the spread of science-related mis/disinformation (Bucchi, 2019).

In this article, to provide insight into the relationship between trust and science in the Internet era, especially in terms of socially salient fields like climate change, I focused on the example of direct communication between expert scientists and an interested and concerned audience through the blogging platform, explaining that what could be described as mainstream climate blogging is indicative of a movement that began before the appearance of blogs, and generally the Web, when scientists recognized the potential for public communication through Usenet, perhaps via other similar online platforms as well.

To examine the character of the knowledge constituting the nucleus of the material communicated through the mainstream climate blogs, I relied upon the concept of "primary source knowledge", initiated by Collins and Evans (2007) as part of their sociological approach to scientific knowledge. Informed by the philosophy and history of science, their perspective has brought about some useful conceptual tools for analyzing and understanding better not only the nature of scientific knowledge but also the relationship between trust and science. My study suggests that, within the mainstream climate blogs' communicative environment, it is not only the understanding of primary source knowledge that can be enhanced, which has admittedly a challenging nature for non-scientists, but also its credibility. In this sense, it can be argued that the analyzed mainstream climate blogs are a both authoritative and trustworthy niche of science communication.

The importance of science communication, as an endeavour, is largely associated with ideas such as democracy, justice, and shared heritage (Davies, 2021). However, many questions concerning equality in science communication and public engagement with scientific knowledge could be raised. For instance, the fact that all thirty-eight interviewees in my study were male users should make us investigate and reflect on the underrepresentation of women in science communication, at least when it comes to science communication online. We also need to examine how a better understanding of science could be possible for citizens with different educational backgrounds than the ones the interviewees of this case study have.

As I have already mentioned, having followed a qualitative approach which is based on interviews with both scientist-bloggers and blog readers, I believe I have offered a more symmetrical analysis of the communication of science through the blogging platform. Moreover, by emphasizing the readers' experiences and perceptions, I have also provided an empirical example of primary source knowledge. Merely suggesting that primary source knowledge can be gained from the Internet misses the diversity of the Internet's environments, the various sociocultural practices unfolding in them, and, accordingly, the different meanings revealed through such practices. I therefore consider that my reinterpretation of primary source knowledge as part of the mainstream climate blogs' communicative framework, including my argumentation about how such knowledge is related to trust in science, have further developed the concept per se, revealing another aspect of its relationship to the Internet and the nature of trust citizens put in science and scientists.

Funding Acknowledgment This research is co-financed by Greece and the European Union (European Social Fund - ESF) through the Operational Programme "Human Resources Development, Education and Lifelong Learning" in the context of the project "Reinforcement of Postdoctoral Researchers - 2nd Cycle" (MIS-5033021), implemented by the State Scholarships Foundation (IKY).

Operational Programme
Human Resources Development,
Education and Lifelong Learning
Co-financed by Greece and the European Union

References

Abate, T. (1995). Let's go ('net) surfing now. *BioScience, 45*(8), 522–524. https://doi.org/10.2307/1312694

Agley, J. (2020). Assessing changes in US public trust in science amid the COVID-19 pandemic. *Public Health, 183*, 122–125. https://doi.org/10.1016/j.puhe.2020.05.004

Ayres, L. (2008). Thematic coding and analysis. In L. M. Given (Ed.), *The SAGE encyclopedia of qualitative research methods* (pp. 868–868). Sage Publications.

Barbalet, J. (2019). Trust: Condition of action or condition of appraisal. *International Sociology, 34*(1), 83–98. https://doi.org/10.1177/0268580918812268

Baumer, E., Mark, S., & Tomlinson, B. (2011). Bloggers and readers blogging together: Collaborative co-creation of political blogs. *Computer Supported Cooperative Work, 20*, 1–36. https://doi.org/10.1007/s10606-010-9132-9

Bucchi, M. (2019). Facing the challenges of science communication 2.0: Quality, credibility and expertise. *EFSA Journal, 17*, e170702. https://doi.org/10.2903/j.efsa.2019.e170702

Bulman-Fleming, S., Wang Edward, T. H., Heuer Gerald, A., Seung-Jin, B., Amengual, C. M., Sam, N., Klamkin Murray, S., Howard, M., Bjorn, P., Michael, H., Straffin Philip, D., Reiner, M., Wardlaw William, P., Trinity University Problem Group, Fukuta Jiro, University of Wyoming Problem Circle, Richard, P., Kenneth, S., Adam, R., & Kedlaya Kiran, S. (1992). Problems. *Mathematics Magazine, 65*(5), 348–356. https://doi.org/10.2307/2691250

Bultitude, K. (2011). The why and how of science communication. In R. Premysl (Eds.), *Science Communication* (pp. 1–18). European Commission.

Collins, H., & Evans, R. (2007). *Rethinking expertise.* The University of Chicago Press.

Collins, H., Luis, R.-G., & Paul, G. (2017). A note concerning primary source knowledge. *Journal of the Association for Information Science and Technology, 68*(5), 1105–1110. https://doi.org/10.1002/asi.23753

Collins, H. (2014). *Are we all scientific experts now?* Polity Press.

Collins, H. (2018). Studies of expertise and experience. *Topoi: An International Review of Philosophy, 37*(1), 67–77. https://doi.org/10.1007/s11245-016-9412-1

Cook, J., Dana, N., Green Sarah, A., Mark, R., Bärbel, W., Rob, P., Robert, W., Peter, J., & Andrew, S. (2013). Quantifying the consensus on anthropogenic global warming in the scientific literature. *Environmental Research Letters*, *8*(2), 024024. https://iopscience.iop.org/article/10.1088/1748-9326/8/2/024024

Cook, J. (2022). Understanding and countering misinformation about climate change. *Research Anthology on Environmental and Societal Impacts of Climate Change*, 1633–1658.

D' Souza, G. E., & Walton III, B. T. (1997). A guide to Internet resources in sustainable development. *Applied Economic Perspectives and Policy, 19*(1), 122–127. https://doi.org/10.2307/1349682

Davies, S. R. (2021). An empirical and conceptual note on science communication's role in society. *Science Communication, 43*(1), 116–133. https://doi.org/10.1177/1075547020971642

Donath, J. (2014). *The social machine: Designs for living online.* The MIT Press.

Eggleton, T. (2013). *A short introduction to climate change.* Cambridge University Press.

Elgesem, D., Lubos, S., & Nicholas, D. (2015). Structure and content of the discourse on climate change in the blogosphere: The big picture. *Environmental Communication, 9*(2), 169–188. https://doi.org/10.1080/17524032.2014.983536

Farmer, G. T. (2015). *Modern climate change science: An overview of today's climate change science.* Springer.

Fisher, D. (2003). Studying social information spaces. In C. Lueg & D. Fisher (Eds.), *From Usenet to CoWebs* (pp. 3–19). Springer.

Fløttum, K., Anje, M. G., Øyvind, G., Nelya, K., & Andrew, S. (2014). Representations of the future in English language blogs on climate change. *Global Environmental Change, 29*, 213–222. https://doi.org/10.1016/j.gloenvcha.2014.10.005

Gardiner, S. M., Simon, C., Dale, J., & Henry, S. (2010). *Climate ethics: Essential readings.* Oxford University Press.

Giddens, A. (2009). *The Politics of climate change.* Polity Press.

Golbeck, J. (2013). *Analyzing the social web.* Morgan Kaufmann Publishers Inc.

Goosse, H. (2015). *Climate system dynamics and modelling.* Cambridge University Press.

Guest, G., MacQueen, K. M., & Namey, E. E. (2012). *Applied thematic analysis.* Sage Publications Inc.

Haerlin, B., & Parr, D. (1999). How to restore public trust in science. *Nature, 400*, 499-499. https://doi.org/10.1038/22867

Hamilton, L. C., & Safford, T. G. (2021). Elite cues and the rapid decline in trust in science agencies on COVID-19. *Sociological Perspectives, 64*(5), 988–1011. https://doi.org/10.1177/07311214211022391

Hendriks, F., Dorothe, K., & Rainer, B. (2016). Trust in science and the science of trust. In B. Bernd (Ed.), *Trust and communication in a digitized world* (pp. 143–159). Springer.

Houghton, J. (2004). *Global warming: The complete briefing* (3rd ed.). Cambridge University Press.

Houston, L., & Capalbo, S. (2021). Economics. In S. Andreas (Ed.), *Introduction to climate science*. Retrieved January 24, 2023, from https://eng.libretexts.org/Bookshelves/Environmental_Engineering_(Sustainability_and_Conservation)/Introduction_to_Climate_Science_(Schmittner_2021)/01%3A_Text/09%3A_Economics

Huber, B., Matthew, B., de Zúñiga, H. G., & James, L. (2019). Fostering public trust in science: The role of social media. *Public Understanding of Science, 28*(7), 759–777https://doi.org/10.1177/0963662519869097

Hulme, M. (2009). *Why we disagree about climate change: Understanding controversy, inaction and opportunity*. Cambridge University Press.

Jasanoff, S. (2011). Cosmopolitan knowledge: Climate science and global civic epistemology. In J. S. Dryzek, R. B. Norgaard, & S. David (Eds.), *The Oxford handbook of climate change and society* (pp. 130−144). Oxford University Press.

Kaiser, J. (1999). NetWatch. *Science, 283*(5399), 139–139. http://www.jstor.org/stable/2897369

Kreps, S. E., & Kriner, D. L. (2020). Model uncertainty, political contestation, and public trust in science: Evidence from the COVID-19 pandemic. *Science Advances, 6*(43), eabd4563. https://doi.org/10.1126/sciadv.abd4563

Kvaløy, B., Henning, F., & Ola, L. (2012). The publics' concern for global warming: A cross-national study of 47 countries. *Journal of Peace Research, 49*(1), 11–22. https://doi.org/10.1177/0022343311425841

Lewandowsky, S., Stritzke Werner, G. K., Freund, A. M., Klaus, O., & Krueger, J. I. (2013). Misinformation, disinformation, and violent conflict: From Iraq and the "War on Terror" to future threats to peace. *American Psychologist, 68*(7), 487–501. https://doi.org/10.1037/a0034515

Lorenzoni, I., Pidgeon, N. F., & O' Connor, R. E. (2005). Dangerous climate change: The role for risk research. *Risk Analysis: An International Journal, 25*(6), 1387–1398.https://doi.org/10.1111/j.1539-6924.2005.00686.x

Mackenzie, D. (2007). Finding the ratchet: The political economy of carbon trading. *London Review of Books, 29*(7). Retrieved January 24, 2023, from https://www.lrb.co.uk/the-paper/v29/n07/donald-mackenzie/the-political-economy-of-carbon-trading

Makulowich, J. S. (1993). The use of electronic communications in environmental health research. *Environmental Health Perspectives, 101*(1), 34–35. https://doi.org/10.1289/ehp.9310134

McGeehin Heilferty, C. (2012). "An internet family": Online communication during childhood cancer. In L. Rebecca Ann (Ed.), *Produsing theory in a digital world: The intersection of audiences and production in contemporary theory* (pp. 159−176). Peter Lang Publishing.

Medin, Douglas L., & Megan, B. (2014). The cultural side of science communication. *Proceedings of the National Academy of Sciences, 111*, 13621–13626. https://doi.org/10.1073/pnas.1317510111

Mehlenbacher, A. (2019). *Science communication online: Engaging experts and publics on the internet*. The Ohio State University Press.

Mostafa, M. M. (2016). Post-materialism, religiosity, political orientation, locus of control and concern for global warming: A multilevel analysis across 40 nations. *Social Indicators Research, 128*(3), 1273–1298. https://doi.org/10.1007/s11205-015-1079-2

Nagel, J., Thomas, D., & Jeffrey, B. (2008). *Workshop on sociological perspectives on global climate change*. Retrieved January 24, 2023, from https://www.asanet.org/wp-content/uploads/savvy/research/NSFClimateChangeWorkshop_120109.pdf

Oppenheimer, M. (2005). Defining dangerous anthropogenic interference: The role of science, the limits of science. *Risk Analysis: An International Journal, 25*(6), 1399–1407. https://doi.org/10.1111/j.1539-6924.2005.00687.x

Oreskes, N., & Conway, E. M. (2010). Defeating the merchants of doubt. *Nature, 465*(7299), 686–687.https://doi.org/10.1038/465686a

Oreskes, N. (2004). The scientific consensus on climate change. *Science, 306*(5702), 1686–1686. https://doi.org/10.1126/science.1103618

Peters, H. P., Sharon, D., Joachim, A., Yin-Yueh, L., & Dominique, B. (2014). Public communication of science 2.0: Is the communication of science via the "new media" online a genuine transformation or old wine in new bottles? *EMBO Reports, 15*(7), 749–753. https://doi.org/10.15252/embr.201438979

Poberezhskaya, M. (2017). Blogging about climate change in Russia: Activism, scepticism and conspiracies. *Environmental Communication, 12*(7), 942–955. https://doi.org/10.1080/17524032.2017.1308406

Resnik, D. B. (2011). Scientific research and the public trust. *Science and Engineering Ethics, 17*(3), 399–409. https://doi.org/10.1007/s11948-010-9210-x

Sarathchandra, D., & Kristin, H. (2020). Trust/distrust judgments and perceptions of climate science: A research note on skeptics' rationalizations. *Public Understanding of Science, 29*(1), 53–60.

Schmittner, A. (2021). *Introduction to Climate Science*. Retrieved January 24, 2023, from https://eng.libretexts.org/Bookshelves/Environmental_Engineering_(Sustainability_and_Conservation)/Introduction_to_Climate_Science_(Schmittner_2021)

Shanahan, M.-C. (2011). Science blogs as boundary layers: Creating and understanding new writer and reader interactions through science blogging. *Journalism, 12*(7), 903–919. https://doi.org/10.1177/1464884911412844

Sharman, A. (2014). Mapping the climate sceptical blogosphere. *Global Environmental Change, 26*, 159–170. https://doi.org/10.1016/j.gloenvcha.2014.03.003

Simpson, T. W. (2012). What is trust? *Pacific Philosophical Quarterly, 93*(4), 550–569. https://doi.org/10.1111/j.1468-0114.2012.01438.x

Somerville, R. C. J., & Hassol, S. J. (2011). Communicating the science of climate change. *Physics Today, 64*(10), 48–53. https://doi.org/10.1063/PT.3.1296

Thorsen, E. (2013). Blogging on the ice: Connecting audiences with climate-change sciences. *International Journal of Media & Cultural Politics, 9*(1), 87–101.

Treen, K. M. d'I., Williams Hywel, T. P., & O'Neill Saffron, J. (2020). Online misinformation about climate change. *Wiley Interdisciplinary Reviews: Climate Change, 11*(5), e665. https://doi.org/10.1002/wcc.665

Van Rensburg, W., & Head, B. W. (2017). Climate change sceptical frames: The case of seven Australian sceptics. *Australian Journal of Politics & History, 63*(1), 112–128. https://doi.org/10.1111/ajph.12318

Weigold, M. F. (2001). Communicating science: A review of the literature. *Science Communication, 23*(2), 164–193. https://doi.org/10.1177/1075547001023002005

Whyte, K. P., & Crease, R. P. (2010). Trust, expertise, and the philosophy of science. *Synthese, 177*(3), 411–425. https://doi.org/10.1007/s11229-010-9786-3

Wintterlin, F., Friederike, H., Mede Niels, G., Rainer, B., Julia, M., & Schäfer, M. S. (2022). Predicting public trust in science: The role of basic orientations toward science, perceived trustworthiness of scientists, and experiences with science. *Frontiers in Communication, 6*. https://doi.org/10.3389/fcomm.2021.822757

Wynne, B. (1995). Public understanding of science. In J. Sheila (Ed.), *Handbook of science and technology studies* (Revised ed., pp. 361–388). Sage Publications Inc.

Wynne, B. (2006). Public engagement as a means of restoring public trust in science—Hitting the notes, but missing the music? *Public Health Genomics, 9*(3), 211–220. https://doi.org/10.1159/000092659

Xiao, X., Porismita, B., & Yan, S. (2021). The dangers of blind trust: Examining the interplay among social media news use, misinformation identification, and news trust on conspiracy beliefs. *Public Understanding of Science, 30*(8), 977–992. https://doi.org/10.1177/0963662521998025

Zoukas, G. (2019). *Climate blogging in a post-truth era: Opportunities for action and interaction. Mainstream scientist-produced climate blogs as a climate science communication niche*. Retrieved January 24, 2023, from http://hdl.handle.net/1842/35540

Emancipatory Data Literacy and the Value of Trust

Birte de Gruisbourne

Abstract Data literacy has been formulated as a future skill for the 21st century. While data literacy is often thought as an individualized competence which tends to responsibilize the individual, this paper suggests an idea of *emancipatory data literacy* as a structural competence and critical attitude towards data and data-driven arguments. This attitude includes a mastery of the tools to trust in data on a well-grounded manner and is conceptualized with a special focus on affect, which makes data literacy a genuinely relational phenomenon. As an epistemological practice data literacy is a genuine part of what it means to talk about or with science. Since data is often understood as one of the major indicators for objectivity, data literacy helps to trust in scientific arguments and in the same way know limits and conditionality of objectivity.

1 Introduction

In the past few years data literacy[1] has become a central issue in higher education in Germany. Data literacy was defined as a so-called future skill (Suessenbach et al., 2021), initiatives and institutions signed a data literacy charter underlining the aim to foster data literacy education for every citizen (Schüller & Koch, n.d.), and the data strategy of the federal German government imagines a "skilled society – self-determined and informed" (The Federal Government, 2021). All these initiatives try to give an answer to current problems we face in data intensive societies by educating individuals to become data literate which is understood as "the ability to

[1] This article originates in a data literacy project I have been involved in. I thank my colleagues Annika Bush, Sonja Dolinsek, Tobias Matzner, and Christian Schulz, since many of my thoughts in this paper came into being in our collective doing of and thinking about data literacy education. I also thank Niklas Corall, Maike Niehaus and Lea Schulz for their comments and Ammu Joshy for her valuable review.

B. de Gruisbourne (✉)
Institut für Medienwissenschaften, Universität Paderborn, Warburger Straße 100, 33098 Paderborn, Germany
e-mail: birte.de.gruisbourne@uni-paderborn.de

M. M. Resch et al. (eds.), *The Science and Art of Simulation*,
https://doi.org/10.1007/978-3-031-68058-8_15

collect, manage, evaluate, and apply data, in a critical manner" (Risdale, 2015). This goal is mostly framed as a rational endeavour: students, employees and citizens shall gain competences to decide responsibly on the basis of data. Data is also read as a sign for objectivity (Gitelman & Jackson, 2013; Rosenberg, 2013) implicating that data literacy enhances objective and scientific judgments. As such data literacy is regularly envisioned as a kind of empowerment through the ability to make informed decisions based on data. Data literacy shall therefore help to overcome the need to blindly trust companies, journalists or politics. This tension between literacy and trust, i.e. rationality and affect, is a typical pattern which can be found (often implicitly) in the data literacy discourse.

In the following I will argue against the strong opposition between trust and literacy. I will first give a brief idea of what I mean, when I talk about trust as an affective relation. After a short discussion of the problem of responsibilization in the context of neo-liberal governance, I will elaborate on what I call *emancipatory data literacy* with a focus on the relation between trust and literacy in data or data-driven arguments. Since data as such often function as an argument or as a demonstration of objectivity, this function has to be analysed and questioned. Given that we live in a data-driven society and that many political, economic or social decisions are based on some or another kind of data, citizens in democratic societies need competences to evaluate these processes. When people are supposed to trust in decisions based on data whatsoever, adequate competences will be needed. For living in a datafied society changes what needs to be understood about socio-technical, mediated or political structures. While data literacy imagined by employers and neo-liberal governance often tends to focus on practical skills to purposefully work with data, the concept of *emancipatory data literacy* is much more structural than skill oriented. Of course, there are many important practical skills to cope with data we should expect from people who actually have to make data-driven decisions like politicians, (data) scientists or journalists. Citizens however may not need to collect and analyse data in order to take part in democratic decision making in a well-informed manner. Yet, some structural competence about data, its concept, function, and power structures, helps to enhance trust in data-driven arguments on reasonable grounds. I argue, that data literacy as a structural competence helps scientists and citizens alike to trust and distrust (their own) data, communicate wisely and trust arguments derived from data.

2 Trust and Affect

Trust is a risky affective relation. When I trust someone or something I put my faith in the other's reliability be it a parachute, a vaccine, a friend's promise, a scientist and their data on climate change or a politician and their decisions. Trust does not need to be an intersubjective relation and the current focus on data literacy and its role for trust needs an account which can exceed intersubjective trust. So, if trust is some affective reliance, it does not need to be reliance in someone but can also mean to trust in something. Trust can be formulated as an attitude of optimism which allows to take the risk to be totally wrong and forms a genuine affective relation all humans engage in.[2] We have some *feeling* that someone is trustworthy, sometimes even against good reason, sometimes backed by arguments. As an affective relation trust depends on the affective disposition of the trustor and the affective arrangement surrounding the decision. Affective arrangements were conceptualized as a working concept for interdisciplinary research on affect and emotion. They are therefore broadly defined as:

> heterogeneous ensembles of diverse materials forming a local layout that operates as a dynamic formation, comprising persons, things, artifacts, spaces, discourses, behaviors, and expressions in a characteristic mode of composition and dynamic relatedness (Slaby et al., 2019).

The affective arrangement can be understood as a framework in order to come to terms with the various factors that frame individual and social affective encounters like those of trust and enables a closer look at the elements which condition them. The affective arrangement is thus the space where trust is negotiated and is part of the structures which form our engagement with the situation. When people trust they trust in resonance with their environment, i.e. with the affective arrangement as an intersubjective relation and institutional setting.[3] In other words: To ask for trustworthiness is in the first step an affective process in a specific social relation the trustor and trustee find themselves in and that *may* be followed by rational evaluation. Typical factors for a person to trust someone's statements are their authority, reputation, title, status or shared values (Origgi, 2008). These factors are inseparable from the surrounding affective arrangement: a manager wearing an expensive suit in a formal setting embodies other indicators for trustworthiness than a naked scientist in the bathroom, although both might have a certain expertise on the subject matter. It is not only their expertise, but the whole arrangement which frames their trustworthiness. Switch suits and space and our evaluation will differ.

Yet, it is not only the arrangement but also the trustor's affective disposition (Mühlhoff, 2019) which influences their expectations. They might be some liberal student who does not belief in outer appearance and therefore trust the naked scientist more or they might otherwise have learned that a serious appearance matters most and trust the manager. Both do not have solely reasonable grounds for their trust,

---

[2] I take the stance of trust as an "attitude of optimism" from Jones (1996) who differentiates between trust in persons and reliance in things. Because of her more or less psychological perspective and her focus on good will, it is impossible to trust a parachute in her view. I do not want to give up the concept of trust in things and especially trust in structures as it is often used in our ordinary language. But she may be right insofar as the optimistic attitude of trust is still indirectly bound to human relationships. Scientific and political structures could be understood as a surrogate for intersubjective relations that are not possible in socially distant encounters. Although trust may be intersubjectively learned and executed, it must still be more than a metaphor when we talk about trust in things or structures.

[3] By using the term institution, I do not only have concrete institutions like schools or civil services in mind, but also other stabilized and stabilizing social structures like scientific peer-review which influence our relations of trust.

but evaluate the whole material setting according to their sedimented and embodied experiences. Like this, trust looks like a genuinely misguided practice we should correct by a rational evaluation in order to eradicate all those biases. Indeed, much educational effort is put into empowering students to become literate enough to decide autonomously, as the idea of a "skilled society – self-determined and informed" the Federal German Government puts forward in its data strategy. I will come back to this individualized and rational subjectivation techniques later.

If people thus trust on the basis of their own sedimented affective experiences and social norms and evaluate the whole system or environment for indicators of trustworthiness, we need to know how to trust well. The trust researcher Gloria Orrigi underlines especially for reputation as an indicator for perceived trustworthiness that it

> seems impossible to make sense of our trust in information on at least reasonable grounds without paying attention to the biased heuristics we use to evaluate the social information and to the systemic distortions of the reputational systems in which the information is embedded. My strong epistemological point here is that reputation is not just a collateral feature that we may consider or not in our epistemic practices. Without access to other people's reputations and evaluations, without a mastery of the tools that crystallize collective evaluations, coming to know would be impossible (Origgi, 2008).

Trust for Orrigi ultimately functions as a mostly epistemic relation and shows that we cannot understand what it means to trust if we do not understand the structurally biased dimension of the tools of trustworthiness. But trust is not just some kind of good faith in a person's goodwill and an intersubjective relation of dependence. The risk, the trustor takes, is to be wrong about the world and while there often exist many different rational criteria to prove the trustor's information, trust tends to ignore fact checking and intentionally depends on the trustor's affective evaluation of the other. The risk of trust hinges on the trustor's (mostly affective) analysis of the trustees trustworthiness which depends not necessarily on expertise, but on differently biasing criteria, like the affective arrangement and disposition. This furthermore implies that trust is not only conditioned by social relations, but also stabilizes these relations by reproducing hierarchies of trustworthiness. Yet, the mastery of the tools Origgi mentions, can also be understood as a possibility to challenge these stabilizing effects. Much research in social epistemology has been done on implicit bias' discriminating effects on all kinds of decisions, many of which became sadly popular in the COVID-19 pandemic. Still, trust is not just one mistake in evolution we as imperfect humans have to deal with, but a central factor in our social life whereby I do not only mean the necessary reduction of complexity which is needed in highly distributed societies. Trust can be understood as an affective skill or competence which should be valued and taken seriously, especially in epistemic settings that are traditionally rationalized, but highly dependent on embodied and affective relations. Since scientific practice in not excluded from that affective dimension,[4] trust is important within science as well as in the relation between the scientific subjectivity and citizenship.

[4] For the specific role of embodiment in paradigmatic epistemic practices like the laboratory see (Knorr Cetina, 1992: 121): "It is a knowledge that draws upon scientists' *bodies* rather than their minds. Consciousness or even intentionality are left out of the picture."

Therefore, I want to understand Origgi's 'mastery of the tools' of trustworthiness as embodied skills to read and speak codes of trustworthiness in social relationships. This mastery is affective, but as a skill or competence it goes beyond somehow drifting with one's affects. Mastery means to be appropriately disposed to evaluate the trust-setting affectively. The affective dispositions which condition one's behaviour of trust are not inaccessible to reason, nor are they encapsulated bodily structures which cannot be challenged. Instead, affective dispositions are embodied and sedimented affective experiences, framed by the current environment and reshaped in every new social and material interaction. The subject is disposed to affect and be affected in their way in every encounter with any environment. Nonetheless their affectivity is not only the stable frame for the encounter but is destabilized in every encounter as well. So, the disposition is bodily as well as situated, individual and social. When trust depends on so many moments of uncertainty, lacks objectivity and tends to biased decision-making, how can we then find a way to do it right? And why should we trust our trust? We somehow need to learn how to trust well. Concerning trust in data and data-driven arguments within and beyond scientific discourse and communication *emancipatory data literacy* education can help as a practice which draws not only on rationalization, but also forms our affective practices with data and data-driven arguments.

3 The Problem of Individual Responsibilization

As already mentioned in the introduction, data literacy education is often thought to be an answer to social and political problems caused by biased data or wrong interpretations of data, fake news, or distrust in scientific facts, e.g. on vaccine safety or climate change. This is problematized as an individual lack of practical skills concerning data, data-driven arguments, and data-driven decision making. As we have seen, trust as an affective relation can enhance these problems, because our affective dispositions and the surrounding affective arrangements are not easily accessible to us. As an answer to these questions data literacy is often thought as a rational remedy, a consciousness which enables us to detect fraud and make self-determined decisions. Although usually conceptualized as a competence, i.e. a bundle of knowledge, skills, and *values*, the data literacy discourse tends to take the affective dimension of education not seriously enough (Ahlborn et al., 2021; Dander, 2018). As a relational phenomenon affectivity sheds light on the socio-political and mediated aspects of trust in data and data-driven arguments.

Still, data literacy describes an individual competence learners gain. This is of course structurally true of every educational framework: education is something an individual learner receives, but many theoretical representations of the educational process only focus on this individualized and rationalized setting and thereby run the risk to hide the affective dimension due to this difficulty of every rational representation (Alsop, 2015; de Gruisbourne & Matzner 2021). In the current data literacy discourse this often implies that the student or data literate citizen is responsible for their decisions and their use of data 'self-determined and informed'. This self-determined and informed subject is then portrayed as a rational and autonomous individual, which mirrors the ideal of neo-liberal subjectivity and thereby touches the debate on responsibilization (Brown, 2017; Lemke, 2001). Stemming from Foucauldian perspectives in criminology responsibilization describes the process of moving responsibility from state actors to individuals as part of neo-liberal governance (Garland, 1996) and has from there on moved to other mostly sociological discourses (Brown, 2021; Karaian, 2014; Phoenix & Kelly, 2013; Shamir, 2008). The criminologist Pat O'Malley defines responsibilization therefore as following:

> Responsibilization is a term developed in the governmentality literature to refer to the process whereby subjects are rendered individually responsible for a task which previously would have been the duty of another – usually a state agency – or would not have been recognized as a responsibility at all. The process is strongly associated with neo-liberal political discourses, where it takes on the implication that the subject being responsibilized has avoided this duty or the responsibility has been taken away from them in welfare state era and managed by an expert or government agency (O'Malley, 2009: 522).

Hence, competence does not necessarily mean that students gain empowering agency, but may also imply that responsibility is assigned to them, although they might be lacking the means and power to realize what the data literate citizen should do according to their skills, knowledge, and values. The tension between responsibilization and empowerment shows data literacy's ambivalence as with competence also comes responsibility. But responsibility need not be responsibilization. To make this tension productive the empowering moments need to be contextualized and enlarged on a socio-political scale (Burkhardt et al., 2021; Matzner, 2017). Data literacy can mean to empower citizens or students to challenge power, but may also form a duty or burden to become literate and act accordingly. As long as literacies (be it data, scientific, alphabetic etc.) are individualized and understood as a remedy against the need to trust, literacy initiatives impend to become responsibilizing. I want to propose that *emancipatory data literacy* can be empowering if it is thought as a structural and relational competence.

4 Data and Data Literacy

To make clear what this idea could practically mean in the area of neo-liberal responsibilization versus empowerment and trust, I will give an overview what data literacy could be and what we[5] consider important skills in datafied societies in uncertain times. So, what are data and data literacy, how are they relevant, and which data literacy is needed?

---

[5] Since some of the following thoughts have been formulated in (Bush et al., 2022), I say we for once. Our working paper includes a data literacy competence framework for higher education which resonates with my theoretical approach in this text.

One standard definition of data literacy is the one from Risdale's Knowledge Synthesis Report which is a meta-analysis of different, mostly US data literacy frameworks: "Data literacy is the ability to collect, manage, evaluate, and apply data, in a critical manner" (Risdale, 2015) or an ability to "speak data" (Bhargava, 2022). Data literacy is to be found in the field of other literacies like: ethical literacy, science literacy, statistical literacy, information literacy, digital literacy, or media literacy with many overlapping elements, whereby data literacy's focus within these fields lies on data and data-driven arguments. I use data-driven arguments as an umbrella term for many kinds of data products, like usual arguments in a public discourse, but also other products which indirectly function as arguments like visualizations or digitizations. Many often cited frameworks start with a broad concept of data, but end with statistics (e.g. Risdale 2015; Schüller et al., 2019). So, in the actual data literacy discourse data mostly means statistics and/or big data, whereas other forms of data like qualitative data or historical artifacts are often neglected though not irrelevant in data-practices. Data literacy as a structural competence needs to integrate these different types of data, their function and limits, especially when it comes to different forms of scientific practice. Although in the following I will be mostly talking about quantitative data these are just one form of data the structural competence of *emancipatory data literacy* can focus on.

In an introduction to her concept of *Creative Data Literacy* Catherine d'Ignazio states that data "has become a currency of power. [...] As a result, knowing how to collect, find, analyze, and communicate with data is of increasing importance in society" (d'Ignazio, 2017). We can derive at least two demands from this: on the one hand, becoming able to speak data enables the data literate person to take part in power, to have and be able to pay an important currency. On the other hand, data literacy needs to take a distant look at data's power status: to avoid the trap of neoliberal responsibilization, a data literate citizen also needs to challenge the role of the currency as such. An emancipatory and relational instead of an individualizing concept of data literacy does not only include to speak data in a technical sense, but challenges its role in socio-technical arrangements. Data's power is closely related to their promise of objectivity, the promise that data may speak for themselves and reveal never imagined truths (Anderson, 2008; d'Ignazio & Klein, 2020). In order to understand these relations, it is important to differentiate data according on their kind and form, but also on their political, historical, socio-technical, or mediated environments. What data are, thus the material of data literacy education, depends on their context and function and needs to stay flexible and adjust to the context of data.

Data can be understood as aggregated abstractions which represent the world and in the same vein effect the world from the moment they are collected (Gitelman & Jackson, 2013).[6] As abstractions data contain decisions which information should become data and which shall be ignored, so that every data analysis is necessarily a cut-out. Data analysis tries to gain information about the world and these data

[6] Gitelman and Jackson (2013) define data's characteristics as abstract, aggregative, and graphically mobilized.

collections and analyses act performatively through citation of past data and meaning-making for present and future purposes. This is most obvious if a data analysis is also predictive, e.g. if someone's creditworthiness depends on (big) data about their social group's average solvency (Chun, 2021; Matzner, 2016). But the fact that some data exist and others do not already forms our world in various ways, because different things come into focus—which sometimes also implies to come into being (Criado Perez, 2019; d'Ignazio & Klein, 2020). In different contexts data can be differently used and materialized, and cannot be separated from their function. Daniel Rosenberg compellingly defines data by their rhetorical function, meaning that a single datum only becomes meaningful as part of an argument, in relation to other data and their contexts. While facts cease to be facts if they are proved wrong, data stay as data (Rosenberg, 2013: 18). This functionality bears on the properties of data. Leonelli (2015) for example defines portability as the central criterion of data, Thatcher (2016) takes market orientation and efficiency as a central point. The first one talks about scientific data in biology, the second focuses on big data in digital capitalism, and both define data through their specific function in epistemic or capitalist practice. Although there are similarities between the portable data as a basis for biological research and the fungible commodity capitalism works with, the perspectives change according to the function data have and the way they are used. What unites these concepts is that data have a function which structures arguments and have a certain power. So, data's structure (abstraction, aggregation, rhetoric and performativity) somehow defines data, but the specific usage varies and makes different promises according to context and function.

With these variations of data in mind, it seems difficult to define data literacy via general practical skills and so a functional differentiation is needed. One first distinction can be made between employability and citizenship as different targets of data literacy education. While the employable elements of data literacy might be best understood as a subject-specific expertise—which can mean specific technical and methodological knowledge and skills a student gains—data literacy for democratic citizens needs to be understood much more structurally. In order to understand the relation of trust and data literacy the citizenship perspective seems more fruitful, although in the practical teaching process these two perspectives are of course intertwined, not least because a student is always already a citizen somewhere.[7] While employees are mostly in the position to decide things based on data, citizens are more often compelled to trust.

If data literacy is supposed to enhance democratic empowerment for citizens in the digital age, statistics may not be the most important skill even if we only look at quantitative data. Often practiced as a mixture of a bit of data science and statistics, data literacy misses to include an idea of the socio-technical and mediating conditions as well as the affective arrangements in which data practices are embedded.

[7] Citizenship as a condition for political participation is of course much more precarious than this example may imply, as it depends on birth and race, but also on class, dis_ability, and/or gender. To counter those inequalities *emancipatory data literacy* as a means of democratic participation should also speak data for those who cannot speak data for themselves.

If data or data products function as arguments or reasons for data-driven decisions it is important not only to be methodologically skilled, but also trained in critical thinking beyond the correct application of rules (be they ethical, methodological or mathematical). If we take data's promise for objectivity as a currency of power seriously, epistemological knowledge and a critical attitude towards these promises might be even more important as it hints at a possible need also for the best skilled data scientist to become data literate in the emancipatory way (Hachmeister et al., 2021).

5 Overcoming Responsibilization by Embracing Trust and Relationality

In their critique of forms of responsibilizing governance in literacy discourses Rahul Bhargava et al. use the term *data inclusion* instead of literacy. They derive their argument from Claude Lévi-Strauss who problematized international literacy programs as a way to subordinate people as proper workers (Bhargava, 2015). It has become clear that data literacy as the future skill cited in the introduction is not only about mature citizenship, but also about employability as many programs openly state.[8] By choosing the term inclusion as an ability to participate the focus shifts from responsibilization towards a society providing an inclusive environment similar to the disability movement's demand for accessibility. This also means that data literacy is not only a competence citizens can gain but also a social and political responsibility. For data literacy to be not only responsibilizing it must include skills to profoundly challenge power asymmetries and objectivity claims, while acknowledging different grades of expertise and therefore the necessity to trust. Education then is still about individual learners and their competences, but individualized as they are, these competences can be framed differently, i.e. relational. Thought as a common and political project, *emancipatory data literacy* can therefore avoid the problem of neo-liberal responsibilization, because it is not the responsible citizen but the social relation which has to provide conditions for citizens to challenge data and data driven arguments and enables the mastery of the tools of informed trust as a further layer in *emancipatory data literacy*.[9]

In their groundbreaking book "Data Feminism" Catherine d'Ignazio and Lauren F. Klein give some examples how empowerment through data literacy education for individuals and critique of power can be integrated and thereby escapes responsibilization through practical empowerment (d'Ignazio & Klein, 2020). In a chapter with the telling title "Teach Data Like an Intersectional Feminist" they show how in a local

[8] The future skill framework (Suessenbach et al., 2021) for example is co-authored by the McKinsey & Company.

[9] Framed like that, Bhargava (2015) might be right that literacy is not the adequate term any more. On the other hand, literacy has always been ambivalent and it may be our task to make this ambivalence productive.

lotto project for high school learners data literacy is at once used to elucidate people about exploitative social relations and teach basic statistics. High school learners in New York did not only learn about the probability of chance-based games, but also about the social structures of the whole lottery game. Besides estimating the chance to win the lottery (basic statistics), they also learned that the lottery's profits go back to the state's budget to fund different social programs. But, as they also learned,

> lottery tickets are not purchased equally across all income brackets or all neighbourhoods. Low-wage workers buy more tickets than their higher-earning counterparts. What's more, the revenue from ticket purchases is not allocated back to those workers or the places they live. Because of this, scholars have argued that the lottery system is a form of regressive taxation – essentially a "poverty tax" – whereby low-income neighbourhoods are "taxed" more because they play more, but do not receive a proportional share of the profit (d'Ignazio & Klein, 2020: 67).

Furthermore, they learned this in practice as they went to and came from areas where people with lower income live and collected data about other people's experiences with the lottery in their own neighbourhood. The goal here was to create a data-driven argument which shows their own opinion as well as an analysis of the data they collected. The authors state that the learners became better in statistics but—what they state as even more important—the learners brought their knowledge back to their neighbourhoods. This approach challenged power by questioning the promises of the lottery and brought it to the people who were concerned. Although practical skills in statistics were needed, what seems more important for the project's success was the awareness about the relevance and force of the learner's own data analysis and the social impact their analysis can have which entails relational and affective moments. Those are usually no part of data literacy education although it made the project successful in the first place.[10]

Furthermore, this can be read as an example for the possible challenges for trust data literacy can bring forward. Besides the fact that lottery games as such work with people's hopes and trustfulness, the data project itself would not have been possible if there had not been an affective basis of trust in the high school students: people trusted them more as they tried to challenge the hope they put in the lottery system, because they were from their neighbourhood. Here Orrigi's criteria of reputation (high school students) and shared values and backgrounds (same neighbourhood) come back in. As this example shows, data literacy education as a form of critique needs to work with these soft factors in order to succeed in its challenge of power. *Emancipatory data literacy* here means not only statistics, but a social practice which is sensitive to qualitative and quantitative data, but also to affective structures of trust. In this project data literacy is inherently bound to an analysis of the affective arrangement the data project is part of and frames its method according to this arrangement and

[10] Although this case is taken from a data literacy education project, it seems fruitful to evaluate these interrelations for the case of citizen science, a practice that is often thought of as a means to enhance trust in science and which needs lots of emotional and communicative labor on the side of citizens *and* reseachers in order to be sucessful. See for exemplary evaluations of the accessory work behind citizen science and trustworthy data collection: (Gilfedder et al., 2019; Sandhaus et al., 2019; Thornton & Leahy, 2012).

the participant's dispositions. This important affective relation, that contributes to the project's success, is trust. Trust, which is not only established through knowledge and data expertise, but by acknowledgement of the socio-affective circumstances. This is true in both directions, the one of the learners and the one of the neighbourhood.

An emancipatory concept of data literacy can be understood as a critical practice within data-driven societies, not small data science for individuals. It is furthermore highly situated in the learner's social situation, but also dependent on the kind of data the learner works with. As an educational process, data literacy is still something individuals go through, but not necessarily as responsibilized individuals. However, *emancipatory data literacy* is not to be thought as a neo-liberal future skill, a cure against the relational, affective and insecure relation of trust, via responsibilization of the individual. Data literacy can be thought as an inclusive practice which teaches empowering practical skills—data awareness (Höper & Schulte, 2021)—as well as it sensitizes learners to data's epistemological limits and goes hand in hand with a critique of data as a currency of power.

6 Conclusion: Emancipatory Data Literacy as a Critical Attitude

If we take seriously that data literacy as a competence entails despite all its rational knowledge and practical skills affective elements, data literacy forms a critical attitude towards the authority of data as such and asks for context and power structures. This critical attitude enables learners to detect if there is something wrong and to know where to search and whom to ask. If we ask for people to (affectively) trust in (rational) data, data driven arguments, and facts data literacy can be one way to find a path through the foggy data forest. This is why we need to understand data literacy also as an ability to cope with the epistemological limits of data-driven arguments and their role in public discourse. Data literacy can sometimes simply be a competence to ask the right questions, because the data literate citizen has a structural knowledge and understands central elements of our data society although they may not be able to collect and analyse data. But they understand what data do and have an idea how arguments based on data function and where possible problems lie.

I understand *emancipatory data literacy* (as other neighbour-literacies) as an epistemic virtue for informed trust and mistrust in data-driven arguments and a critical attitude towards our data society as a whole. A data literate individual does not only ask: "Is this data visualization adequate to represent the problem?", but also "what is missing? Which problematic social structures are reproduced? What does this chart do beyond representation? And whom can I ask?" A data literate citizen is also someone who embodies the competence to trust, based on a basic ability to speak a bit of data, and challenge power. Someone who knows their limits and if necessary, forms associations with others who have the required expertise, who is furthermore able to communicate between different fields be they scientific disciplines or social groups. And while data literacy stands on reasonable grounds the data literate citizen is also affectively bound to trust, because they understand the structure of data and data-driven arguments. Data literacy is to be understood as a structural knowledge of how data-driven arguments function regarding their performativity, objectivity claims and role in power dynamics. As a political awareness it does not only ask for truth, but also for power and discrimination. This includes the ability to challenge or prove reputation and authority, in order to trust in a well-grounded manner.

Emancipatory data literacy is one element of the practices that make up trust in science. For scientific findings are always based one some kind of research data. Furthermore, science communication uses data visualizations or data stories as one important tool to show evidence and prove objectivity and truth. To understand the affective dimension of every objectivity claim, the tools of trust and trustworthiness can help to better understand which factors engender trust and may help citizens to find information when and how to trust, and to communicate that information wisely. As *emancipatory data literacy* includes not only skills and knowledge, but also values, it can be described as a critical attitude, is therefore always already closely related to trust and with its epistemological perspective especially to trust in science.

References

Ahlborn, J., Verständig, D., & Stricker, J. (2021). Embracing Unfinishedness. Kreative Zugänge Zu Data Literacy. *Medienimpulse, 59*(3), 1–42. https://doi.org/10.21243/mi-03-21-18

Alsop, S. (2015). Encountering science education's capacity to affect and be affected. *Cultural Studies of Science Education, 2016*(11), 551–565. https://doi.org/10.1007/s11422-015-9692-6

Anderson, C. (2008). The end of theory: The data deluge makes the scientific method obsolete. *Wired* (blog). https://www.wired.com/2008/06/pb-theory/

Brown, B. (2021). Responsibilization and recovery: Shifting responsibilities on the journey through mental health care to social engagement. *Social Theory & Health, 19*, 92–109. https://doi.org/10.1057/s41285-019-00097-x

Brown, W. (2017). *Undoing the demos: Neo-liberalism's stealth revolution.* Zone Books.

Burkhardt, M., Grashöfer, K., Miyazaki, S., & Weich. 2021. Welche Daten? Welche Literacy? Ein Kommentar zur Data-Literacy-Charta des Stifterverbandes. *Medienimpulse, 59*(4), 1–12. https://doi.org/10.21243/mi-04-21-13

Bush, A., de Gruisbourne, B., Schulz, C., & Matzner, T. (2022). Data Literacy Kompetenzrahmen für Hochschulen. Workingpaper 2021, January. https://www.campus-owl.eu/fileadmin/campus-owl/dalis/documents/Kompetenzrahmen_Workingpaper210921.pdf

Bhargava, R. (2015). Beyond data literacy: Reinventing community engagement and empowerment in the age of data. *Data Pop Alliance White Paper Series.*

Bhargava, R. (2022). Speaking data. *Datatherapy.org* (blog). March 16. https://datatherapy.org/2014/07/09/speaking-data/

Chun, W. H K. (2021). *Discriminating data. Correlation, neigbourhoods, and the new politics old recognition.* MIT Press

Criado Perez, C. (2019). *Invisible women. Exposing data bias in a world designed for men.* Penguin Random House

Dander, V. (2018). Medienpädagogik im Lichte | Im Schatten digitaler Daten. *MedienPädagogik: Zeitschrift Für Theorie Und Praxis Der Medienbildung, Zurück in Die Zukunft Der Medienpädagogik. «Subjekt», «Bildung» Und «Medien*Kritik» Im Lichte | Im Schatten Digitaler Daten.* https://doi.org/10.21240/mpaed/diss.vd.01.X. März: 1–134

Garland, D. (1996). The limits of the sovereign state: strategies of crime control in contemporary society. *The British Journal of Criminology*, 36 (4 (Autumn 1996)), 445–471.

Gilfedder, M., Robinson, C. J., Watson, J. E. M., Campbell, T. G., Sullivan, B. L., & Possingham, H. P. (2019). Brokering trust in citizen science. *Society & Natural Resources, 32*(3), 292–302. https://doi.org/10.1080/08941920.2018.1518507

Gitelman, L., & Jackson, V. (2013). Introduction. In *"Raw Data" is an Oxymoron* (pp. 1–14). MIT Press.

Gruisbourne, B. de, & Matzner, T. (2021). Herausforderungen digitaler Lehre. Eine Perspektive der Care-Ethik. In I. Neiske, J. Ostenhushenrich, N. Schaper, U. Trier, & Vöing, N. (Eds.), *Hochschule Auf Abstand. Ein Multiperspektivischer Zugang Zur Digitalen Lehre.* Bielefeld: transcript.

Hachmeister, N., Weiß, K., Theiß, J., & Decker, R. (2021). Balancing plurality and educational essence: Higher education between data-competent professionals and data self-empowered citizens. *Data.* https://doi.org/10.3390/data6020010

Höper, L., & Schulte, C. (2021). Datenbewusstsein: Aufmerksamkeit für die eigenen Daten. In Humbert, L. (Ed.), *Nformatik - Bildung von Lehrkräften in Allen Phasen.* Lecture Notes in Informatics (LNI). Gesellschaft für Informatik.

d'Ignazio, C. (2017). Creative data literacy. Bridging the gap between the data-haves and data-have nots. *Information Design Journal, 23*(1), 6–18.

d'Ignazio, C., & Klein, L. F. (2020). *Data feminism.* MIT Press.

Jones, K. (1996). Trust as an affective attitude. *Ethics, 107*(1), 4–25.

Karaian, L. (2014). Policing 'Sexting': Responsibilization, respectability and sexual subjectivity in child protection/crime prevention responses to teenagers' digital sexual expression. *Theoretical Criminology, 18*(3), 282–299. https://doi.org/10.1177/1362480613504331

Knorr Cetina, K. (1992). The couch, the cathedral, and the laboratory: On the relationship between experiment and laboratory in science. In *Science as practice and culture.* Chicago University Press.

Lemke, T. (2001). 'The birth of bio-politics': Michel Foucault's lecture at the Collège de France on neo-liberal governmentality. *Economy and Society, 30*(2), 190–207. https://doi.org/10.1080/03085140120042271

Leonelli, S. (2015). What counts as scientific data? A relational framework. *Philosophy of Science, 82*(December), 810–821.

Matzner, T. (2016). Beyond data as representation: The performativity of big data in surveillance. *Surveillance & Society, 14*(2), 197–210.

Matzner, T. (2017). Data science education as contribution to media ethics. In *Paderborn symposium on data science education at school level 2017: The collected extended abstracts* (pp. 28–32). Paderborn. https://www.telekom-stiftung.de/sites/default/files/files/PaderbornSymposiumDataScience2017_0.pdf

Mühlhoff, R. (2019). Affective disposition. In J. Slaby & C. von Scheve (Eds.), *Affective societies—Key concepts* (pp. 119–130). Routledge.

O'Malley, P. (2009). Repsonsibilization. In A. Wakefield & J. Fleming (Eds.), *The sage dictionary of policing* (pp. 522–644). SAGE Publications Ltd.

Origgi, G. (2008). Trust, authority and epistemic responsibility. *Theoria, Universidad Del Bío-Bío, 23/1*(61), 35–44.

Phoenix, J., & Kelly, L. (2013). 'You have to do it for yourself': Responsibilization in youth justice and young people's situated knowledge of youth justice practice. *The British Journal of Criminology* 53 (3 (May 2013)), 419–437. https://doi.org/10.1093/bjc/azs

Risdale, C. (2015). Strategies and best practices for data literacy education. Knowledge Synthesis Report.

Rosenberg, D. (2013). Data before the fact. *Raw Data Is an Oxymoron*, 15–40.
Sandhaus, S., Kaufmann, D., & Ramirez-Andreotta, M. (2019). Public participation, trust and data sharing: Gardens as hubs for citizen science and environmental health literacy efforts. *International Journal of Science Education, Part B: Communication and Public Engagement, 9*(1), 54–71. https://doi.org/10.1080/21548455.2018.1542752
Schüller, K., Busch, P., & Hindinger, C. (2019). Future Skills: Ein Framework für Data Literacy—Kompetenzrahmen und Forschungsbericht. Arbeitspapier Nr.27. *Hochschulforum Digitalisierung.* https://doi.org/10.5281/zenodo.3349865
Schüller, K., & Koch, H. (n.d.) Charter Data Literacy (Engl.). https://www.stifterverband.org/sites/default/files/data-literacy-charter.pdf
Shamir, R. (2008). The age of responsibilization: On market embedded morality. *Economy and Society, 37*(1), 1–19. https://doi.org/10.1080/03085140701760833
Slaby, J., Mühlhoff, R., & Wüschner, P. (2019). Affective arrangements. *Emotion Review, 11*(1), 3–12. https://doi.org/10.1177/1754073917722214
Suessenbach, F., Winde, M., Klier, J., & Kirchherr, J. (2021). Future skills 2021. 21 Kompetenzen für eine Welt im Wandel. Edited by Stifterverband für die Deutsche Wissenschaft e.V. https://www.stifterverband.org/medien/future-skills-2021
Thatcher, J. (2016). Data colonialism through accumulation by dispossession: New metaohors for daily data. *Society and Space, 34*(6), 990–1006.
The Federal Government. (2021). Data Strategy of the Federal German Government. An Innovation Strategy for Social Progress and Sustainable Growth. https://www.bundesregierung.de/resource/blob/992814/1950610/fb03f669401c3953fef8245c3cc2a5bf/datenstrategie-der-bundesregierung-englisch-download-bpa-data.pdf?download=1
Thornton, T., & Leahy, J. (2012). Trust in citizen science research: A case study of the groundwater education through water evaluation & testing program. *Journal of the American Water Resources Association (JAWRA),* 48 (5): 1032–40. https://doi.org/10.1111/j.1752-1688.2012.00670.x

Only a Theory? Substantive and Methodological Strategies for Regaining Trust in Science

Andrija Šoć and Monika Jovanović

Abstract Even after decades of emphasizing the importance of competent, impartial, and widely understandable communication of scientific claims, one can still often hear the phrase 'but it's just a theory'. Trust in science seems easily shaken, hard to acquire, and almost impossible to regain. The aim we have in this paper is to discuss the problem of trust in science and to outline a viable solution to it. Our approach unfolds in three successive stages. First, we will go through the empirical research pertaining to the issue. Second, we will take a more nuanced look at what the phrase 'trust in science' actually means. Third, we will point to some commonly suggested substantive strategies for restoring trust in science and argue that they will not be successful if they are not accompanied by methodological strategies aimed at a thorough understanding of the phenomenon of trust in science.

1 Trust and Its Correlates

The importance of trust in science has never been more obvious, nor more overtly called for, than in the past two years. However, polls show that trust in science is on the wane, as it has been for the past few decades. Understanding this phenomenon requires us to go beyond the general notion, or a broad demand for trust in science *simpliciter*. As we will claim, one of the key misunderstandings about trust in science lies in thinking, by both scientists and the general public, that the only item of interest regarding science is its results. In fact, the central problem is conveying the distinctive character of the scientific process itself, be it the question of methodology, obtaining results that do not match the predictions, misinterpreting data, etc.

What the general public can see as weaknesses of the scientific method—the lack of ready-made answers, attaching probabilities and uncertainties to an expressed

A. Šoć (✉) · M. Jovanović
Faculty of Philosophy, University of Belgrade, Belgrade, Serbia
e-mail: andrija.soc@f.bg.ac.rs

M. Jovanović
e-mail: mojovano@f.bg.ac.rs

M. M. Resch et al. (eds.), *The Science and Art of Simulation*,
https://doi.org/10.1007/978-3-031-68058-8_16

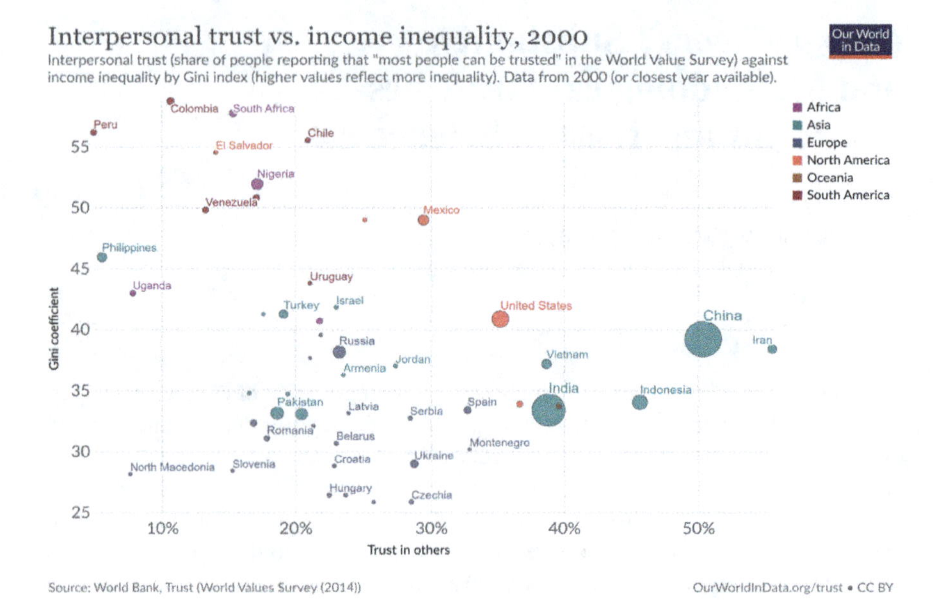

Fig. 1 Interpersonal trust versus income inequality. https://ourworldindata.org/grapher/interpersonal-trust-vs-income-inequality *Source* World Bank, World Value Survey (2000)

view, or frequent missteps in a wide variety of research projects—actually represents the natural process of discovering the elements of a wide range of complex phenomena. All this is clear enough to a scientist or a philosopher of science. However, why haven't others caught up to the satisfying degree? And how can that situation be remedied? Answering this question is paramount since a lack of trust in science, as has been evident in the past few years, can endanger public health, the environment, and overall quality of life.

One common approach to understanding trust in general, be it interpersonal, political or trust in science, is to explore whether such a phenomenon has any correlation with a set of relevant societal parameters.[1] There are two important types of such parameters—economic and political. Two of the most important economic parameters in correlation with the levels of trust—the Gini coefficient (GC) and the GDP *per capita.* First, let's look at a survey that measures trust against the GC (Fig. 1).

Even though the economic state of a country, which GC captures sufficiently representatively, might seem to reliably track the level of trust, we can see below that there is only weak correlation. For instance, countries such as Hungary and the Czech Republic, which score low on the equality portion of the chart, exhibit less trust than countries such as Vietnam or India. Perhaps expectedly, countries with extremely high Gini coefficient display a lack of interpersonal trust (Colombia and

[1] Trusting in other people is, it seems, a *sine qua non* of trust in science. Since we learn about scientific results from reporters, journalists, scientists, it stands to reason that if one doesn't place trust in people, placing trust in science will be that much harder.

South Africa are at the very top with respect to GC), but they also carry the baggage of decades-long conflicts and armed struggle. In addition, we can observe that almost no country scores higher than 50% regarding interpersonal trust, and only China and Iran score barely above that level. In so far as GC represents an informative picture of the state of a country's economic output and power, it is not reliably predictive of trust. As one might hold that GC isn't a sufficiently comprehensive parameter to correlate with the levels of trust, we will move on to exploring the second major correlate, GDP *per capita.* The following chart, created by 'World Value Survey', contains relevant data (Fig. 2).

As we can see, although there are examples of both measurements being expectedly high (Norway and Netherlands), or expectedly low (Philippines, Colombia, Ghana), there are countries with the disconnect between high GDP per capita and low interpersonal trust (Germany, United States, Singapore, Kuwait). There are very few countries in which over 50% of people agree with the trust statement ('most people can be trusted') and some have as low a GDP per capita as countries in which less than 10% of the people agree with the statement (compare, say, Vietnam and China with the Philippines). Thus, it seems that the economic power of a country,

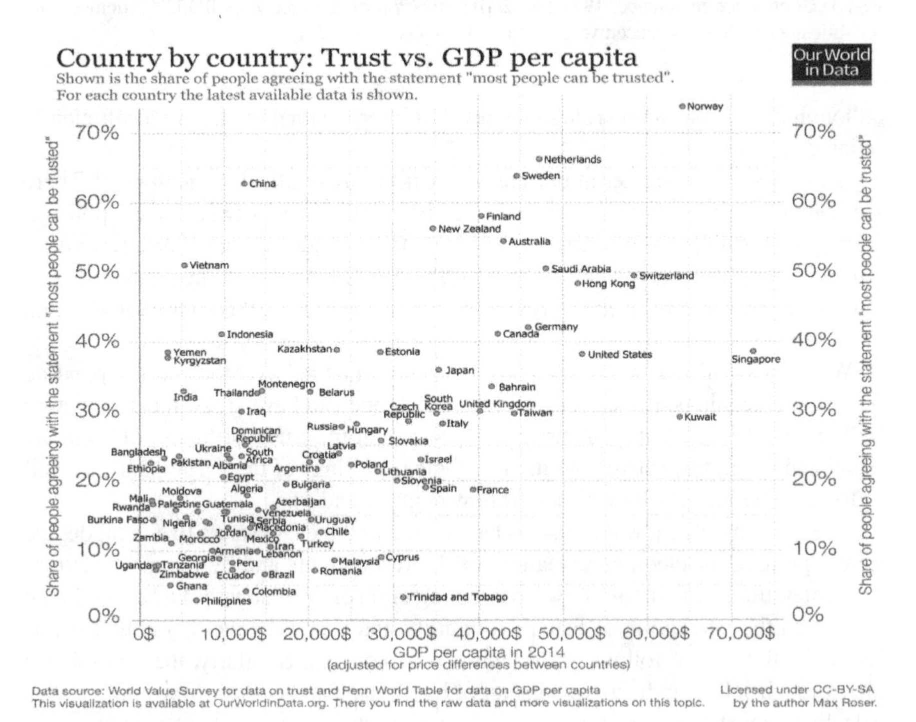

Fig. 2 Country by country: trust versus GDP per capita. https://ourworldindata.org/grapher/share-agreeing-most-people-can-be-trusted-vs-gdp-per-capita?xScale=linear *Source* World Value Survey, Penn World Table (2014)

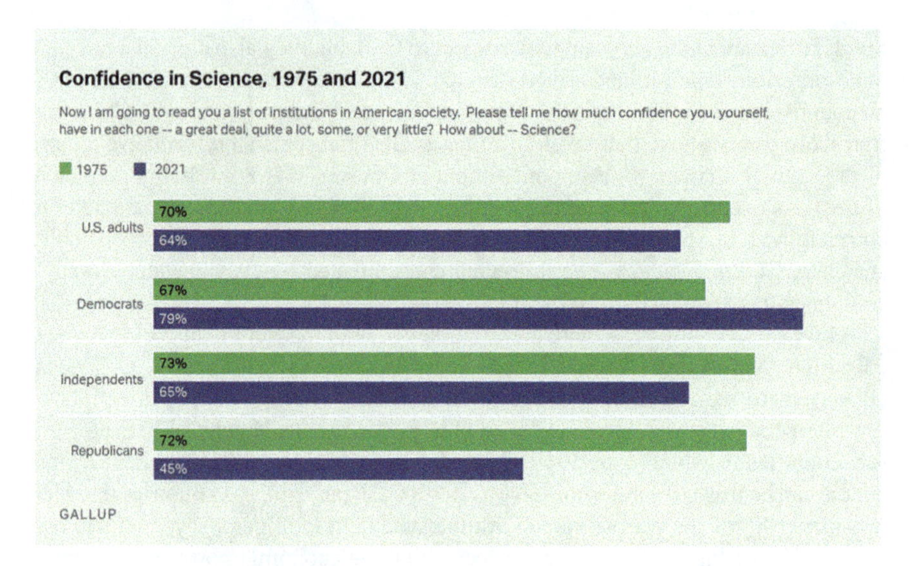

Fig. 3 Confidence in science, 1975 and 2001. https://news.gallup.com/poll/352397/democratic-republican-confidence-sciencediverges.aspx *Source* Gallup (2021)

although it may play some role in the level of interpersonal trust, isn't a sufficiently reliable explanatory factor.

Another set of correlates important for the issue are political or ideological. There are specific polls and surveys examining the relationship between one's political leaning and one's trust in science. One of the polls, conducted in the past few years, which seems particularly relevant, is the following Gallup poll, which makes it clear that there is a large drop in the reported trust in science in the conservative camp within the USA (Fig. 3).

While this poll illustrates how trust in science changed over the years depending on political leanings of the surveyed US adults, one can have an even better picture by looking at two additional surveys which match the trustworthiness of Anthony Fauci and CDC, respectively, to the ideological leanings of the participants, as well as to the news sources they closely follow (Figs. 4 and 5).

The two surveys show the trust in Dr. Fauci and the CDC varies to a great degree by the political or ideological leaning of the respondents and by the news sources they entertain. The first poll shows that the selection of news sources had a significant impact on the perception of Fauci's reliability and trustworthiness, and that it was lower for those who follow conservative news sources. Similarly, the second poll provides a brief but indicative snapshot of how trust changed over time and presents a decline almost across the board. It is again clear that the sharpest drop in both cases occurred within the group of participants who identify as Republicans.

One interpretation of the data above could be that, for some reason, those who are more inclined to adhere to the left-wing are prepared to trust in science more than those who gravitate toward conservatism. In support of this, a recent poll on the

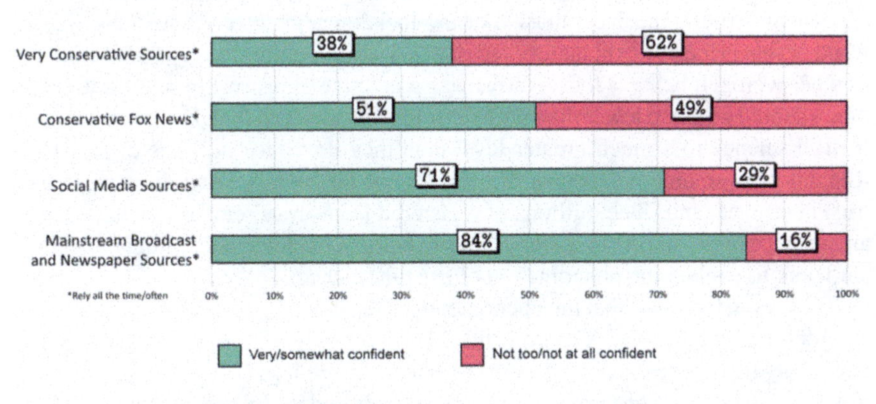

Fig. 4 Confidence in Dr. Anthony Fauci. https://www.annenbergpublicpolicycenter.org/public-trust-in-cdc-fda-and-fauci-holds-steady-survey-shows/ *Source* Annenberg Public Policy Center survey, June 2–22, June (2021)

Figure 16

Trust In The CDC And Dr. Anthony Fauci Has Declined Since April

Percent who say they have **a great deal** or **a fair amount** of trust in each of the following to provide reliable information on coronavirus:

● September 2020 ● April 2020

Centers for Disease Control and Prevention

Total 67% 83%

Democrats 74% 86%

Independents 70% 81%

Republicans 60% 90%

Dr. Anthony Fauci

Total 68% 78%

Democrats 80% 86%

Independents 71% 76%

Republicans 48% 77%

SOURCE: KFF Health Tracking Polls. See toplines for full question wording.

KFF

Fig. 5 Trust in the CDC and Dr. Anthony Fauci. https://www.kff.org/health-misinformation-and trust/report/kff-health-tracking-poll-september-2020/ *Source* KFF Health Tracking Polls (2020)

trust in the scientific community conducted by 'General Social Survey' in January 2022 shows that the gap between the citizens who identify as Democrats and those who identify as Republicans is the widest it has been in the past 49 years (GSS started conducting the poll in 1973). Currently, the gap is 68% to 32% (Borenstein & Fingerhuth, 2022). As Marcia McNutt, the president of the National Academy of

Sciences, comments on the data (ibid.), the reason for this gap is the Republicans' conservative sensibility, as well as their ideological opposition toward any form of mandates (be it vaccines, masks, or the lockdowns). However, we may wonder whether that is the key explanatory factor. If it were, wouldn't the gap be even wider? In other words, if being a conservative was predictive of distrust in science, then the figure should be much lower than 32%. Similarly, liberally inclined Democrats would trust in science to a much greater level than they do. Since millions of registered Republicans remain confident in the viability of the vaccines and do express their trust in science, and since millions of registered Democrats remain vaccine-hesitant and express their distrust in science, we can hardly conclude that political leaning is sufficient to explain the phenomenon of trust in science.

The reason for this is that the phenomenon is too complex to be explained by some set of correlative factors. A more fruitful approach might be to abandon the vague notion of 'trust in science' and examine what science itself encompasses and how trust itself involves a spectrum of multifaceted attitudes. To better understand how there is a mismatch between attempting to adequately understand trust in science and the way it is researched on the most general level, we can look at following survey (Fig. 6), conducted by Pew Research Center.

The contrasting views within this table demonstrate high levels of disagreement between professional scientists and the public not on general, but on very specific terms. Some of the widest gaps occur on the question of food safety, vaccine mandates, and climate change. There is almost no issue on which the scientists and the public are aligned and where they are, the issue at hand is not nearly as significant and far-reaching as the ones where there is a mismatch in attitudes.

The pertinent question to ask now is what the decrease in trust means. Some of the recent polls indicate that the generic survey questions on trust in science get largely affirmative answers. As we have seen in the Gallup poll shown earlier (Fig. 3), the US adults, who displayed great disapproval in the virtual consensus held by professional scientists in different areas still display an overall 64% confidence in science. But what exactly are they confident in, if not in the issues on which scientists fully agree? It seems, then, that to understand the phenomenon of trust in science, we must ask more nuanced questions and treat it as a complex whole that consists of different aspects, each of which could have multiple relations to others. We move to this topic in the next section.

2 Types of Trust and Complexity of Science

In this section, we want to explore the complexity of the phenomenon of trust in science. One of the key respects in which it is important to take this complexity into account is in the analysis of different types of surveys. On the one hand, surveys on trust in science often use the generalized notion of science. On the other hand, because science is a multifaceted phenomenon, different participants might focus on different aspects of science. This, in turn, would imply that the same response might have different meaning for different participants within the same group of people. Thus, how one interprets what trust means will determine how one will respond to a survey question.[2] However, if those conceptions vary, then there can be no uniform analysis of a survey result. Moreover, the research that relies on overly generalized notions cannot properly explain the phenomena these notions pertain to. We will illustrate this problem by examining a recent paper that distinguishes between trust in science and critical evaluation (O'Brien et al., 2021). The authors of the study aim to:

> identify two critical determinants of vulnerability to pseudoscience. First, participants who trust science are more likely to believe and disseminate false claims that contain scientific references than false claims that do not. Second, reminding participants of the value of critical evaluation reduces belief in false claims, whereas reminders of the value of trusting science do not. We conclude that trust in science, although desirable in many ways, makes people vulnerable to pseudoscience (O'Brien et al., 2021, 1).

Across four experiments, the study explored how people who report a high level of trust in science react to pseudoscientific claims as well as how emphasizing the need to trust in science affects the decline of trust in contrast to the effect of emphasizing the need for critical evaluation. What the researchers found was that the latter plays a greater role in regaining trust in science and in protecting oneself from pseudoscientific claims than the former:

> Our four experiments and meta-analysis demonstrated that people, and in particular people with higher trust in science (Experiments 1–3), are vulnerable to misinformation that contains pseudoscientific content. Among participants who reported high trust in science, the mere presence of scientific labels in the article facilitated belief in the misinformation and increased the probability of dissemination. Thus, this research highlights that trust in science ironically increases vulnerability to pseudoscience, a finding that conflicts with campaigns that promote broad trust in science as an antidote to misinformation (O'Brien et al., 10).

Even though authors correctly underline that merely promoting broad trust in science may not be as effective as usually thought, it seems that by proposing for critical evaluation to be explored independently from calls for trust, they fail to recognize that trust can only be based on critical evaluation. If it is not, then the reported trust isn't genuine. If by 'trust in science' we mean simply placing faith in whatever is presented under that title, then such an attitude is, contrary to the spirit of science, highly dogmatic. Such a dogmatic trust—trust not based on any well-founded reasons—should be contrasted with genuine trust or, as it might be called, enlightened trust. Thus, one could make a distinction between enlightened trust, dogmatic trust, and distrust.[3] An attitude of enlightened trust would, unlike that of dogmatic trust need to be already based on critical evaluation, whereas distrust,

---

[2] This phenomenon is explored, for example, by Davidov (2009).

[3] A different partition of trust-related attitudes often emphasizes a distinction between trust, mistrust and distrust. For instance, in two recent papers (See Jennings et al., 2021a, 2021b), the reserachers similarly make such a distinction. Their use of 'mistrust' follows several definitions, among others Lenard (2008) and Citrin and Stoker (2018). Lenard (2008, 313) defines mistrust as 'a cautious attitude towards others; a mistrustful person will approach interactions with others with a careful and questioning mindset' whereas distrust denotes 'a suspicious or cynical attitude towards others'.

similarly to dogmatic trust, is roughly accompanied by the lack of consideration of a particular topic or problem and is more focused on a person, group or institution which originates or disseminates the view that one distrusts.[4] However, the reported susceptibility to pseudoscientific claims shows that the self-reported trust was merely dogmatic.

Based on the previous paragraph, we may conclude that the study by O'Brien et al. has two important upshots. First, it correctly identifies the need for critical evaluation in the public reception of scientific claims. Secondly, it incorrectly disassociates critical evaluation from trust in science. In order to restore trust in science, it is paramount to first avoid conflating dogmatic and enlightened trust. This can be accomplished only if research methods (including, but not only, opinion polls, surveys and questionnaires) are modified in a way that can help us differentiate between the two. Furthermore, to understand such an attitude when it is directed specifically at science, these methods also have to elicit and demonstrate their subjects' awareness of the complexity of science.

While science is comprised of different aspects, trust in science is usually reduced to trust in the effects of science and, more specifically, in the scientific results achieved within various areas of scientific research. However, equating science with scientific results is too narrow. Even worse, the reason why pseudoscientific claims tend to get traction in light of the supposed scientific failures lies in the common insistence that one should trust science because it works. This, however, leaves an open question of how to view science when it doesn't seem to work, or if it fails to find a successful solution to a problem. In other words, once scientists express doubt or demonstrate uncertainty, a dogmatic pseudoscientific perspective can easily be put forward in order to gain public trust.[5] If that happens at the time of a crisis, it can have detrimental consequences.[6]

Citrin and Stoker (2018, 50) make the distinction that whereas trust refers to a belief in the trustworthiness of others, 'mistrust reflects doubt or skepticism about the trustworthiness of the other, while distrust reflects a settled belief that the other is untrustworthy'. As can be seen from these defintions, mistrust is viewed as something akin to skepticism. One possible qualm we might raise regarding the views above is that genuine, or enlightened, trust ought to be accompanied by a questioning mindset without leading to unfounded skepticism. A distinction that better captures this should thus highlight two superficially similar, but profoundly different trust-related attitudes such as dogmatic trust and enlightened trust, rather than two types of attitudes that are on the other end of the spectrum. A full division of trust-attitudes would have to capture the entire spectrum, so as to cover (at least) dogmatic trust, enlightened trust, mistrust and distrust. For a comprehensive definition of the concept of trust in terms of a general attitude, see, e.g. Šoć (2021a, 2021b).

[4] In adopting such a view of distrust, we are inclined to accept the definition put forward by Citrin and Stoker (see previous footnote), although we recognize that more in-depth research is required in order to fully take into account the entire spectrum of trust-related attitudes.

[5] Pseudoscience comes in different forms and mirrors different areas of genuine scientific disciplines. For pseudoscientific claims regarding climate change, see e.g. Torcello (2016). See also: Kitcher (1982), Pennock (2010), Smith (2012).

[6] See an extensive review of pertaining issues at Escandon et al. (2021). See also: Jovanović (2021).

When it comes to different aspects of science, they are well known within the scientific circles.[7] The crucial aspect is certainly the complex scientific methodology that includes various forms of experiments and studies, statistical analysis, modeling, use of informational technology, and more, none of which are featured within pseudoscience.[8] Aside from the application of rigorous methodological strictures, there are also important epistemic aspects of a proper scientific approach. These include the standard of falsifiability,[9] Bayesian belief-update tendencies, and the demand for supporting theories with sound evidence.

All these aspects stand in contrast with pseudoscientific claims, which are immune to any updates by way of constant construction of ad hoc modifications and promotion of unfalsifiable hypotheses and often resort to anecdotal evidence. If, for instance, someone advances the view that the thesis of intelligent design, rather than the theory of evolution, properly explains the complexity of life, they are also able to accommodate any sort of counterexample with an ad hoc modification.[10] And whereas the theory of evolution relies on concrete evidence, yields testable predictions, and has clear falsification standards, the ID claim doesn't have the aforementioned features of a proper scientific theory.

However, a common maneuver by the proponents of pseudoscientific claims is to treat their claims as at least either equal to or even superior to scientific views. One can frequently hear that scientific theory is 'only a theory'. Such a move, based on an incorrect understanding of the technical meaning of 'theory' represents a clear attempt to diminish science's significance, explanatory power, and claim to objectivity. The biggest problem, however, arises when fundamental misunderstanding of how science works carries over into a crisis that requires a novel scientific solution. The public health crises such as the coronavirus pandemic or the ever-increasing problem of climate change demand the application of strategies that the public can perceive as being overly radical, too unsafe, or too costly. If someone doubts the research that points toward implementation of such measures or incorrectly expects science to provide solutions either quickly and completely or not at all, they will be firmly against this and will be inclined to vote for the party or the politician who is ready to proclaim that the research is wrong and that the money required for dealing with public health crises or environmental issues will not be invested in such causes. For example, the investment in renewable energy sources or reduction of off-shore drilling is perceived as a threat to the economy of a country.[11] The development

[7] An extensive overview can be found in Humphreys (2016).

[8] There are different ways to demarcate between science and pseudoscience. A useful overview is provided by Hansson (2021, ch. 4). Here we don't aim to provide an essentialist definition of science (see Dupre, 1993 for an alternative, Wittgensteinian suggestion). Rather, we are attempting to enumerate some crucial aspects of science in order to gain a better understanding the meaning of different attitudes toward science.

[9] The classic statement can be found in Popper (1959).

[10] One of the clear distinctive marks of pseudoscience is its resistance to counterarguments. See, for instance, Thagard (1978). One of the common forms of pseudoscientific thinking are, for instance, conspiracy theories.

[11] See, for example, Stanley (2015, 16, 90), and McKinnon (2016, 210).

of vaccines is perceived by a sizeable number of people in many countries either as an unnecessary waste of time and money or as a direct danger to the health of individuals.[12] Even though the scientific consensus tells us that spending money on renewable energy sources is cheaper than letting climate change endanger the environment, or that the vaccines are safe and effective, the skepticism persists.

Thus, when O'Brien et al. (2021, 11–12) conclude that 'our results suggest that advocacy for trusting science must go beyond scientific labels, to focus on specific issues, critical evaluation, and the presence of consensus among several scientists as a source or claims of scientific studies in support of a claim', they seem to both underestimate the entrenchment of unreasonable skepticism and idealize the level of trust in scientists themselves. After all, the research data we mentioned earlier (Fig. 6) demonstrates the discord between the public and the scientific community. It also demonstrates the futility of simply insisting on the scientific consensus when the public disagrees with scientists on at least a dozen major issues. Where there is a genuine issue to be discussed in cases when experts disagree,[13] here the problem is even more fundamental, as the public fails to put the trust, not in one or the other expert, but in the large majority of experts working on key problems. The implications are far-reaching and the effect of such an attitude on failure to get vaccinations and thus endangering public health has been visible during 2021 (and has continued into 2022 to a degree, although the situation has been improving).

The key problem, then, is not merely that people distrust scientific claims, or that they, as we mentioned before, exhibit misplaced, or dogmatic support for scientific claims, thereby reducing science to the level of pseudoscience. It is, rather, that the public is prone to misunderstanding what it is that is distinctive of science as an exploratory and explanatory endeavor and what it is that makes science unique. Again, research methodology, mathematical and computational tools, falsifiability, Bayesian reasoning, testability, and predictive power are all aspects of a scientist's approach to any problem. To trust in science is to trust not in some, but in all of them, as well as in scientists themselves. And when something turns out to go against these elements—if a scientist turns out to be corrupt or unreliable, if a theory yields wrong predictions, if scientists state that they do not yet know something, or that a study can't be replicated,[14] the proper response wouldn't be to dismiss the entire endeavor but to carefully look at where things went wrong. As the concept of trust in science is multifaceted, so is the concept of distrust. This makes them difficult to be precisely researched by employing generalized approaches. As we noted at the beginning of this section, the same answer to a survey question can mean different things. One person may trust in science in the sense of trusting in the expertise of scientists. Another may trust in research methodology in general but believe that

---

[12] See, for instance, https://healthcareready.org/wp-content/uploads/2021/06/HcR-2021-Domestic-Preparedness-Poll_Key-Findings.pdf, 21.

[13] Expert disagreement can in some instances present a challenge for non-experts (see: Goldman, 2001), but there are views that it is also conducive to creating a consensus among experts. See, for instance, Dellsén (2018). See also a case study on expert disagreement by de Cruz and de Smedt (2013).

[14] See more on the issues of reproducibility in Baker (2016).

Opinion Differences Between Public and AAAS Scientists

% of each group saying the following

	Among AAAS members surveyed			
Biomedical sciences	**All AAAS members surveyed**	**Working Ph.D. Scientists**	**Active Research Scientists**	**U.S. adults**
Safe to eat genetically modified foods	88	90	91	37
Favor use of animals in research	89	92	92	47
Safe to eat foods grown with pesticides	68	72	71	28
Humans have evolved over time	98	99	99	65
Childhood vaccines such as MMR should be required	86	87	87	68
Climate, energy, space sciences				
Climate change is mostly due to human activity	87	88	90	50
Growing world population will be a major problem	82	83	83	59
Favor building more nuclear power plants	65	65	66	45
Favor more offshore oil drilling	32	31	30	52
Astronauts essential for future of U.S. space program	47	47	46	59
Favor increased used of bioengineered fuel	78	80	79	68
Favor increased use of fracking	31	29	28	39
Space station has been a good investment for U.S.	68	65	64	64
N	3,748	1,627	1,246	2,002

Survey of U.S. adults Aug. 15-25, 2014. AAAS survey Sept. 11-Oct. 13, 2014. "Working Ph.D. Scientists" are those employed full time who have a doctorate degree in a medical, natural or physical science; "Active Research Scientists" are "Working Ph.D. Scientists" who also report having received a research grant within the past five years.

PEW RESEARCH CENTER

Fig. 6 Opinion differences between public and AAAS scientists. *Source* Pew Research Center

scientists are corrupt or incompetent. The distrust may be due to how science fails to provide certainty of answers regarding some acute issues.

Trust in science, then, was revealed to be the trust in the applicability of science or its results, but not in its methodology or its Bayesian features. Since science cannot be disassociated from any of its distinctive features, failure to recognize the significance of any of them represent the failure to truly trust in science. If so, the question is then not only how to reformulate the studies of the level of trust in science, but also how to help the public recognize the multifaceted nature of the scientific process and to gain an accurate insight into what its true strengths are, especially in contrast to any proposed alternatives to the scientific mode of research. In the next section, we turn to these questions.

3 The Road to Enlightened Trust

Before outlining some possible suggestions for improved scientific communication, we need to isolate key communicative challenges the public and the scientists are commonly faced with.[15] The first of these challenges is the phenomenon of echo chambers.[16] On a more general level, we can describe it as a cluster of mechanisms that reinforce dogmatic attitudes toward a subject. These mechanisms feature behavioral feedback loops. One such loop is represented by a choice to follow news sources that adhere to a single ideology or perspective. Another is reflected in a choice to discuss issues only with the individuals who already share some specific view. Even more perniciously, some loops aren't directly reinforced by our behavior but are only partly influenced by it. One such mechanism is AI-based algorithms featured in the way social media filters a high number of different information sources and displays to end-users only those that align with the previous choice of information sources (Vaidhyanathan, 2018).

Another challenge is the frequency with which the media and the public can equate studies that were peer-reviewed and those which weren't. The websites which enable preprints to be posted serve a meaningful role in accessing important insights in an open arena. However, reliable scientific communicators must be cautious about presenting the preprints in this light and share the results with all the appropriate reservations before they are peer-reviewed.[17] Part of the problem in recognizing this need is the overt sensationalism of the traditional media, which tends to unfairly increase the public's expectations from scientific endeavors while also misrepresenting how such endeavor actually unfolds.

However, the road toward turning those sentiments into an attitude of genuine trust, the trust based on information and reasons, is not difficult to find, nor to traverse. The struggle against dogmatism has been present in both science and philosophy since their very beginnings, even though both have suffered long periods of stagnation due to the failure to address the issue successfully in different eras. Modern philosophy ushered an age where scientific thinking could finally rely on the authority of reason, not the reason of authority and different authors had different suggestions for how to enhance the critical mindset. Here we want to point out John Stuart Mill's poignant words and see how they presciently help propose an actionable solution to the aforementioned challenges:

> The steady habit of correcting and completing [one's] own opinion by collating it with those of others, so far from causing doubt and hesitation in carrying it into practice, is the only stable

[15] One encouraging trend is that the role of scientific communication has seemingly come to the forefront. See, for example Davis (2022). However, it will take much more than simply describing the failure of communication or connecting it only to political or ideological agendas to arrive at actionable corrective communication strategies.

[16] See, for instance, Parsell (2008) and Nguyen (2020).

[17] The problem with preprints still isn't sufficiently recognized and ought to garner more research in order to be fully recognized as a serious problem. For a recent view along these lines see Mullins (2021).

> foundation for a just reliance on it: for, being cognisant of all that can, at least obviously, be said against him, and having taken up his position against all gainsayers – knowing that he has sought for objections and difficulties, instead of avoiding them, and has shut out no light which can be thrown upon the subject from any quarter – he has a right to think his judgment better than that of any person, or any multitude, who have not gone through a similar process (Mill, 2003, 103).

What Mill describes here is a multi-step process that includes a Bayesian way of thinking in the form of constantly correcting one's own opinion, eliminating echo chambers by constantly comparing one's view to the opposing side, and, equally importantly, strengthening it by submitting it to the harshest rational criticism possible. To close the gap between the public and the working scientists, the public has to incorporate these elements into their way of thinking. Since scientists already practice this, the only way to achieve such an outcome is for scientists themselves to become educators.

Scientific communication, thus, must primarily increase in quality by scientists themselves becoming the conduit between what science is and how science is perceived, between what scientific results mean and how these results are understood. This can come in a form of frequent articles in widely-read daily or weekly publications, public lectures, clear and engaging social media posts, and the like. Such contributions, aside from accurately discussing how science genuinely leads to significant discoveries, must also constantly highlight the multifaceted nature of scientific research, contrasting it explicitly to all the flaws that are too often misrepresented as the strength of pseudoscience. Over time, and ingrained through school curricula, such a process should lead to the closing of the gap between the professional scientists and the public. Only then can science be fully harnessed to the greatest possible degree, and any ensuing crises will have a greater chance of being successfully overcome.

These solutions for improving the level of scientific communication in some sense merely reiterate what has become a general call for the need to trust in science. Suggestions similar to those given in the previous paragraph are put forward by Parikh (2021), who argues that science and scientists have to earn the public's trust by employing a more widely understandable jargon and by contextualizing the process of scientific discovery. At the end of the article, Parikh states that 'fortunately, we start from a solid foundation', because majority of people do agree that science and technology make our lives better.[18] However, as we have seen, the issue with such an appeal to broad trust is that more specific research questions show the public doesn't believe in the positive impact of the GMO foods, in the need for mandating vaccinations, etc. What, we may ask, are the scientific contributions that make our lives better—those which supposedly drive the public's trust in science—if not reducing

[18] Similar point is argued in a *Time* article by Oreskes (2019b), as well as in Oreskes (2019a). See more at https://time.com/5709691/why-trust-science. Furthermore, a project at Harvard aims to leverage 'data science, science and technology studies, and related disciplines to analyze the breakdowns in public trust, and to ask what steps could be taken to promote better mutual understanding'. https://datascience.harvard.edu/trust-science. See also Rolin (2020).

the problems of hunger and poverty, or reducing dangers from epidemics of infectious diseases?

The issue, then, with this and similar attempts to find solutions for the problem of trust in science (some of which we mentioned in the footnote above) is that they view it independently of problems of how to properly research this complex phenomenon. The substantive issues are not recognized as being inextricably connected to methodological issues. While there is a general agreement on which strategies to apply in order to improve scientific communication, if they are as broad as the generalized notion of 'trust in science', they will also be as vague and as ineffective as science itself seems to those who initially distrust it. Thus, it is important to make a distinction between substantive strategies, aimed at restoring trust in science, and methodological strategies, aimed at better understanding the spectrum of trust-related attitudes toward science (both enlightened and dogmatic trust, but also mistrust and distrust).

In contrast to commonly held views, we want to emphasize that applying substantive strategies must go hand in hand with the improvement of research methodology. Without first using better methodology to understand the phenomenon of trust, how can there be an effective application of substantive communication strategies at all? The most one could reasonably hope to achieve is strengthening an already present dogmatic trust and reinforcing the enlightened trust, where it exists. This would, ironically, be another instance of an echo chamber, a mutual agreement between those who are already prone to think along the same lines.

To counter such an effect, our first suggestion is for the notion of trust in science to be subdivided into its different components and surveyed in parallel to the generalized notion of trust. The aspects of science we discussed earlier should be carefully delineated and presented in different experimental settings. Polls, surveys, and questionnaires are certainly an important component of developing a detailed understanding of how the public's perception of science and its trust in it evolves, rises, falls, or stays steady. However, they have to be more finely-tuned to the complexity of both the attitude towards science and to the complexity of science itself. Only if we isolate different components of this complex attitude can we find which factors are explanatorily important and which are merely spurious. An important component of that research ought to be longitudinal studies, which could track how one's attitude toward science changes with the development of debates in different areas of science, while also trying to connect it to levels of misinformation present in different types of media. Additionally, our second suggestion is for different discussions between those who belong on the entire spectrum of trust to be conducted in order for researchers to see to what degree the initial distrust or dogmatic trust are amenable, and under what conditions.[19] Only with insights gained from such research is it reasonable to formulate different substantive strategies for approaching and educating general public.

---

[19] The type of discussion we have in mind has been extensively researched in the form of deliberative discussion and applied to various educational settings. See, e.g. Šoć (2021a, 2021b) for one possible view of how such discussions can be fruitful in an educataional setting. Some of the most prominent sources that describe deliberative discussions in experimental settings are, e.g. Fishkin (2009) and Steiner (2012).

The success of any approach cannot, of course, be guaranteed in advance. But if we are right, what is certain is that the crude notion of generalized trust in science would at least be supplanted with a more detailed, particularized notion of what science is. In so far as it more accurately reflects the object of trust, different avenues of researching it would certainly help us gain a better view of the subjects of trust—the individual members in any society who are simultaneously the most affected by how science works and who also have the most to gain by accurately reflecting on what science is.

Acknowledgements This paper was written as a part of the research project "Humans and Society in Times of Crisis", which is funded by the Faculty of Philosophy, University of Belgrade. The realization of this research was financially supported by the Ministry of Science, Technological Development, and Innovations of the Republic of Serbia, as a part of financing scientific research at the University of Belgrade, Faculty of Philosophy (contract number 451-03-47/2023-01/ 200163). We would like to thank editors and reviewers for insightful comments and valuable suggestions.

References

Baker, M. (2016). 1,500 scientists lift the lid on reproducibility. *Nature, 533*, 452–454. https://doi.org/10.1038/533452a

Borenstein, S., & Fingerhut, H. (2022). Americans' trust in science now deeply polarized, poll shows. *AP News*, January 26. Retrieved January 29, 2022, from https://apnews.com/article/coronavirus-pandemic-science-health-covid-19-pandemic-4e99139d995581319dffab4107627a5e

Citrin, J., & Stoker, L. (2018). Political trust in a cynical age. *Annual Review of Political Science, 21*, 49–70. https://doi.org/10.1146/annurev-polisci-050316-092550

Davidov, E. (2009). Measurement equivalence of nationalism and constructive patriotism in the ISSP: 34 countries in a comparative perspective. *Political Analysis, 17*, 64–82. https://doi.org/10.1093/pan/mpn014

Davis, J. (2022). Why are Americans confused about COVID? Blame it on poor communication. *USA Today*, January 26. https://www.usatoday.com/story/opinion/2022/01/26/cdc-covid-messaging-problems-fauci-rochelle-walensky/6563468001/?gnt-cfr=1

De Cruz, H., & De Smedt, J. (2013). The value of epistemic disagreement in scientific practice: The case of Homo Floresiensis. *Studies in History and Philosophy of Science Part A, 44*, 169–177. https://doi.org/10.1016/j.shpsa.2013.02.002

Dellsén, F. (2018). When expert disagreement supports the consensus. *Australasian Journal of Philosophy, 96*, 142–156. https://doi.org/10.1080/00048402.2017.1298636

Dupré, J. (1993). *The disorder of things: Metaphysical foundations of the disunity of science.* Harvard University Press.

Escandón, K., Rasmussen, A. L., Bogoch, I. I., et al. 2021. COVID-19 false dichotomies and a comprehensive review of the evidence regarding public health, COVID-19 symptomatology, SARS-CoV-2 transmission, mask wearing, and reinfection. *BMC Infect Disease, 21*(710). https://doi.org/10.1186/s12879-021-06357-4

Fishkin, J. (2009). *When the people speak: Deliberative democracy & the public consultation.* Oxford University Press.

Goldman, A. (2001). Experts: Which ones should you trust? *Philosophy and Phenomenological Research, 63*, 85–110. https://doi.org/10.2307/3071090

Hansson, S. O. (2021). Science and pseudo-science. In E. N. Zalta (Ed.), *The Stanford encyclopedia of philosophy*. Retrieved January 29, 2022, from https://plato.stanford.edu/archives/fall2021/entries/pseudo-science

Humphreys, P. (Ed.). (2016). *The Oxford handbook of philosophy of science*. Oxford University Press. https://doi.org/10.1093/oxfordhb/9780199368815.001.0001

Jennings, W., Stoker, G., Bunting, H., Valgarðsson, V., Gaskell, J., Devine, D., McKay, L., & Mills, M. C. (2021a). Lack of trust, conspiracy beliefs, and social media use predict COVID-19 vaccine hesitancy. *Vaccines, 9*, 593. https://doi.org/10.3390/vaccines9060593

Jennings, W., Stoker, G., Valgarðsson, V., Devine, D., & Gaskell, J. (2021b). How trust, mistrust and distrust shape the governance of the COVID-19 crisis. *Journal of European Public Policy, 28*, 1174–1196. https://doi.org/10.1080/13501763.2021.1942151

Jovanović, M. 2021. Krize, misaoni eksperimenti i fikcija: moralne intuicije između teorije i prakse [Crises, thought experiments and fiction: Moral intuitions between theory and practice, in Serbian]. In N. Cekić (Ed.), *Etika i istina u doba krize* [Ethics and truth in a time of crisis] (pp. 271–282). JDP Službeni glasnik.

Kitcher, P. (1982). *Abusing science: The case against creationism*. MIT Press.

Lenard, P. T. (2008). Trust your compatriots, but count your change: The roles of trust, mistrust and distrust in democracy. *Political Studies, 56*, 312–332. https://doi.org/10.1111/j.1467-9248.2007.00693.x

McKinnon, C. (2016). Should we tolerate climate change denial? *Midwest Studies in Philosophy, 40*, 205–216. https://doi.org/10.1111/misp.12056

Mill, J. S. (2003). On liberty. In M. Warnock (Ed.), *Utilitarianism and on liberty*. Blackwell.

Mullins, M. (2021). Opinion: The problem with preprints. The Scientist, November 21. Retrieved January 27, 2022, from https://www.the-scientist.com/critic-at-large/opinion-the-problem-with-preprints-69309

Nguyen, C. T. (2020). Echo chambers and epistemic bubbles. *Episteme, 17*, 141–161. https://doi.org/10.1017/epi.2018.32

O'Brien, T. C., Palmer, R., & Albarracin, D. (2021). Misplaced trust: When trust in science fosters belief in pseudoscience and the benefits of critical evaluation. *Journal of Experimental Social Psychology, 96*, 1–13. https://doi.org/10.1016/j.jesp.2021.104184

Oreskes, N. (2019a). *Why trust science?* Princeton University Press.

Oreskes, N. (2019b). Science isn't always perfect—but we should still trust it. *Time*, October 24. https://time.com/5709691/why-trust-science

Parikh, S. (2021). Why we must rebuild trust in science. *The PEW Charitable Trusts*, February 9. https://www.pewtrusts.org/en/trend/archive/winter-2021/why-we-must-rebuild-trust-in-science

Parsell, M. (2008). Pernicious virtual communities: Identity, polarisation and the Web 2.0. *Ethics and Information Technology, 10*, 41–56. https://doi.org/10.1007/s10676-008-9153-y

Pennock, R. T. (2010). The postmodern sin of intelligent design creationism. *Science and Education, 19*, 757–778. https://doi.org/10.1007/s11191-010-9232-4

Popper, K. R. (1959). *The logic of scientific discovery*. K. R. Popper (Trans.). Basic Books.

Rolin, K. (2020). Trust in science. In J. Simon (Ed.), *The Routledge handbook of trust and philosophy* (pp. 354–366). Routledge.

Smith, K. (2012). Homeopathy is unscientific and unethical. *Bioethics, 26*, 508–512. https://doi.org/10.1111/j.1467-8519.2011.01956.x

Stanley, J. (2015). *How propaganda works*. Princeton University Press.

Šoć, A. (2021). Ka obuhvatnom pojmu poverenja [Toward a comprehensive concept of trust, in Serbian]. *Kritika, 2*, 35–56. https://doi.org/10.5281/zenodo.4784241

Šoć, A. (2021). Deliberative education and quality of deliberation: Toward a critical dialogue and resolving deep disagreements. In I. Cvejić, P. Krstić, N. Lacković, O. Nikolić (Eds.), *Liberating education: What from, what for?* (pp. 123–145). Institute for Philosophy and Social Theory, University of Belgrade (Novi Sad: Sajnos).

Steiner, J. (2012). *Foundations of deliberative democracy*. Cambridge University Press.

Thagard, P. R. (1978). Why astrology is a pseudoscience. *Philosophy of Science Association, 1*, 223–234. https://doi.org/10.1086/psaprocbienmeetp.1978.1.192639
Torcello, L. (2016). The ethics of belief, cognition, and climate change pseudoskepticism: Implications for public discourse. *Topics in Cognitive Science, 8*, 19–48. https://doi.org/10.1111/tops.12179
Vaidhyanathan, S. (2018). *Antisocial media: How Facebook disconnects us and undermines democracy*. Oxford University Press.

Undermining Trust in Science: No Fraud Required

Yujia Song and Maciej Balajewicz

Abstract In this paper, we caution against a sort of naive trust in science, which assumes that normative standards of scientific inquiry are upheld except in rare cases of fraud or research misconduct. Since scientists are subject to the norms and practices of the institutions that support their work, we need to look beyond *individuals'* behavior to examine the *institutional* contexts that shape the practice of science. This paper probes into two central aspects of the scientific institution as they play out in the U.S.—peer review and grant funding. We argue that problems more pervasive and intractable than outright fraud threaten to undermine trust in science. The push for positive results by journals and funding agencies alike, coupled with institutional pressures to publish, entails a serious misalignment of goals for scientists—the kind of efficiency needed to maximize positive outputs comes at the expense of the time, effort, and discipline that good science calls for.

1 Introduction

Is it obvious that we should trust science? As cliché as it might sound, the answer depends on what "science" means. If we mean an enterprise that aims at truth about the empirical world (or at least the best model of it), its practitioners working diligently toward that goal by employing the scientific method and exercising epistemic virtues such as intellectual humility, honesty, and open-mindedness, then we clearly have good reason to trust science. The interesting question, however, is how well science *as it is done* approaches this Platonic ideal. Since real scientists are also *employees* (in academia or industry), their community embedded in and supported by organizations such as universities, government agencies, and corporations, the question cannot be settled by looking at individual instances of research misconduct alone—we must

Y. Song (✉)
Philosophy Department, Salisbury University, 1101 Camden Avenue, Salisbury, MD 21801, USA
e-mail: ysong@salisbury.edu

M. Balajewicz
Independent Researcher, Boulder, USA
e-mail: maciej@maciejbalajewicz.com

M. M. Resch et al. (eds.), *The Science and Art of Simulation*,
https://doi.org/10.1007/978-3-031-68058-8_17

also examine the institutional contexts that shape and regulate the process of scientific inquiry.

This paper probes into two central aspects of the scientific institution as they play out in the U.S.—peer review and grant funding—though we believe every stage in the scientific professional "pipeline," from the recruitment and training of graduate students to the career milestones of researchers (e.g., hiring, promotion, etc.), needs a thorough, critical examination if we truly care about the integrity of science. We focus on peer review because it forms the core of the "self-correcting mechanism" of science. Although grants fall outside of "science" in the abstract sense, they are integral to the actual practice of science. Given how much scientists (in and outside academia) rely on government funding to conduct research, it is surprising how little attention has been paid to the role of funding practices in shaping how scientists work.

We argue that problems more pervasive and intractable than outright fraud threaten to undermine the trustworthiness of the peer review and grant funding systems. The push for positive results by journals and funding agencies alike, coupled with institutional pressures to publish, entails a serious misalignment of goals for scientists—the kind of efficiency needed to maximize positive outputs comes at the expense of the time, effort, and discipline that good science calls for. Thus, we caution against naive trust in science not because individual scientists may fall short of the rational ideals of science—sometimes even intentionally—but because institutional pressures can often constrain their attempts to realize those ideals.

2 Insider's Perspective

Before we proceed, some preliminary remarks about our methodology are in order. Since we are concerned with science as it *is* done rather than as it *appears* to or *should* be done, it is essential to consult empirical facts about its practice rather than abstract models based on popular imagination or philosophical theorizing. Given the primacy of insider knowledge for understanding a practice, we draw on descriptions of first-personal experience to explore the inner workings of peer review and grant funding. This insider's perspective is augmented with metascientific studies that examine patterns and trends in publications and funded projects.

Although first-personal accounts lack the objectivity that empirical studies promise, they are indispensable for our inquiry given the paucity of research into the day-to-day experience of scientists. However comprehensive the metascientific studies are, they only reveal the *products* of peer review and grant funding systems—the actual *processes* of decision-making, including *motivations* and *contexts*, at various stages remain hidden behind the data. When authors address these issues in their studies, they typically fall back on a sort of common knowledge among "insiders." For example, one of the most influential papers on the topic (Edwards & Roy, 2017) states, "While there is virtually no research exploring the impact of perverse incentives on scientific productivity, most in academia would acknowledge a collective shift in our behavior over the years (Table 1), emphasizing quantity at the expense of quality (53)." The table, which lists ten "growing perverse incentives in academia" along with their *intended* effects and *actual* effects on individual researchers and universities is attributed to personal communication with John Regehr, a computer science professor (ibid., 52).[1] In other words, the causal relationships presented as facts in the table do not come from systematic empirical studies; the table is slightly modified from Regehr's personal blog post (Regehr, 2011), based presumably on his own experience and observations.

Notably, the adverse effects Regehr describes are "collective" (in Edwards and Roy's words), not individual; they are normalized behavior, not blatant violations of research ethics. This observation is shared by veteran researchers and newcomers alike. A PhD student in computer science goes so far as to claim, "Explicit academic fraud is, of course, the natural extension of the sort of mundane, day-to-day fraud that most academics in our community commit on a regular basis" (Buckman, 2021). Buckman gives some examples of "mundane fraud":

> Trying that shiny new algorithm out on a couple dozen seeds, and then only reporting the best few. Running a big hyperparameter sweep on your proposed approach but using the defaults for the baseline. Cherry-picking examples where your model looks good, or cherry-picking whole datasets to test on, where you've confirmed your model's advantage. Making up new problem settings, new datasets, new objectives in order to claim victory on an empty playing field. Proclaiming that your work is a "promising first step" in your introduction, despite being fully aware that nobody will ever build on it.

We quote Buckman's critique at length to show the gap between the insider's perspective and outsider's.[2] Buckman may be overly cynical, but as we show below, there is wide agreement among "insiders" that broader systemic issues do more harm to science than "explicit academic fraud."

3 Peer Review

Incidences of academic fraud are alarming against the background assumption that peer review *usually* works. We are comforted by the thought that despite the media attention such cases receive, only a small number of irresponsible researchers are

[1] A 2019 report by the National Academies of Sciences, Engineering, and Medicine (NASEM), titled *Reproducibility and Replicability in Science* (NASEM, 2019) in turn cites this paper as a major source on the topic of misaligned incentives.

[2] Ironically, a formal investigation by science "insiders" like the NASEM report plays down such personal observations, blaming instead the news media for distorting the public's perception of science (NASEM, 2019). As journalists pursue "rare, unexpected, or novel events and topics," the authors write, they tend to focus on "single-study, breakthrough results" (158). The report suggests that the (mis)representation in the media may even go beyond selective bias to intentional inaccuracy: "There is some evidence that media stories contain exaggerations or make causal statements or inferences that are not warranted when reporting on scientific studies (ibid.)."

violating the norms of conduct and a small number of editors and reviewers are not sufficiently careful in carrying out their respective duties. However, this assumption turns out to be rather questionable. As we look into the roles of the editor, reviewer, and author, it is instructive to keep in mind the norms that *should* govern the peer review process such as those suggested by Longino (2002, 206) for effective "critical scrutiny" of research within the scientific community: "(1) the availability of venues for and (2) responsiveness to criticism, (3) public standards (themselves subject to critical interrogation), and (4) tempered equality of intellectual authority."

3.1 The Editor

What gets reviewed is already filtered through the overarching goal of positive results: journals discourage negative results.[3] This remains true even after the Reproducibility Project (Open Science Collaboration, 2015) brought the replication crisis in psychology to public attention in 2015. A 2017 study finds that only 3% of 1151 psychology journals were found to permit replication studies in their instructors to authors (Martin & Clarke, 2017). This means that the replication crisis is not merely one of *product*, but one of *process.* It is unlikely to go away as long as journals do not change their practice; yet the gatekeepers of these journals are themselves scientists, subject to external pressures on scientific research to produce short-term "deliverables."

If we consider replication a form of criticism, we can see that the first of the four conditions suggested by Longino, availability of "venue" for criticism, is already compromised even before peer review starts. Longino herself acknowledges that "a complex set of processes in the institutions of contemporary science… works against satisfaction of this requirement (2002, 129)." Factors she highlights, including "[t]he limitations of space, the relation of scientific research to production and commerce whose consequence is privatization of information and ideas, and an understanding of research as the generation of positive results (ibid.)," inevitably shape peer review from the top-down.

As for the review process itself, it may come as a surprise to an outsider that many journals do not use double-blind review, where the author's and reviewer's identities are withheld from each other. For example, *Nature* journals started offering double-blind review as an *option* for authors in March 2015, after a trial run at *Nature Geoscience* and *Nature Climate Change* from June 2013 ("Nature Journals Offer Double-Blind Review", 2015). Ironically, the editorial admits that "[t]he uptake of the double-blind method has been much lower than the enthusiasm expressed in surveys suggested it would be. No more than one-fifth of monthly submissions to these

---

[3] According to authors of a study that simulated the effects of this bias, "The model suggests that a fixation in top-tier journals on significant or positive findings tends to drive trustworthiness of published science down, and is more likely to select for false positives and fraudulent results" (Grimes et al., 2018, 11).

journals are choosing the double-blind route." Two years after the change, things didn't look any better. A 2017 Nature Publishing Group study shows that between March 2015 and February 2017, only 12% of submissions to 25 Nature-branded journals chose the double-blind option (Enserink, 2017). It is not clear why most authors still prefer the single-blind option when given the choice. Mark Burgman, editor of *Conservation Biology*, seems to suggest that authors fear a negative reaction for going against the grain: "There's the idea that if you go double blind, you have something to hide" (ibid.). Alternatively, their preference may simply reflect the harsh reality that double-blind review *is* more effective—as the study (ibid.) finds, papers that opt for single-blind review are far more likely to be desk-rejected (8% compared to 23%) and less likely to be accepted after peer review (25% compared to 44%).

Further complicating the matter is the fact that formal double-blinding does not ensure anonymity as much as one would like to think. Specialization within a field, expanded networks of collaborators, and pre-publication discussions of one's work are some of the reasons why complete anonymity is often impossible ("Nature Journals Offer Double-Blind Review", 2015). For example, Burgman states that reviewers at *Conservation Biology* "who make a guess get it right about half of the time" (Enserink, 2017). Since such compromises of anonymity result from legitimate—and often unavoidable—interactions in the scientific community, the *Nature* editorial cites them as reasons for making double-blinding optional for all of its journals.

While we agree that the double-blind system is not perfect, it is important to recognize that the *way* the peer review system is set up—whether double-blind is the default or "opt-in"—has profound implications for the behavior of the reviewers and authors. Given the enormous pressure to publish, plus the fact that double-blind review *lowers* one's chance of acceptance (as it should), there is little incentive for most scientists to choose this option. For *Nature* to say, "We expect that some authors will choose it because of concern about biases, others purely *on principle*" [emphasis added] (ibid.) is thus at best naively optimistic and at worst downright irresponsible.

Another weakness of the peer review process is the lack of adequate preparation and resources for editors to enforce publishing ethics. Although this paper does not focus on fraud and other research misconduct, we want to note that a robust accountability mechanism is crucial for upholding the integrity of the publication process. Anecdotal evidence casts serious doubt on the efficacy of such mechanisms. Joseph Hilgard, a psychology professor, describes in a blog post the institutional barriers he encountered when trying to report scientific misconduct (Hilgard, 2021). What he thought was "a rather obvious case of unreliable data and possible research misconduct" turned out to be a protracted uphill battle with various journals. While some editors were willing to investigate the case, others seemed confused about how to proceed, and one editor-in-chief explicitly refused to enforce their journal's own code of ethics. The biggest irony of this two-year saga was that not only has the author continued to publish more articles, but every time Hilgard pointed out his problems, he made sure to avoid them in the next article. As Hilgard realized, "it is very easy to generate unreliable data, and it is very difficult to get it retracted."

3.2 The Reviewer

The difficulty Hilgard encountered highlights the centrality of trust *within* the scientific community: journal editors and authors trust the reviewers to follow standards of impartial and critical evaluation, and reviewers trust authors to honestly report their methodology and data. In the case of reviewers, single-blind review allows biases based on the author's identity and institution (Okike et al., 2016). Even with double-blind review, conscious or unconscious biases may still persist as the author's identity can often be revealed in other ways. As a science writer notes:

> Say you have a prejudice against people from certain countries, or newcomers, or feminists. There could be markers of authors being from that group which no amount of redacting citation lists and "our previous work has shown" can remove – including jargon or equipment that they use, writing style, quality of graphics, who they're studying, and more. Entire subfields can be markers for gender, for example (Bastian, 2017).

We may question the extent to which such personal biases are tolerated and even fostered in the scientific community, but some biases are *imposed* by the community. One such bias is a preference for "what is hot right now"—popular areas of research, methods, or theories. According to a recent study that examines "1.8 billion citations among 90 million papers across 241 subjects," an increased number of publications only reinforces, rather than challenges, accepted ideas within a field (Chu & Evans, 2021)." The authors go on to explain this apparent paradox: "Scholars in fields where many papers are published annually face difficulty getting published, read, and cited unless their work references already widely cited articles (ibid.)."

An *STAT* in-depth report on Alzheimer's research confirms this finding (Begley, 2019):

> In more than two dozen interviews, scientists whose ideas fell outside the dogma recounted how, for decades, believers in the dominant hypothesis suppressed research on alternative ideas: They influenced what studies got published in top journals, which scientists got funded, who got tenure, and who got speaking slots at reputation-buffing scientific conferences.

One of the researchers interviewed, Ruth Itzhaki, was among those who questioned the dominant view that amyloid was the cause of Alzheimer's. She had trouble getting her work published as top journals often cited previous rejections as grounds for rejection. Journal rejections in turn were used to discredit her funding application (ibid.).

What Itzhaki and other skeptics in her field experienced may be unusually egregious, but collective biases for certain kinds of research make it harder to uphold the norm of equality Longino imagines necessary for critical discourse in science. The third condition, "public standards" for evaluation, is difficult to satisfy as well. There is evidence that review quality of conference submissions can be surprisingly inconsistent. In 2014, the Neural Information Processing Systems Conference gave 10% of submissions to two separate program committees to review. Of the papers accepted by one program committee, almost 60% were rejected by the other. As a computer scientist sees it, this suggests that "the fate of many papers is determined by the specifics of the reviewers selected and not just the inherent value of the work itself" (Littman, 2021).[4]

One obvious reason for such inconsistency in review outcomes is a lack of time, partly because the reviewers themselves are too busy with other responsibilities[5] and partly because conferences and journals receive explosive volumes of submissions. Moreover, reviewers have little incentive to spend too much time on the job. Contrary to what Longino suggests—criticism must receive public recognition and carry almost as much weight as original research—reviewers are not rewarded for their work. Although they may be minimally recognized (on their CV) as a reviewer for certain journals, that is independent from the *quality* of their reviews.

3.3 The Author

The sheer amount of scientific output and pressure to publish incentivizes scientists to find ways to "stand out" from the crowd. Yet, as we have seen, the need to "stand out" ironically promotes conservatism over innovation as deviation from the "canon" (i.e., top-cited papers) risks lower chances of publication. Moreover, since scientists have limited time and resources, good research is far from optimal in achieving that goal. According to a recent study (Smaldino & McElreath, 2016) that simulates the evolutionary effects of an incentive structure that rewards positive findings and publications quantity, such a structure will "in the absence of countervailing forces, select for methods that produce the greatest number of publishable results. This, in turn, will lead to the natural selection of poor methods and increasingly high false discovery rates."

Another common strategy authors use is to give the *appearance* of significance. According to an analysis of positive and negative words in paper abstracts (Vinkers et al., 2015), the occurrence of terms like "innovative" and "groundbreaking" increased by 25 times over the 1974–2014 period while "promising" and "novel" increased by 30 and 40 times respectively. Usage of the word "unprecedented" increased by 50 times.

The upshot of the overwhelming demand for quantity—and the various strategies of outright or "mundane" fraud adopted in response to it—is that researchers committed to conscientious, good-quality work are not only inadequately rewarded, but *punished* for doing the right thing. From their simulation study, Smaldino and McElreath (2016) find that the existing incentive structure selects for lower effort, even with unrealistically favorable conditions for replication and harsh punishments

[4] It is worth noting that Littman brings up this information in an article about blatant academic misconduct—"collusion rings" in computer science research, where researchers collude to review each other's submissions positively, in violation of the ethics of peer review.

[5] For example, as we discuss in §4.2, grant recipients are found to spend disproportionately large amount of their time not on actual research, but on administrative tasks.

for false positive results. Similarly, another simulation study finds that such an incentive structure renders "diligent researchers" (those who take replication seriously and strive to minimize false positive submissions) much less competitive for publication and funding compared to their "careless" or "unethical" counterparts (Grimes et al., 2018). As the authors point out, "diligent researchers are unfairly affected by careless or unethical conduct, with avoidable false positives or unethical publications garnering disproportionate reward at their expense (11)."

To make things worse, the ever-increasing dependence of science on resources (in terms of funding, equipment, researchers, etc.) means that those who take time to do good work risk not just a low publication record, but the ability to maintain one's position in the scientific community at all. As Smaldino and McElreath (2016) put it,

> Scrupulous research on difficult problems may require years of intense work before yielding coherent, publishable results. If shallower work generating more publications is favoured, then researchers interested in pursuing complex questions may find themselves without jobs, perhaps to the detriment of the scientific community more broadly.

To reiterate, the problems we want to highlight with peer review are not so much about individuals' failures (in their roles as the editor, reviewer, or author) exposed in cases of outright fraud or misconduct. Rather, the problems are built into the peer review system and are therefore more insidious than cases of explicit fraud.

4 Grant Funding

For grant funding, we will focus on three aspects of the system: the goals implicit in the system, the selection process, and the main players involved in the process. While we find some similarities here with peer reviews for journals (and conferences), one important difference is that the players in this system are not just academics. The involvement of funding agencies entails that applicants are beholden to expectations set not just by academia, but by those who decide what to fund and for how much.

4.1 Goals

Regarding which projects get funded, we see increasing emphasis on short-term, "high impact" results. William G. Kaelin, Jr., the 2019 Nobel Prize recipient lamented in a *Washington Post* op-ed (Kaelin, 2020):

> Today, federal research funding is increasingly linked to potential impact, or deliverables, and basic scientists are increasingly asked to certify what they would be doing with their third, fourth and fifth years of funding, as though the outcomes of their experiments were already knowable. But the most important scientific discoveries often began with an unexpected experimental result that a scientist then pursued.

The demand for "deliverables" produces two seemingly contradictory results similar to what we have seen with journal publications. On the one hand, research shows that the grant system is biased against genuine innovation (Guthrie et al., 2018). On the other hand, like journals, funding agencies also seem to favor original research over replication studies.[6] But the contradiction is only apparent, for the sort of "original research" more likely to get funded builds on widely accepted models or theories rather than departs from them. This sort of conservatism, as opposed to (true, not apparent) innovation or replication, is most likely to yield positive results in the short term.

The emphasis on application further contributes to the problem. A quick look at the federal research and development funding report (Sargent, 2022) reveals:

1. While the federal government is the largest contributor to funding of basic research (41% in 2019, most recent available data), basic research accounts for 27% of the federal R&D spending in 2021 (compared to 28% for applied research, 42% for development).
2. Within basic research, 50% of the funding came from the Department of Health and Human Services and 14% from the National Science Foundation.
3. The Department of Defense remains the top federal funding agency on R&D, followed by the Department of Health and Human Services, accounting for 40.1% and 27.6% respectively of the estimated spending in 2021.

The funding breakdown suggests that basic research, which an outsider might take "science" to be, receives a relatively small portion of funding compared to research geared towards specific objectives or applications. The largest funding agencies for "science" are interested in very practical matters of life and death, not intellectual goals of truth or validity. The 2021 budget also highlights certain targeted funding areas, or "Science and Technology Supporting the 'Industries of the Future'," including artificial intelligence (AI), quantum information science (QIS), 5G/advanced communications, biotechnology, and advanced manufacturing. The directions set by funding agencies contribute to the bias towards "hot topics" discussed earlier.

4.2 Process

The review process of the grant system shares similar flaws in reliability and fairness as journal peer reviews. Although grants are meant to support top scientists, studies have found mixed evidence correlating funding status with research quality. A study of the most highly cited scientists (primary authors of papers with at least 1000 citations) eligible for NIH funding between 2001 and 2012 finds that study-section

[6] According to a survey by *Vox* (Belluz et al., 2016): "Jon-Patrick Allem, a social scientist at the Keck School of Medicine of USC, noted that funding agencies prefer to support projects that find new information instead of confirming old results."

members (grant application reviewers) comprise a small fraction (0.8%) even though the vast majority of them (88.9%) have current NIH funding (Nicholson & Ioannidis, 2012). In the 2018 grant peer review study (Guthrie et al., 2018), researchers find clear evidence that peer review is "at best only a weak predictor of future research performance" (6) and that "ratings vary considerably between reviewers" (7).

As with journal peer reviews, grant peer reviews suffer from reviewer bias. Notably, given the way grant review processes are set up, a certain kind of "proximity" bias is particularly salient. The 2012 study of NIH funding finds that "the grants of study-section members were more similar to other currently funded NIH grants than were non-members' grants" (Nicholson & Ioannidis, 2012). According to the authors, "This could suggest that study-section members fund work that is more similar to their own, or that they are chosen to serve as study-section members because of similarities between their own and funded grants" (ibid.). If it is the former, it would mean a form of reviewer bias that has also been identified in other studies (Guthrie et al., 2018). If it is the latter, then the bias is built into the funding process, which effectively rewards conservatism over innovation. As researchers point out, "We feel that by allowing grant holders to serve as grant reviewers, a conflict of interest becomes inescapable. Exceptional creative ideas may have difficulty surviving in such a networked system. Scientists who think creatively may be discouraged by the funding process and outcomes…" (Nicholson & Ioannidis, 2012).

Another problem of the grant funding process is the excessive burden it places on the applicants. According to a 2018 survey of over 10,000 principal investigators (PIs) of federally-funded projects, PIs spent an average of 44.3% of their research time on administrative and other requirements associated with applying for and managing the grants (Schneider, 2020). This was an increase of about 30 min per week from the 2005 and 2012 surveys, which estimated the value to be 42.3%. The report also notes an increase of respondents, from about one-third in 2012 to almost half in 2018, expressing concern about graduate students being discouraged from pursuing academic research careers (ibid.). However, this is hardly the only kind of negative impact on graduate students since they are in fact the ones responsible for much of the research, if not the grant application or management itself.

4.3 Players

Unlike journals, funding agencies do not all rely on peer review to award grants. Instead of a panel of experts in the relevant research area, program managers are employed to review grant applications. The problems with peer review we have seen above are only exacerbated by the power dynamics inherent in such a system: when a single program manager gets to make funding decisions, they potentially wield considerable power over researchers in entire fields. Yet, given the scarcity of research on this aspect of grant funding systems, we can only get a sense of how things work from personal anecdotes shared online, often anonymously. For example, in a discussion thread on scientific funding in the U.S., a comment read (Salzberg, 2015)[7]:

> DARPA [the Defense Advanced Research Projects Agency, a research and development agency of the U.S. Department of Defense] is also known (notoriously, one could say) for operating a good-old-boys system, where you have to butter up the DARPA program managers to get any funding.

In another thread on a similar topic, an anonymous user wrote (TheWaywardProf, 2021):

> Various orgs like the Office of Naval Research, Air Force Research Labs, etc. fund a LOT of basic science. Each one has several long-standing "program managers" who are former scientists that have an agenda for the sorts of things they want to fund. A lot of this funding is dependent on how excited you can get the PM. Think of this like pitching an Angel investor. If you get in good with a PM they can fund you at a decent rate for a long time!

Besides the potential for corruption, the existence of powerful program managers makes clear that "fellow scientists" are not the only people deciding which projects get funded. As we touched on in §4.1, other powerful players, many working outside the scientific community, set the "agenda" for grant funding. From the outsider's perspective, it is easy to neglect the significance of grants for a career in science. For the insiders though, grants not only pave the way for career advancement, but can often determine whether one *has* a career in science at all. This does not justify academic fraud, but it hardly makes for a work environment conducive to realizing the ideals of science.

5 Limitations and Conclusion

5.1 Limitations

It is difficult to gauge the extent and severity of the impact of such flaws in peer review and grant funding systems on the quality of research. As other researchers have pointed out, reliable empirical evidence is hard to come by. For example, the NASEM report acknowledges the lack of adequate evidence for evaluating the degrees of non-reproducibility and non-replicability (NASEM, 2019, 82). It also notes that surveys of researchers about questionable practices are methodologically defective to serve as "reliable source for a quantitative assessment (97)."

As for funding practices, researchers are well-aware of the irony in the lack of empirical basis for preferring the current model to alternatives. Authors of the 2018 grant peer review study reported that "no funding agencies have made significant use of alternative systems" (Guthrie et al., 2018).

---

[7] The commenter, Steven Salzberg, is the Bloomberg Distinguished Professor of Biomedical Engineering, Computer Science, and Biostatistics at Johns Hopkins University.

Yet, first-personal accounts from blogs, online forums, and news articles show that the problems we have discussed are well-known within the scientific community. In the same blog post that criticizes "mundane fraud," Jacob Buckman faults a culture that condones and facilitates such behavior (Buckman, 2021):

> When standards are low, it's to no individual's advantage to hold themselves to a higher bar. Newcomers to the field see these things, they learn, and they imitate. Often, they are directly encouraged by mentors. A graduate student who publishes three papers a year is every professor's dream, so strategies for maximal paper output become lab culture. And when virtually every lab endorses certain behaviors, they become integral to the research standards of the field.
>
> But worst of all: because everybody is complicit in this subtle fraud, nobody is willing to acknowledge its existence.

As Buckman points out, the cost of whistle-blowing is high. Ultra specialization means that those who work in the same area are likely to know each other; by speaking out against a colleague or her own team, one risks ostracism from other colleagues and teams, i.e., career suicide. In contrast, the benefits of conformity are considerable. Of course one could object that all this is only personal opinion held by a small number of people; a claim as sweeping as Buckman's that "mundane fraud" is an everyday, regular occurrence among scientists should not be taken as incontrovertible until the evidence is in. In the spirit of science, we would certainly welcome more evidence one way or the other. Indeed, the scarcity of robust research on this topic may well be indirect evidence for a view like Buckman's.

5.2 Conclusion

The trustworthiness of "science" is not a "given." The *real* scientific "institution" encompasses organizations like universities, government agencies, and corporations. *Real* scientists, who must navigate this institution successfully for their careers, are beholden to its norms and practices. *Real* scientific success is measured not so much by the quality of one's work as by the quantity of publications, citations, and grants one receives. With scientific research practices compromised by misaligned incentives, no fraud is required to undermine the trustworthiness of (actual) science.[8]

As we have seen, this is not news to the practitioners themselves. Although the knowledge of their own complicity is itself grave cause for concern, it also offers hope for change. For "outsiders," our critique shouldn't be taken as reason for *mistrusting* science. Instead, a more nuanced understanding of how science actually works is but a starting point for reforming it. As for the insiders, what is lacking is not the understanding, but a culture where it can be openly acknowledged and discussed.

[8] Although the NASEM report (NASEM, 2019) lists "publication bias" and "misaligned incentives" as two of six sources of non-replication that are "unhelpful" to scientific progress (as opposed to "helpful" sources that are built into the nature of scientific inquiry), there is no indication that they merit special attention compared to (innocent) errors, methodological flaws, or outright fraud.

While it is beyond the scope of this paper to discuss possible solutions, the bottom line is that change must be structural. As we celebrate Katalin Karikó's "meteoric" rise to fame during the COVID-19 pandemic—in the words of an interviewer for the *Proceedings of the National Academy of Sciences* (Nair, 2021)—we must also consider why she hit rock bottom in academia in the first place. In the interview, despite Karikó's acknowledgement of her inability to secure major grants during her academic career, the interviewer's response unwittingly affirms the media portrayal of her experience as the triumph of single-minded dedication: "The story of your career in science is a tale of success against the odds. Do you have a message for early-career scientists who might be struggling to get their unconventional ideas validated?"

References

Bastian, H. (2017). The fractured logic of blinded peer review in journals. PLOS Blogs: Absolutely Maybe. Retrieved January 03, 2023, from https://absolutelymaybe.plos.org/2017/10/31/the-fractured-logic-of-blinded-peer-review-in-journals/

Begley, S. (2019). The maddening saga of how an Alzheimer's 'cabal' thwarted progress toward a cure for decades. STAT, June 25. https://www.statnews.com/2019/06/25/alzheimers-cabal-thwarted-progress-toward-cure/

Belluz, J., Plumer, B., & Resnick, B. (2016). The 7 biggest problems facing science, according to 270 scientists. Vox, September 7. http://www.vox.com/2016/7/14/12016710/science-challeges-research-funding-peer-review-process

Buckman, J. (2021). Please commit more blatant academic fraud. The Buckman Zone. Retrieved January 03, 2023, from https://jacobbuckman.com/2021-05-29-please-commit-more-blatant-academic-fraud/

Chu, J. S. G, & Evans, J. A. (2021). Slowed canonical progress in large fields of science. *Proceedings of the National Academy of Sciences, 118* (41).

Edwards, M. A., & Roy, S. (2017). Academic research in the 21st century: Maintaining scientific integrity in a climate of perverse incentives and hypercompetition. *Environmental Engineering Science, 34*(1), 51–61.

Enserink, M. (2017). Few authors choose anonymous peer review, massive study of nature journals shows. Science, September 22. https://www.science.org/content/article/few-authors-choose-anonymous-peer-review-massive-study-nature-journals-shows

Grimes, D. R., Bauch, C. T., & Ioannidis, J. P. A. (2018). Modelling science trustworthiness under publish or perish pressure. *Royal Society Open Science, 5*(1), 171511.

Guthrie, S., Ghiga, I., & Wooding, S. (2018). What do we know about grant peer review in the health sciences? *F1000 Research, 6.*

Hilgard, J. (2021). I tried to report scientific misconduct. How did it go? Crystal Prison Zone. Retrieved January 03, 2023, from http://crystalprisonzone.blogspot.com/2021/01/i-tried-to-report-scientific-misconduct.html

Kaelin, W. G. (2020). Why we can't cure cancer with a moonshot. *The Washington Post*, February 11.

Littman, M. L. (2021). Collusion rings threaten the integrity of computer science research. *Communications of the ACM, 64*(6), 43–44.

Longino, H. E. (2002). *The fate of knowledge*. Princeton University Press.

Martin, G. N., & Clarke, R. M. (2017). Are psychology journals anti-replication? A snapshot of editorial practices. *Frontiers in Psychology, 8*, 523.

Nair, P. (2021). QnAs with Katalin Karikó. *Proceedings of the National Academy of Sciences, 118*(51).

National Academies of Sciences, Engineering, and Medicine. (2019). *Reproducibility and replicability in science.* National Academies Press.

Nature journals offer double-blind review. 2015. *Nature, 518*(274). https://doi.org/10.1038/518274b.

Nicholson, J. M., & Ioannidis, J. (2012). Conform and be funded. *Nature, 492*(7427), 34–36.

Okike, K., Hug, K. T., Kocher, M. S., & Leopold, S. S. (2016). Single-blind vs double-blind peer review in the setting of author prestige. *JAMA, 316*(12), 1315–1316.

Open Science Collaboration. (2015). Estimating the reproducibility of psychological science. *Science, 349*(6251), aac4716.

Regehr, J. (2011). Perverse incentives in academia. Embedded in Academia. Retrieved January 03, 2023, from https://blog.regehr.org/archives/632

Salzberg, S. (2015). Comment on "a bad idea from congress about how to fund science." *Hacker News.* https://news.ycombinator.com/item?id=9955086.

Sargent, J. F. (2022). Federal research and development (r&d) funding: FY2022. *CRS Report* 46869.

Schneider, S. L. (2020). *2018 Faculty workload survey research report: Primary findings.* Federal Demonstration Partnership. https://thefdp.org/wp-content/uploads/FDP-FWS-2018-Primary-Report.pdf

Smaldino, P. E., & McElreath, R. (2016). The natural selection of bad science. *Royal Society Open Science, 3*(9), 160384.

TheWaywardProf. (2021). Comment on "Our brutal science system almost cost us a pioneer of mRNA vaccines." *Hacker News.* https://news.ycombinator.com/item?id=26126669.

Vinkers, C. H., Tijdink, J. K., & Otte, W. M. (2015). Use of positive and negative words in scientific PubMed abstracts between 1974 and 2014: Retrospective analysis. *BMJ, 351*, h6467. https://doi.org/10.1136/bmj.h6467

www.ingramcontent.com/pod-product-compliance
Lightning Source LLC
Chambersburg PA
CBHW070736301224
19657CB00003BA/20

9 783031 680571